Sugarcane

Agricultural Production, Bioenergy, and Ethanol

Sugarcane

Agricultural Production, Bioenergy, and Ethanol

Edited by

Fernando Santos
Universidade Estadual do Rio Grande do Sul, Porto Alegre, RS, Brazil

Aluízio Borém
Universidade Federal de Viçosa, Viçosa, MG, Brazil

Celso Caldas
Central Analítica LTDA, Maceio, AL, Brazil

AMSTERDAM • BOSTON • HEIDELBERG • LONDON
NEW YORK • OXFORD • PARIS • SAN DIEGO
SAN FRANCISCO • SINGAPORE • SYDNEY • TOKYO
Academic Press is an imprint of Elsevier

Academic Press is an imprint of Elsevier
125, London Wall, EC2Y 5AS.
525 B Street, Suite 1800, San Diego, CA 92101-4495, USA
225 Wyman Street, Waltham, MA 02451, USA
The Boulevard, Langford Lane, Kidlington, Oxford OX5 1GB, UK

Notices
Knowledge and best practice in this field are constantly changing. As new research and experience broaden
our understanding, changes in research methods, professional practices, or medical treatment may become
necessary.

Practitioners and researchers must always rely on their own experience and knowledge in evaluating
and using any information, methods, compounds, or experiments described herein. In using such information
or methods they should be mindful of their own safety and the safety of others, including parties for whom
they have a professional responsibility.

To the fullest extent of the law, neither the Publisher nor the authors, contributors, or editors, assume
any liability for any injury and/or damage to persons or property as a matter of products liability, negligence
or otherwise, or from any use or operation of any methods, products, instructions, or ideas contained in the
material herein.

ISBN: 978-0-12-802239-9

British Library Cataloguing-in-Publication Data
A catalogue record for this book is available from the British Library.

Library of Congress Cataloging-in-Publication Data
A catalog record for this book is available from the Library of Congress.

For Information on all Academic Press publications
visit our website at http://store.elsevier.com/

Typeset by MPS Limited, Chennai, India
www.adi-mps.com

Printed and bound in the USA

Working together
to grow libraries in
developing countries

www.elsevier.com • www.bookaid.org

Contents

Chapter 3: Planting **35**

Victor Francisco Araújo de Medeiros Barbosa

Tomaz Caetano Cannavam Ripoli and Marco Lorenzzo Cunali Ripoli

Chapter 11:   Breeding Program and Cultivar Recommendations 241

Márcio Henrique Pereira Barbosa and Luís Cláudio Inácio da Silveira

Chapter 12:   Molecular Biology and Biotechnology................................257

Aluízio Borém, Jorge A. Doe and Valdir Diola

Chapter 15: Ethanol Fermentation 311

João Nunes de Vasconcelos

Chapter 17: Industrial Waste Recovery 365

*Sarita Cândida Rabelo, Aline Carvalho da Costa and
Carlos Eduardo Vaz Rossell*

Chapter 18: Sugarcane Bioenergy 383

Sizuo Matsuoka, José Bressiani, Walter Maccheroni and Ivo Fouto

List of Contributors

Wellington Sacco Altran Raízen — Unidade Gasa, Andradina — SP, Brazil

Márcio Henrique Pereira Barbosa Federal University of Viçosa, Viçosa, MG, Brazil

Thales Velho Barreto Velho Barreto Inc. and ECO, Technology and Industrial Equipment Inc, Brazil

Aluízio Borém Universidade Federal de Viçosa, Viçosa, MG, Brazil

José Bressiani Canaviallis, Maceio, AL, Brazil

Celso Caldas Central Analítica LTDA, Maceio, AL, Brazil

Claudio Soares Cavalcante CSC Engenharia, Campinas, São Paulo, SP, Brazil

Antônio Carlos Duarte Coelho Federal University of Pernambuco, Recife, PE, Brazil

Aline Carvalho da Costa School of Chemical Engineering, State University of Campinas, Campinas, SP, Brazil

Antônio Alberto da Silva Universidade Federal de Viçosa, Viçosa, MG, Brazil

Alexandre Ferreira da Silva Embrapa Milho e Sorgo, Sete Lagoas, MG, Brazil

Luís Cláudio Inácio da Silveira Federal University of Viçosa, Viçosa, MG, Brazil

Fernando Medeiros de Albuquerque F. Medeiros Consultoria, Recife, PE, Brazil

Leonardo Angelo de Aquino Federal University of Viçosa, Viçosa, MG, Brazil

Francisco de Assis Dutra Melo Universidade Federal Rural de Pernambuco, Recife, Pernambuco, PE, Brazil

Maria Bernadete S. de Campos Universidade Federal de São Carlos, UFSCAR, São João Del Rei — MG, Brazil

Pedro Henrique de Cerqueira Luz Universidade de São Paulo, Pirassununga — SP — FZEA — USP

Victor Francisco Araújo de Medeiros Barbosa Universidade Federal de Viçosa, Viçosa, MG, Brazil

Rubens Alves de Oliveira Federal University of Viçosa, Viçosa, MG, Brazil

Willians Xavier de Oliveira Fundação Getúlio Vargas, São Paulo, SP, Brazil

Sérgio de Oliveira Procópio Embrapa Tabuleiros Costeiros, Aracaju, SE, Brazil

João Nunes de Vasconcelos Federal University of Alagoas, Maceió, AL, Brazil

Ronaldo Medeiros dos Santos University of Brasilia, Brasília, DF, Viçosa, MG, Brazil

Valdir Diola Department of Plant Physiology, Universidade Federal de Viçosa, Brazil; **In memoriam**

Jorge A. Doe Texas A&M University, College Station, TX, USA

Luiz Carlos C.B. Ferraz Universidade de São Paulo, ESALQ, São Paulo, SP, Brazil

Evander Alves Ferreira Universidade Federal dos Vales do Jequitinhonha e Mucuri, Diamantina, MG, Brazil

Ivo Fouto AGN Bioenergy, São Paulo, SP, Brazil

Leandro Galon Universidade Federal da Fronteira Sul, Erechim, RS, Brazil

José Marinaldo Gleriani Universidade Federal de Viçosa, Viçosa, MG, Brazil

Daniella Macedo Imaflora, Piracicaba, SP, Brazil

Newton Macedo Imaflora, Piracicaba, SP, Brazil

Walter Maccheroni AGN Bioenergy, São Paulo, SP, Brazil

Fernando Bomfim Margarido Santelisa Vale, Brazil

Paulo do Carmo Martins Federal University of Juiz de Fora, Juiz de Fora, MG, Brazil

Sizuo Matsuoka Vignis S.A., Santo Antonio de Posse, State of São Paulo, Brazil

Wilson R.T. Novaretti Federal University of Juiz de Fora, Juiz de Fora, MG, Brazil

Angélica Maria Patarroyo Universidade Federal de Minas Gerais, Belo Horizonte, MG, Brazil

Sarita Cândida Rabelo Brazilian Bioethanol Science and Technology Laboratory—CTBE/CNPEM, Campinas, SP, Brazil

Márcio Mota Ramos Federal University of Viçosa, Viçosa, MG, Brazil

Marco Lorenzzo Cunali Ripoli John Deere – Latin America, Indaiatuba, SP, Brazil

Tomaz Caetano Cannavam Ripoli Universidade de São Paulo – Escola Superior de Agricultura "Luiz de Queiroz", São Paulo, SP, Brazil; **In memoriam**

Fernando Santos Universidade Estadual do Rio Grande do Sul, Porto Alegre, RS, Brazil

Carlos Alberto Alves Varella Federal Rural University of Rio de Janeiro, Rio de Janeiro, RJ, Brazil

Carlos Eduardo Vaz Rossell Brazilian Bioethanol Science and Technology Laboratory—CTBE/CNPEM, Campinas, SP, Brazil

Godofredo Cesar Vitti Universidade de São Paulo, Piracicaba – SP – ESALQ – USP

Foreword

Sugarcane crops have been established in Brazil since nearly five centuries ago. At that time, sugar cane brandy and brown sugar were special products.

For almost a century, Brazil has had cars fueled by ethanol and has been an important player in world sugar production and exports.

However, in the last 35 years the industry experienced its most impressive leap in production and productivity, based on absolutely spectacular technological progress.

The Proálcool Program, the largest program on energy alternatives resulting from the seventies' "oil shocks", conferred a new face to the sugarcane production chain.

Soon after, the introduction of the Program for Sugarcane Payment by Sucrose Contents produced one of the greatest technological revolutions of 20th century agribusiness: new varieties were developed, as well as different cropping practices, planting and harvesting dates, fertilizer formulas, everything changed; mechanization evolved greatly, and techniques were vigorously implemented in all segments of the agricultural industry.

All such advancements made Brazil become, in a sustainable and highly competitive manner, the world's largest exporter of sugar and ethanol.

The prospects for the future are even more promising: the so-called "green economy", so widely used by the world's greatest leaders, opens up amazing opportunities for agroenergy, whether for biofuels, bioelectricity, or for bagasse pellets as an alternative to firewood in the fireplaces of cold countries.

All of these topics are well addressed in this timely and notable book. At this moment in history, when global warming represents a major challenge, the role of the sugarcane production chain transcends national boundaries. However, there is a concerning aspect: the lack of coordination among sector policies, both in the public and private spheres.

So far, we have not defined the amount of ethanol we should or want to produce, in what period of time, and for which market — internal or external?

There are no long-term contract models.

We have not defined who will be in charge of logistics, storage, production contracts, or final product certification.

There is no liaison between the areas of technological development and human resources training.

Nothing is defined as to the future of hydrated ethanol.

Nothing is organized on the issue food × energy, a ridiculous theme that still gets media attention because of minor interests of other sectors.

The production system, so well described by Barbosa Lima Sobrinho in the Sugarcane Production Statute in the forties, turned to dust with IAA's end.

Sugarcane suppliers, who "deliver" their production to plants, and do not sell it, are in an extremely awkward position in terms of links in the production chain, since they cannot choose who they sell to: they can only sell to industries nearby their agricultural area. This renders the production chain uneven. In addition, since the end of IAA, the process lacks arbitration, even though Consecana is a good beginning.

Finally, in such a promising segment for Brazil, at such an important time, this lack of coordination may inhibit the country's progress, even as regards leading a global geopolitical shift, with the exportation of technology to poor tropical countries in Latin America, Africa, and Asia for them to produce agroenergy associated with food.

For all this, we welcome the publishing of this enlightening book, written by some of the foremost experts in each of the topics addressed.

Dr. Roberto Rodrigues
Former Secretary of Agriculture

Preface

Sugarcane is native to the warm temperate to tropical regions of South Asia, and is used for sugar, ethanol and spirit production. Sugarcane is the world's largest crop by production quantity. In 2014, the FAO estimated that it was cultivated on about 29.0 million hectares, in more than 90 countries, with a worldwide harvest of 1.84 billion tons. Brazil was the largest producer of sugar cane in the world. The next five major producers, in decreasing amounts of production, were India, China, Thailand, Pakistan and Mexico.

Cane accounts for over 80% of sugar produced; most of the rest is made from sugar beets. Sugarcane predominantly grows in the tropical and subtropical regions, and sugar beet predominantly grows in colder temperate regions of the world. In India, between the sixth and fourth centuries BC, the Persians, followed by the Greeks, discovered the famous "reeds that produce honey without bees". They adopted and then spread sugar and sugarcane agriculture. A few merchants began to trade in sugar — a luxury and an expensive spice until the 18th century. Before the 18th century, cultivation of sugarcane was largely confined to India. Sugarcane plantations, like cotton farms, were a major driver of large human migrations in the 19th and early 20th centuries, influencing the ethnic mix, political conflicts and cultural evolution of various Caribbean, South American, Indian Ocean and Pacific island nations.

Sugarcane became an even more important crop with the importance of bioenergy in today's society. Bioenergy is renewable energy made available from materials derived from biological sources and sugarcane is currently the major source of biofuel.

The Brazilian sugarcane industry employs modern agronomic management practices to enhance productivity and protect the environment. In fact, Brazil is the leader in sugarcane production and research.

Written by experts in each topic addressed, the intention is that this book will be used by new and advanced students, as well as serving as a reference book for those interested in the sugarcane crop and processing. Instructors are encouraged to select specific chapters to meet classroom needs. Readers will also benefit from the list of references that accompany each chapter.

The Editors

CHAPTER 1

Agricultural Planning

Fernando Bomfim Margarido[1] **and Fernando Santos**[2]

[1]*Santelisa Vale, Brazil* [2]*Universidade Estadual do Rio Grande do Sul, Porto Alegre, RS, Brazil*

Introduction

Nowadays, management involves less risk than it used to. However, the responsibility involved is much greater, considering the technological processes surrounding an administrative decision. According to the classical definition, it could be said that managing means planning, organizing, directing and controlling. On the basis of this definition, planning means deciding in advance what should be done for a particular purpose to be achieved, namely, to maximize agricultural and industrial yield and, thus, profits. That is the starting point for good management.

The sugar and ethanol sector in Brazil is going through one of its best periods. There has been significant change in the sector's dynamics resulting in reduced competitiveness among industrial units, expansion of cultivated areas and adjustments in strategies adopted by companies. This chapter addresses planning through technical expertise aimed at operational practices. It is, therefore, a simplified view of planning.

1.1 Planning

The main role of agricultural managers is to foment the activity. In simple terms, fomenting the agricultural activity means guaranteeing the supply of raw materials for the industry, which involves, in the case of sugarcane culture, agricultural production, soil conservation and preparation, planting, crop practices for cane-plants, harvesting, crop practices for the ratoon and supplying mills with raw material during the harvest period. Such supply relates not only to the total quantity of cane to be crushed over the harvest period, but also the constant hourly supply, involving the concept of logistics throughout the plantation, observing machinery size and personnel availability. The agricultural production system is relevant for strategic planning in industrial units, so as to anticipate production, storage and marketing of final products.

According to Pinazza (1985), high productivity levels derive from four basic types of factors: physical, structural, institutional and development factors. Physical factors represent the edaphic and climatic conditions of a region and agricultural production. Institutional factors involve government action by means of implementation of agricultural policies. Development factors are related to the research system, and to what extent knowledge generates increased productivity. Structural factors refer to the management system adopted, and have a decisive influence on the strategic and operational performance of mills and distilleries.

Agricultural planning observes industrial planning, therefore, the starting point is the amount one plans to process along the next three growing seasons. It is important to take into account that the agricultural sector requires planning at least 2 years ahead, since it is necessary to arrange partnership agreements, prepare the soil and wait for harvest time — on average, the first cut is carried out 1½ years after planting.

1.1.1 Planning for Planting

In agricultural planning, it is important to know the productive potential of the region vis-à-vis climate, soil quality and resources available for production (use of vinasse, irrigation and fertilization). This information is especially necessary for the introduction of a new unit. When a unit is already in operation, one can look at productivity history over the last 5 or 6 years. Historical data older than 10 years are not pertinent, since varieties will not behave in the same way after such a period.

The technical area is very important as, at this point, it is necessary to survey the amount of arable land available at the various properties, their productive potential, the opportunities to purchase raw materials in regional markets, the options of land renting or production partnerships; in addition, the technical area should analyze the edaphic zoning (per production environment), topography (feasibility of mechanical harvesting), climate characteristics in the region (temperature, rainfall, light, photoperiod, water balance, frost) and the region's road system, anticipating the flow of production. In some cases, these factors make it unfeasible to locate a production unit in a particular area, for example, where high toll fees would increase transportation costs, or areas with a ban on sugarcane burning (as of 2012, in the State of São Paulo) on slopes with over 12% gradient or with the presence of stones. It should be noted that, for sugarcane production in the past, soil fertility was the sole determinant of land value, but currently, topography and presence of obstacles in the area are also determining factors.

Table 1.1 shows an example of a balanced sugarcane plantation, considering theoretical average productivity of the site and areas with equal size in each category of cutting.

Table 1.2 considers that first cut sugarcane has been used for planting and that 1 ha produces seedlings to plant 7 ha. It should be noted that, in this case, the area where the first cut took place was smaller.

Table 1.1: Balanced sugarcane production system.

Cut	Productivity (t/ha)	Area (ha)	Production (t)
Cane-plant	—	4100.00	—
1° cut	120.00	4100.00	492,000.00
2° cut	100.00	4100.00	410,000.00
3° cut	92.00	4100.00	377,200.00
4° cut	81.00	4100.00	332,100.00
5° cut	73.00	4100.00	299,300.00
Other cuts	66.00	2050.00	135,300.00
Total		**26,650.00**	**2,045,900.00**

Productivity data used refer to average productivity in the north of the state of São Paulo, in the Alta Mogiana Region. To analyze a particular region, it is important to consider local productivity.

Table 1.2: Balanced sugarcane production system considering the production of seedlings.

Cut	Productivity (t/ha)	Area (ha)	Production (t)
Cane-plant	—	4100.00	—
1° cut	120.00	3514.29	421,714.29
2° cut	100.00	4100.00	410,000.00
3° cut	92.00	4100.00	377,200.00
4° cut	81.00	4100.00	332,100.00
5° cut	73.00	4100.00	299,300.00
Other cuts	66.00	2050.00	135,300.00
Total		**26,064.29**	**1,975,614.29**

One can observe that the first cut area decreases, since part of it (1/7 on average) is used to produce seedlings for cane-plant planting.

For a better picture of agricultural planning, we will use as an example the construction of a new industrial unit with overall capacity to crush 2,000,000 t of sugarcane, and daily crushing capacity of 12,000 t. In this case, several factors should be considered in planning, such as physical, edaphic and climate conditions in the region, planting system, crop practices and harvesting.

Tables 1.3 to 1.11 refer to a planting plan aimed at crushing 2,000,000 t within 5 years. In this case, initial planting is large (7500 ha), decreasing slightly in the second and third years (5000 ha) and stabilizing in the fourth year (4100 ha). The technical manager in charge of planning can easily use an Excel spreadsheet to make projections, change planting areas and productivity to obtain yearly production values.

The time factor is very important in agricultural planning. In the example presented, it is observed that, soon after the definition of a location for a particular industrial unit, planting should start, so that, as soon as the construction of the industrial plant is completed, the sugarcane is ready to be crushed. In the example, in the first year of operation 800,000 t of

Table 1.3: Planning for the first year of sugarcane production.

Cut	Productivity (t/ha)	Area (ha)	Production (t)
Cane-plant	—	7500.00	—
1° cut	120.00	—	—
2° cut	100.00	—	—
3° cut	92.00	—	—
4° cut	81.00	—	—
5° cut	73.00	—	—
Other cuts	66.00	—	—
Total		**7500.00**	—

In the first year, it is important to plant more than the future equilibrium point. Since the industry needs to crush a larger amount in the first year, it is important to plan according to the yearly evolution of the amount to be crushed.

Table 1.4: Planning for the second year of sugarcane production.

Cut	Productivity (t/ha)	Area (ha)	Production (t)
Cane-plant	—	5000.00	—
1° cut	120.00	6785.71	814,285.71
2° cut	100.00	—	—
3° cut	92.00	—	—
4° cut	81.00	—	—
5° cut	73.00	—	—
Other cuts	66.00	—	—
Total		**11,785.71**	**814,285.71**

In the second year, it is possible to decrease the planting area to 5000 ha, yet the area to be planted is still larger than the area of equilibrium (around 4100 ha).

Table 1.5: Planning for the third year of sugarcane production.

Cut	Productivity (t/ha)	Area (ha)	Production (t)
Cane-plant	—	5000.00	—
1° cut	120.00	4285.71	514,285.71
2° cut	100.00	7500.00	750,000.00
3° cut	92.00	—	—
4° cut	81.00	—	—
5° cut	73.00	—	—
Other cuts	66.00	—	—
Total		**16,785.71**	**1,264,285.71**

As of the fourth year of planning onwards, the planting area stabilizes around 4100 ha, and the yield of sugarcane plantations is maintained over time.

Table 1.6: Planning for the fourth year of sugarcane production.

Cut	Productivity (t/ha)	Area (ha)	Production (t)
Cane-plant	—	4100.00	—
1° cut	120.00	4414.29	529,714.29
2° cut	100.00	5000.00	500,000.00
3° cut	92.00	7500.00	690,000.00
4° cut	81.00	—	—
5° cut	73.00	—	—
Other cuts	66.00	—	—
Total		**21,014.29**	**1,719,714.29**

It is important to point out that after large decreases in planting or in renovation, there is a significant increase in total production in the next year, but a drastic reduction in the second year, because of two factors: (i) part of the first cut cane (1/7), which is used for seedlings, is not used for sowing, and therefore, it is added to the next growing season; (ii) because of renovation itself, which if it is not carried out, increases the cutting area in the following year.

Table 1.7: Planning for the fifth year of sugarcane production.

Cut	Productivity (t/ha)	Area (ha)	Production (t)
Cane-plant	—	4100.00	—
1° cut	120.00	3514.29	421,714.29
2° cut	100.00	5000.00	500,000.00
3° cut	92.00	5000.00	460,000.00
4° cut	81.00	7500.00	607,500.00
5° cut	73.00	—	—
Other cuts	66.00	—	—
Total		**25,114.29**	**1,989,214.29**

After the fifth year, it is possible to keep the area of equilibrium and reach the planned production.

Table 1.8: Planning for the sixth year of sugarcane production.

Cut	Productivity (t/ha)	Area (ha)	Production (t)
Cane-plant	—	4100.00	—
1° cut	120.00	3514.29	421,714.29
2° cut	100.00	4100.00	410,000.00
3° cut	92.00	5000.00	460,000.00
4° cut	81.00	5000.00	405,000.00
5° cut	73.00	7500.00	547,500.00
Other cuts	66.00	—	—
Total		**29,214.29**	**2,244,214.29**

It is observed that, in the sixth year, if the planting area is maintained at 4100 ha, production starts to go over 2,000,000 t. In this case, one can either review the plan so as to reduce the planting area or stick to planning, but then one should start the season ahead of schedule, sell sugarcane or leave it to be harvested the next season.

Table 1.9: Planning for the seventh year of sugarcane production.

Cut	Productivity (t/ha)	Area (ha)	Production (t)
Cane-plant	—	4100.00	—
1° cut	120.00	3514.29	421,714.29
2° cut	100.00	4100.00	410,000.00
3° cut	92.00	4100.00	377,200.00
4° cut	81.00	5000.00	405,000.00
5° cut	73.00	5000.00	365,000.00
Other cuts	66.00	3750.00	247,500.00
Total		**29,564.29**	**2,226,414.29**

Table 1.10: Planning for the eighth year of sugarcane production.

Cut	Productivity (t/ha)	Area (ha)	Production (t)
Cane-plant	—	4100.00	—
1° cut	120.00	3514.29	421,714.29
2° cut	100.00	4100.00	410,000.00
3° cut	92.00	4100.00	377,200.00
4° cut	81.00	4100.00	332,100.00
5° cut	73.00	5000.00	365,000.00
Other cuts	66.00	2500.00	165,000.00
Total		**27,414.29**	**2,071,014.29**

If the planting area is maintained at 4100 ha, there is a tendency that production will be balanced at 2,000,000 t again.

Table 1.11: Planning for the ninth year of sugarcane production.

Cut	Productivity (t/ha)	Area (ha)	Production (t)
Cane-plant	—	4100.00	—
1° cut	120.00	3514.29	421,714.29
2° cut	100.00	4100.00	410,000.00
3° cut	92.00	4100.00	377,200.00
4° cut	81.00	4100.00	332,100.00
5° cut	73.00	4100.00	299,300.00
Other cuts	66.00	2500.00	165,000.00
Total		**26,514.29**	**2,005,314.29**

The stabilization of sugarcane plantations occurs after 6 years of planting 4100 ha.

sugarcane would be crushed, reaching 1,250,000 t in the second year and 1,700,000 in the third year, stabilizing around 2,000,000 t from the fourth year onwards. Readers may be wondering about the fifth and sixth years, when production is 10% higher than that required for crushing. In this case, either the crushing starts earlier or, depending on the region, production is sold to other mills. This kind of situation can be avoided by planting less than 4100 ha in the previous year.

1.1.2 Planning of Varieties

After defining the planting area, one should choose the variety to be planted. The choice of plant variety is a technical and administrative decision. The technical area determines the production environment where the variety will be planted, given the type of soil (through specific analysis) and climate in the region. In the administrative field, it should be noted that the sugarcane harvest time is a function of the variety's cycle.

In the choice of variety according to the cycle (early, middle, late), several factors should be taken into account, such as:

- End of partnership contract;
- Harvesting period of nearby sugarcane plantations;
- Location of the area (avoid late canes in fire hazard areas);
- Possibility of irrigation;
- Production environment (early canes have been preferred in D and E production environments, and late maturity canes in environments A and B. Classification by production environments is a function of productive potential and environment E is the one with higher productive potential); and
- Type of harvesting (mechanical or manual).

In terms of harvest season, canes harvested in April, May and June are considered early canes; in July, August and September they are considered mid-cycle; and those harvested in October and November are considered late (Table 1.12).

Regarding sugarcane maturity, the curves of early, mid-cycle or late sugarcanes are very similar. Typically, the point of greatest richness occurs in late August or early September, just before the rainy season (central—south region), but what determines sugarcane earliness is the fact that the variety is richest in that period vis-à-vis other varieties, i.e., it reaches maturity before other varieties. Late sugarcane keeps maturity longer after the onset of the rainy season and does not "foam".

Generally, the planting of approximately 40% early canes, 30% mid-cycle and 30% late canes is recommended; however, at planting time, it is necessary to assess the amount planted by suppliers and make adjustments accordingly.

Table 1.13 presents the recommendation for management of the main sugarcane varieties planted in the state of São Paulo. This type of table facilitates the display of variety options to choose from.

Table 1.12: Sugarcane harvest times and cycles.

April/May/June Early	July/August/September Mid-cycle	October/November Late

Table 1.13: Characteristics of the main sugarcane varieties planted in the state of São Paulo and management recommendations.

Variety	Advantage	Production Environment					Harvest Time								Restrictions
		A	B	C	D	E	Apr	May	Jun	Jul	Aug	Sep	Oct	Nov	
SP77-5181	Rich	X	X				X	X	X							Not suitable to mechanical harvesting; demanding as regards soil and moisture; in the dry season, the internodes are short; susceptible to borer
SP79-1011	Excellent ratoon	X	X							X	X	X				Takes long to shade between lines; susceptible to rust and borer; may break tips due to strong winds.
SP80-1816	Excellent ratoon	X	X	X					X	X	X	X				Very demanding as regards soil, tips break easily due to winds; very susceptible to spittlebug, and the cane-plant is sparse.
SP80-1842	Good ratoon	X	X	X					X	X	X	X				Takes long to shade between lines: it is susceptible to spitlebug; very prone to tipping and the buds have bugs; and under straw, there is thinning and reduced tillering.
SP80-3280	Good ratoon	X	X	X					X	X	X	X				Demands fertile and humid soil; under these conditions it production is good.
SP81-3250	Rich and productive	X	X	X					X	X	X	X				In low fertility land and with mechanical harvesting, productivity and longevity are reduced; sometimes affected by yellow leaf virus; susceptible to spittlebug.
SP83-2847	Rusticity	X	X	X					X	X	X	X				Highly tolerant to low-fertility soils; very poor and very flourishing; susceptible to smut.
SP83-5073	Rich	X	X	X					X	X	X	X				Very demanding as regards soil, heat and moisture; resistant to borer and tolerant to spittlebug; should only be grown under very good conditions.
SP84-1431	Rich and productive	X	X	X					X	X	X	X				Rust causes productivity losses; in low fertility soils it is very thin; tipping occurs; when harvested early, it overcomes rust.
SP84-2025	Productive	X	X	X					X	X	X	X				It is called half variety, because the demands, maturity and productivity are average; it is sometimes affected by yellow leaf virus.
SP85-3877	Rich	X	X	X					X	X	X	X				Does not present good ratoon after mechanical harvesting; very demanding.
SP85-5077	Productive			X	X						X	X	X			Is very late and poor; grows slowly; needs to be harvested in the middle of the season, in the first cut.
SP86-42	Good ratoon, productive	X	X	X	X			X	X	X	X	X	X			It is very thin and does not tolerate very low fertility soils, very susceptible to borer.
SP86-155	Rich	X	X				X	X	X	X	X	X	X			It is very thin and does not tolerate trampling, as it becomes even thinner, making the crop sparse; presents much loss in mechanical harvesting.
SP87-365	Productive	X	X	X					X	X	X	X				It has low resistance to dry periods and is highly susceptible to borer; very demanding as regards soils; production of cane-plant is not good.
SP89-1115	Rich	X	X				X	X	X	X	X	X				Very demanding as regards soils; production of cane-plant is not good.
SP90-1638	Good ratoon, productive	X	X								X	X		X	Very demanding as regards soils; low tolerance to dry periods.

Source: IDEA News (2004).

It is worth noting that varieties respond differently depending on the way they are managed and to the region where they grow. For example, in some areas rust can be moderate, and the harvest period can change its incidence in the same region. The use of irrigation can also change the performance of varieties that have shooting problems and so on.

1.1.3 Planning for Harvesting

This planning is very important in agriculture, as industry depends on it and it ensures maximum quality of sugarcane crops.

In planning, the quantification of human resources (sugarcane cutters, tractor drivers, drivers, etc.) and material resources (trucks, winches, harvesters, transshipment, service trucks, firemen and tractors) is of utmost importance. Correct quantification determines the success of agricultural supply, and the profit of operations. The harvesting activity is a crucial determinant in the total cost of raw materials and consequently in the final cost of sugar and ethanol. Therefore, correct quantification determines low-cost operations. It is known that a piece of equipment generates more profit depending on its operational production, and that idle equipments mean cost without revenue. Thus, the better the structure that is available, the greater the profitability. The size of a queue of trucks or the lack of sugarcane with which to load them during the season tells us a lot about service efficiency and profitability of the activity. When experienced managers see a queue of trucks to unload sugarcane at a certain industry, they will first question whether there was an interruption at the mill, and then check whether service frontlines are close; if they are, they will conclude that there is an excess of trucks for sugarcane transportation. This is to say that their trucks are operating below capacity.

Quantification should not be overestimated, so as to create idle staff and machinery and increasing costs, nor underestimated to the point of causing undersupply problems to the industry. Considering it is a broad and complex subject, it will not be covered in this chapter, however it deserves mentioning.

The harvesting of sugarcane is carried out in a sequence of three stages: cutting, loading and transporting, and is associated with early, middle and late maturity cycles, taking into account an average harvesting interval day by day, so as to maintain an hourly supply to industries.

For example, a mill with capacity to crush 500 t of sugarcane per hour should receive 500 t per hour. It seems obvious, but if supply is 250 t, the industry will process only 250 t per hour, working below capacity. If supply is 750 t per hour, there will be a queue of trucks to unload; in this case the mill does not stop for lack of sugarcane, but there will be idleness in the harvesting structure, thus, resulting in increased cost.

Industries usually operate with sugarcane harvesting fronts ranging from around 1500 t to 2500 t per day. These numbers may vary, so that an industry grinding 12,000 t per day has about six fronts and the type of harvesting also varies (mechanical or manual). For the quantification of equipments, one should always use averages. For example, to calculate the number of trucks, the average distance from the mill should be considered. Assuming that the average distance is 25 km, harvesting fronts cannot be located at an average greater than 30 km, otherwise inevitably mills will be undersupplied, unless part of the route takes paved roads and despite the distance, travel time is offset by road quality, or truck breakdown for that day is foreseen to be lower than average. The opposite may also occur if fronts are located less than 20 km away: trucks will queue to deliver sugarcane at the mill. One could compare the location of harvesting fronts to a game of chess to be played at every movement of the enemy, or at each event, as within planning there are areas where production exceeds the estimate and areas where this does not occur. There are also events such as fires, pests, diseases, frost; i.e., factors beyond control that require adjustments to be made to the original planning.

Another important factor in planning is that, after the harvest, the sequence of crop practices is defined automatically.

1.2 Final Remarks

Above all, planning means making a plan of what should be done and how it should be done, based on a forecast, in order to obtain the best possible results for companies. Planning plays an important role in farming activities and it has taken on paramount relevance due to the expansion of areas planted with sugarcane, the influence of increased production, and the need to work to a budget.

Finally, it should be noted that the cost of raw materials in a sugar and ethanol industry represents around two-thirds of final product costs (sugar and alcohol). This number reflects the importance of the agricultural sector in the organization.

Bibliography

Abbitt, B., Morton, M., 1980. Florida's sugarcane industry: progress to date. Citrus Veg. Mag. 43, 10, 12–13, 26, 28.

Alvarez, J., Deren, C.W., Glaz, B. 2003. Sugarcane selection for sucrose and tonnage using economic criteria. Proceedings of the Sugar Cane International Conference. November–December 6–10.

Batalha, M.O. (coord.) 2007. Gestão agroindustrial: GEPAI: Grupo de estudos e pesquisas agroindustriais. third ed. São Paulo: Atlas.

Campos, M.C.C., Junior, J.M., Pereira, G.T., Souza, Z.M., Montanari, R., 2009. Planejamento agrícola e implantação de sistema de cultivo de cana-de-açúcar com auxílio de técnicas geoestatísticas. Revista Brasileira de Engenharia Agrícola e Ambiental. 13 (3), 297–304.

Macedo, I.C. (org.). 2005. A Energia da cana-de-açúcar. Doze estudos sobre a agroindústria da cana-de-açúcar no Brasil e a sua sustentabilidade. São Paulo: Berlendis & Vertecchia.

Paiva, R.P.O., Morabito, R., 2007. Um modelo de otimização para o planejamento agregado da produção em usinas de açúcar e álcool. Gest. Prod. 14 (1), 25−41.

Picoli, M.C.A., Rudorff, B.F.T., Zuben, F.J.V. 2007. Estimativa da produtividade agrícola da cana-de-açúcar: estudo de caso da Usina Catanduva. In: Anais XIII Simpósio Brasileiro de Sensoriamento Remoto, Florianópolis, Brasil, 21−26 abril 2007, INPE, p. 331−333.

Pinazza, A.H., 1985. Implicações da gerência agrécola nas usinas e destilarias. Brasil Açucareiro. 103, 26−27.

Robison, L.J., Barry, P., 1996. Present value models and investment analysis. The Academic Page, Northport, AL.

Santos, F.A. 2008. Análise de trilha dos principais constituintes orgânicos e inorgânicos sobre a cor do caldo em cultivares de cana-de-açúcar. Dissertação (Mestrado), Universidade Federal de Viçosa, Viçosa, MG. 64p.

Segato, S.V. et al. (Org.). 2006. Gerência agrícola em destilarias de álcool. Instituto do Açúcar e Álcool, Planalsucar, 1982. Atualização em produção de cana-de-açúcar. Piracicaba.

CHAPTER 2

Physiology

Fernando Santos[1] and Valdir Diola[2],[†]
[1]*Universidade Estadual do Rio Grande do Sul, Porto Alegre, RS, Brazil* [2]*Department of Plant Physiology, Universidade Federal de Viçosa, Brazil*

Introduction

Sugarcane, *Saccharum* spp., is a plant belonging to the family Poaceae and class Monocotyledones. The main species emerged in Oceania (New Guinea) and Asia (India and China). Varieties grown in Brazil and in the world are multi-species hybrids. The main characteristics of this family are spike-like inflorescence, internode stalk, leaves with silica flakes on edge and open sheath. The plant in its native form is perennial, has an erect habit and is slightly decumbent at the initial stage of development. In subsequent stages, it undergoes self-shadowing tiller selection. The height growth continues until the occurrence of any limitation in water supply, of low temperatures or even flowering. Due to the lack of resistance to low temperatures, the crop is best suited in a range of latitude 35°N to 30°S and at altitudes ranging from sea level to 1000 m (Rodrigues, 1995).

It is one of the most important crops in the tropical world, generating hundreds of thousands of direct and indirect jobs. Sugarcane is a major source of income and development. It is the primary raw material for the manufacture of sugar, ethanol and spirits. It is also used as a forage plant in its fresh form.

There are several products made from this plant, as there are many sugarcane compounds, which can be commercially exploited. Currently, sucrose is the most valuable compound, because it is the source for its main products, sugar and ethanol. The average yield of this crop is 53 t/ha of stalks with sucrose levels from 10% to 18%, and 11% to 16% fiber. The plant has C_4 photosynthetic apparatus, so it is highly efficient in converting radiant energy into chemical energy in photosynthetic rates estimated at up to 100 mg CO_2 fixed per dm^2 leaf area per hour. The high rate of biomass accumulation is due to intense photosynthetic activity throughout the growing season and high leaf area index (LAI) of the plant.

This chapter aims to present briefly the main physiological processes responsible for crop growth and development and to show how they are involved in the metabolic processes of the

---

[†] In memoriam

main compounds responsible for sugarcane products commercially exploited. The study of plant physiology covers a much wider field than we will cover, ranging from the expression of specific genes to complex metabolic processes; this would require further studies. We will briefly detail the procedures for obtaining and assimilating carbon and the synthesis of sucrose, which is the main compound of interest in Agronomics currently. Additionally we will circumstantially present the processes of plant growth and development, with emphasis on the two physiological stages of great agricultural importance: flowering and maturation.

2.1 Photosynthesis

The photosynthetic apparatus is located in the chloroplasts, specifically in specialized membranes called thylakoids. These membrane-bound structures are found in high-density — grana thylakoid — and low-density — stroma thylakoid (also called intergrana thylakoids or lamellae) — and are composed of an external matrix, the stroma, and an internal matrix, the lumen. Photosynthesis takes place in the thylakoids because of the presence of photosynthetic pigments, i.e., chlorophylls, which absorb light in the range of 400–700 nm. This spectrum band, which is used by plants as a source of energy for their metabolic activities, is commonly identified as Photosynthetically Active Radiation (PAR), the unit for which is μmol of photons/m^2/s.

The photosynthetic process can be represented by a simplified equation of reduction in which CO_2 receives electrons and CH is reduced. The H_2O is oxidized, releasing O_2, since light promotes the oxidation of water:

$$12H_2O + 6CO_2 \xrightarrow{\text{light + chlorophyll}} C_6H_{12}O_6 + 6H_2O + 6O_2$$

Photosynthesis refers to a series of reactions, which involves light absorption, energy conversion, electron transfer and multiple processes. Enzymes are involved in these processes converting CO_2 and water into sugars.

There are two stages in this process: light reactions — producing O_2, ATP and NADPH and C compounds synthesized from the radiant energy — and dark reactions — the carbon reduction cycle (Calvin cycle), which consumes ATP NADPH and produces carbohydrates. The two phases occur in different regions. The first one occurs in the thylakoid membranes and the second, in the stroma, both measured by enzymes.

2.1.1 Absorption of Light Energy and Water Oxidation

Light energy excites pigments (chlorophylls) and is absorbed. The double bonds of the chlorophyll in the excited state increase its energy level and electrons intersect the energy of photons. The excited electron must return to ground state, losing energy as heat, fluorescence, inductive energy transfer, electron loss or dissipation and utilization of energy. In photosynthesis, the excited electron is donated to a receptor molecule, triggering redox reactions. From the reaction centers (RC), a dimer of chlorophyll in the excited state transfers the electron to the receptor molecule, which results in a process of charge separation. This constitutes the primary event of photosynthesis through light-mediated induction, which promotes the flow of electrons to the photochemical process. Ultimately, these electrons participate in the reduction of $NADP^+$ to NADPH.

The excitement of the RC of photosystem II (PSII) (P680*) generates a strong oxidant (P680+) and promotes an event of extraction of electrons from water, with the consequent formation of O_2. The process of photo-oxidation of water is catalyzed and mediated by the oxygen evolving complex (OEC). The OEC is located on the side of thylakoid membranes, facing the lumen (Figure 2.1). This involves the oxidation of two water molecules, releasing four protons and four electrons. Thus, for each O_2 released, the RC P680 needs to be excited four times, i.e., to absorb the energy of four photons. Each OEC is home to a group of four manganese ions, which act as accumulators of positive charges. Each absorbed photon removes an electron from RC P680, which is immediately replaced by an electron

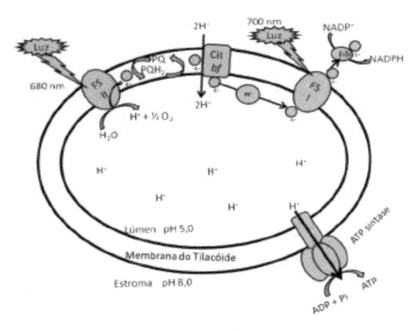

Figure 2.1
Schematic model of the thylakoid membranes, showing the coupling in electron transport and photophosphorylation. The energy stored in the proton gradient generated by the flow of electrons is used for the formation of ATP from ADP and Pi. *Source: Adapted from Kerbauy (2004).*

taken from the cluster of manganese ions of the OEC. The loss of four electrons in a row causes the manganous center to go from state S^0 to S^{4+}, which is the oxidant component that reacts with water, thus restoring the oxidation state of the manganous center to S^0:

$$2H_2O + S^{4+} \rightarrow S^0 + 4H^+ + O_2$$

2.1.2 Photosynthetic Electron Flow and Oxidation of Water

Supramolecular complexes involved in photosynthesis are *photosystem I* (PSI), *photosystem II* (PSII), *cytochrome b6f complex* (Cit b6f) and *ATP synthase complex*. The interconnection between the photosynthetic complexes involved in electron flow is mediated by *mobile carriers* which move within the lipid matrix, such as *plastoquinone* (PQ); within the thylakoids, as *plastocyanin* (PC); or within the stroma, such as *ferredoxin* (Fd). The photosynthetic electron flow between photosystems generates a H^+ proton gradient across the thylakoid membranes. The H^+ gradient provides the momentum in ATP synthesis. In other words, the proton gradient engages *ATP synthase* in the process of storing energy during the photosynthetic electron flow (Figure 2.1).

2.1.3 Photophosphorylation

The synthesis of ATP in chloroplasts, promoted by light, is called *photophosphorylation*. It is driven by the proton motive force generated during the flow of electrons from the light stage. The protons flow through the *ATP synthase* enzyme complex, which crosses the lipid matrix of membranes. The flow of H^+ through the *ATP synthase* complex in favor of the H^+ gradient, is responsible for changes in the configuration of the CF_1 subunit. These changes are necessary for the synthesis of ATP. The ATP synthesized during the photochemical process, besides supporting CO_2 fixation, is used in numerous metabolic pathways that exist within the chloroplasts. As an example, part of the assimilation of NO^{3-}, NH^{4+} and the amino acid synthesis uses the reducing power and ATP generated during the photochemical stage.

2.2 Carbon Metabolism

2.2.1 CO_2 Fixation and C_4 System Sugarcane

Sugarcane, a C_4 metabolism plant, has two distinct enzymes specialized to leaf anatomy that help in fixing CO_2:Rubisco and PEP carboxylase. The primary carboxylation enzyme is PEP carboxylase, which is located in the leaf mesophyll cells and carboxylates CO_2 absorbed from the air via stomata. The resulting phosphoenolpyruvic acid then forms oxaloacetic acid (OAA). Depending on the plant species, the oxaloacetic acid is converted into malate or aspartate and then, by diffusion, is transported to the bundle-sheath cells of leaves, where it is decarboxylated, releasing CO_2 and pyruvic acid. The CO_2 released is again fixed by the enzyme ribulose-1,5-bisphosphate carboxylase. The pyruvic acid, by diffusion, returns to the mesophyll cells, where it is phosphorylated, consuming 2 ATPs, regenerating the enzyme PEP carboxylase, and restarting the entire cycle (Figure 2.2).

The environmental conditions are important, considering the photosynthesis in C_4 plants. For example, under conditions of high temperatures, high light and temporary water deficit, C_4 plants are more efficient than C_3 plants and can accumulate twice as much biomass per leaf area. This is because the PEP-carboxylase, present only in C_4 plants, has high affinity for CO_2, works specifically with carboxylase and does not saturate at high light intensity.

Sugarcane, like other C_4 plants, has mechanisms to improve the photosynthetic efficiency, reducing water loss in hot—dry environments. Biochemical and anatomical changes result in better conversion efficiency of trapped CO_2, thereby reducing the conductance and thus

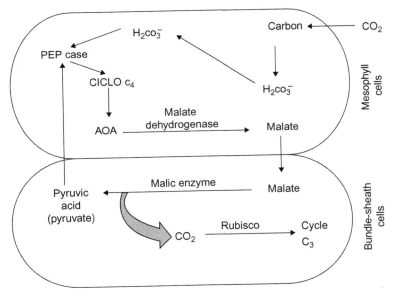

Figure 2.2

General aspects of C_4 pathway in sugarcane. CO_2 enters the mesophyll cell and is converted to HCO^{3-} in the cytosol. This bicarbonate ion reacts with PEP (phosphoenolpyruvate) to form a C_4 acid (oxaloacetate), which is converted into a second C_4 acid (malate) and then transported to a sheath cell. There, C_4 is decarboxylated and CO_2 released is fixed by Rubisco and converted into carbohydrate by the Calvin cycle. The product of decarboxylation (C_3 acids) is transported to the mesophyll cell to regenerate the PEP.

saving water without reducing rates of carbon fixation. C_4 plants concentrate the CO_2 in the bundle sheath cells, where ribulose bisphosphate carboxylase/oxygenase (Rubisco) is found, suppressing almost completely and effectively the oxygenase activity of that enzyme and eliminating photorespiration almost completely.

2.2.2 C_3 Carbon Reactions (Calvin−Benson Cycle)

After C_4 stage, the CO_2 is incorporated into a C_5, the ribulose-1,5-bisphosphate (RuBP), and split into two C_3 compounds. These compounds are the first stable product in the conversion of multiple steps of CO_2 into carbohydrates, the 3-phosphoglycerate (3-PGA). The enzyme that catalyzes this reaction is Rubisco, one of the most important of the three enzymes in the Calvin cycle. As the first enzyme involved in the conversion of CO_2 into carbohydrates, Rubisco plays a critical role in the biochemistry of the chloroplast. Consistent with its role, Rubisco is the most abundant soluble protein in the chloroplast and is possibly one of the most abundant in the biosphere. This enzyme catalyzes two types of reactions, carboxylation and oxygenation. The carboxylation of RuBP yields two molecules of 3-PGA, the first stable intermediate in C_3 photosynthesis. The oxygenation of RuBP results in a molecule of 3-PGA and one of 2-phosphoglycolate.

The Calvin cycle consists of 13 steps, in which the energy produced in photosynthesis in three phases − carboxylation, reduction and regeneration − is used. The carboxylation phase consists of one reaction: the carboxylation of RuBP to produce two molecules of 3-PGA. In the reduction phase, there are two steps. The first one converts PGA into triose phosphate (glyceraldehyde 3-phosphate (3 PGA)). ATP and NADPH (coming from photosynthesis) are used in this phase of the cycle. In the last step, the reactions regenerate RuBP. In this process, an additional ATP is consumed during the conversion of ribulose 5-phosphate into RuBP. All 13 enzymes required in the Calvin cycle are located in the stroma, and 10 of the 13 enzymes are involved in the regeneration phase of the cycle. In addition to Rubisco, the most important enzymes in the cycle are sedoheptulose-1,7-bisphosphatase, which dephosphrylates a sugar bisphosphate to yield a sugar monophosphate, and phosphofructokinase-5-phosphate, which phosphorylates ribulose to form RuBP, regenerating the initial CO_2.

The addition of three molecules of CO_2 and three molecules of sugar C_5 RuBP yields six simple triose molecules each one of which, when phosphorylated and reduced, results in a 3-carbon sugar, or 3-PGA. Five 3-PGAs and three molecules of ATP are used to regenerate the three molecules of RuBP. The molecule of PGA remaining is the net product of carbon fixation and can be used to build carbohydrates (hexoses and sugar) or other cellular components. The energetic demand for the synthesis of a hexose is six CO_2, nine ATP molecules and six NADPH molecules. This energy comes from photosynthesis.

2.2.3 Efficiency of CO_2 Utilization and Characteristics of Photosynthesis in Sugarcane

Photorespiration is dependent on the available CO_2 and the presence of free O_2 in the cell. High external O_2 concentrations and low CO_2 concentrations do not alter the photosynthetic rate, because the compensation point of photosynthesis operates with very low values of CO_2. In some cases, the exchange can be zero. The compensation points of CO_2 with increasing concentration of oxygen suggest a competition between oxygen and CO_2 during photosynthesis. Furthermore, photosynthetic active tissues release a pool of CO_2 immediately after the illumination ceases.

C_4 plants, such as sugarcane, have high rates of photorespiration, but are constrained by relatively sophisticated biochemical mechanisms, since they concentrate CO_2 at the site of carbon fixation. The compensation point of CO_2 for a C_3 plant ($20-100 \, \mu l/l$) is higher than for C_4 plants ($0-5 \, \mu l/l$) and is associated with the presence of photorespiration in C_3 plants.

The characteristics of the cultivars influence the photosynthetic efficiency of sugarcane, as well as climate variations that prevail during the development of the crop. Photosynthesis is negatively correlated with leaf width and positively with leaf thickness. A more vertical position of the leaf in the stalk translates into greater photosynthetic efficiency and usually into high population density, which increases the efficiency of light utilization. Photosynthesis also varies with leaf age, reaching values of C_4 fixation only in recently expanded leaves. Older leaves and very young leaves, on the other hand, perform photosynthesis at levels similar to those of C_3 plants.

The processes of bioconversion of energy in sugarcane crops are more affected by some environmental parameters such as light (intensity and amount), CO_2 concentration, water availability, nutrients and temperature. The increase in irradiance raises the photosynthetic rate, saturation occurring above $0.9 \, cal/cm^2/min$.

The increase of CO_2 in the atmosphere augments the photosynthetic capacity; moderate wind speed increases photosynthesis by boosting the availability of CO_2 for plants. Low wind speed leads to depression in photosynthesis, around noon.

2.3 Synthesis and Storage of Starch and Sucrose

Sucrose is the main form of carbohydrate translocated through the plant by the phloem. Starch is a stable and insoluble reserve of carbohydrates, and the most compressed form of energy conservation. Both starch and sucrose are synthesized from triose phosphate generated by the Calvin cycle.

2.3.1 Synthesis of Starch

The site of starch synthesis in leaves is the chloroplast. Starch is synthesized from the triose phosphate via fructose-1,6-bisphosphate. Glucose-1-phosphate intermediate is converted into ADP-glucose via an ADP-glucose pyrophosphorylase (Figure 2.3) in a reaction that requires ATP and produces pyrophosphate (PPi or $H_2P_2O_7^{2-}$). As in many biosynthetic reactions, pyrophosphate is hydrolyzed into two molecules of orthophosphate (Pi) by a specific inorganic pyrophosphatase, triggering the reaction toward the synthesis of ADP-glucose. Finally, the portion of glucose from ADP-glucose is transferred to the non-reducer end (carbon 4) of the terminal glucose of a starch chain in growth, thus completing the sequence of reactions.

2.3.2 Synthesis of Sucrose

Enzyme assays show that sucrose synthesis occurs in the cytosol from triose phosphate, by a similar route to the route of synthesis of starch, i.e., via fructose-l,6-bisphosphate and

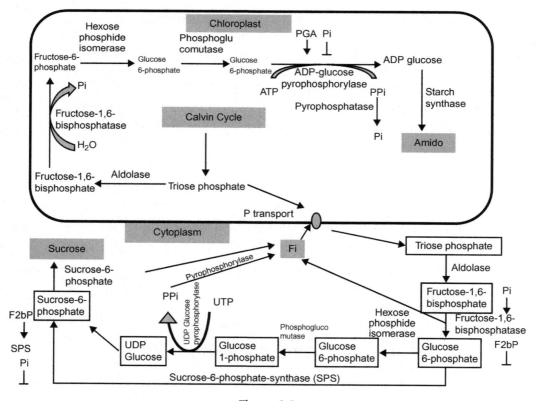

Figure 2.3
Scheme demonstrating the metabolic process of sucrose synthesis and storage of starch.
↓ activation, ⊥ inhibition. *Source: Adapted from Taiz and Zieger (2004).*

glucose-1-phosphate (Figure 2.3). In the synthesis of sucrose, glucose-1-phosphate is converted into UDP-glucose via a UDP-glucose-specific phosphorylase, similar to ADP-glucose pyrophosphorylase of chloroplasts. At this stage, two consecutive reactions complete the synthesis of sucrose. First, sucrose-6-phosphate-synthase catalyzes the reaction of UDP-glucose with fructose-6-phosphate to produce sucrose-6-phosphate and UDP. Second, the sucrose-6-phosphate-phosphatase (phospho-hydrolase) removes the phosphate from sucrose-6-phosphate, producing sucrose. The latter reaction, which is essentially irreversible, moves the former reaction toward the synthesis of sucrose (Figure 2.3).

As in the synthesis of starch, the pyrophosphate formed in the reaction catalyzed by UDP-glucose pyrophosphorylase is hydrolyzed, but not immediately, as in the chloroplasts. Due to the absence of inorganic pyrophosphatase, other enzymes in transphosphorylation reactions may use the pyrophosphate. An example is fructose-6-phosphate-phosphotransferase, an enzyme that catalyzes a reaction similar to the one catalyzed by phosphofructokinase. The difference is that the pyrophosphate replaces ATP as phosphoryl donor. A comparison of reactions shows that the conversion of triose phosphate into glucose-1-phosphate on the routes leading to the synthesis of starch and sucrose has several steps in common. However, these routes using enzyme isoforms (different forms of enzymes that catalyze the same reaction) are specific to the chloroplast or cytosol.

Except for the fructose-1,6-bisphosphatase from cytosol, the synthesis of sucrose is regulated so as to result in sucrose phosphate synthase, an allosteric enzyme activated by glucose-6-phosphate and inhibited by orthophosphate. The enzyme is inactivated in the dark by phosphorylation of a specific serine residue via a protein kinase, and activated by dephosphorylation in the light, via a protein phosphatase. Glucose-6-phosphate inhibits the kinase, while the Pi inhibits the phosphatase (Figure 2.3). Sucrose-6-phosphate synthase (SPS) and sucrose-6-phosphatase (SP) exist as a supramolecular complex, with higher enzyme activity than the isolated activities of constituent enzymes. This non-covalent interaction of the two enzymes involved in the last two steps of sucrose synthesis points to a new regulatory characteristic of carbohydrate metabolism in plants.

The syntheses of starch and sucrose are competing processes that occur in the chloroplast and cytosol, respectively. When the concentration of cytosolic Pi is high, the chloroplast triose phosphate is exported to the cytosol via a Pi carrier, in exchange for Pi, and sucrose is synthesized. When the concentration of cytosolic Pi is low, the triose phosphate is retained within the chloroplast and starch is synthesized.

However, when it comes to sugarcane, the process of starch synthesis does not compete with the synthesis of sucrose as much as the synthesis of fiber. The concentration of fibers is 8−14% in most cultivars, while the starch content can reach only 0.05%. Apart from the difference in concentration, starch is a compound essential to plant life, acting as a potential buffer reserve, which can be interconverted into sugars. By the way, fiber is the form that accumulates sugars in the most unavailable form.

2.3.3 Transport and Allocation of Sucrose

The transport of sucrose synthesized in leaf mesophyll cells to phloem involves translocation across the plasmalemma, in the cell wall, involving a "sucrose carrier", which works in combination with potassium transport, dependent on metabolic energy. The loading of sucrose to the phloem companion cells is performed by a system of co-transport of hydrogen ions, which induce the formation of the electrochemical gradient necessary for the generation of energy in the membrane ATPase system. The mechanism works better in low concentrations of sucrose in the cell wall, resulting in transport to the phloem against the concentration gradient. Whenever the concentration of sucrose in the apoplast reach levels incompatible with the functioning of the sucrose carriers, the enzyme acid invertase, present in the cell wall, is activated, acting on the hydrolysis reaction and turning sucrose into hexoses. Those hexoses are transported back to the mesophyll cells and again converted into sucrose (Figure 2.4).

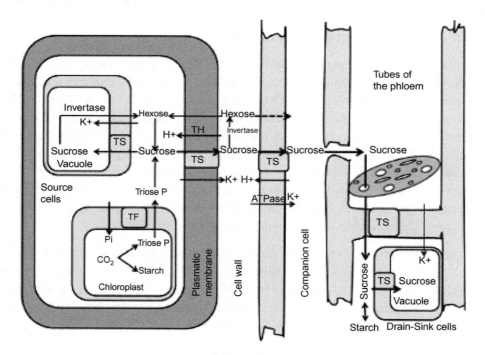

Figure 2.4

Mechanisms of control of source-sink relationships for loading of phloem and transport of sucrose in the plant. (TS — sucrose carrier, TH — H carrier; TF — P carrier).
Source: Adapted from Magalhães (1987).

Sucrose recycling, between the apoplast and symplast, keeps sugar concentration in the cell wall, aiming at the efficient functioning of sucrose carriers. Once inside the companion cells, the sucrose is transferred to the tubes of the phloem preferably via plasmodesmata, passively. The whole process occurs in the "source", i.e., the place of production of carbohydrates, from where they will be translocated to the places of consumption or storage, i.e., the "sinks".

Sucrose follows through the phloem to the sink. The flow of the phloem is formed by the turgor pressure gradient between cells of phloem-source and phloem-sink. This pressure difference is established by the continuous loading of the phloem at the source and discontinuous phloem unloading at the sink (Magalhães, 1987). The phloem loading increases the concentration of sucrose, reducing the osmotic potential, causing the water intake and increasing turgor. This causes reversible deformation of sucrose carrier proteins, thus preventing the complete filling of the vessels (Figure 2.4). The move, by mass flow, reaches the sink-cell, undergoing active unloading into the vacuole of a parenchyma cell in the stalk.

When leaving the phloem, sucrose undergoes breakage and rearrangements before being stored in the vacuole. Conversions are initiated in the external spaces of parenchymatous tissue, where sucrose is converted into glucose and fructose by the action of invertase. The hexoses enter the cytoplasm of parenchyma cells of the stalk by a diffusion process. In the cytoplasm, the reactions are more complex, because hexoses are very reactive and undergo rapid interconversions and phosphorylations. Several enzymes participate in these reactions, such as hexokinases (phosphorylation of glucose and fructose); phosphohexose isomerase (interconversion of glucose-6-P and fructose-6-P); UDPG fructose-6-P transglucosidase (synthesis of sucrose and sucrose-P); numerous non-specific phosphatases; and one sucrose-P; apart from the auxins, which control the system (Figure 2.5).

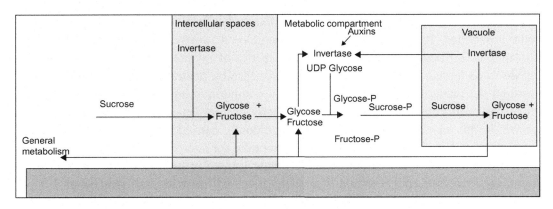

Figure 2.5
Schematic representation of the cycle of sucrose and hexoses and sucrose accumulation in conductive and parenchymal tissue. *Source: Adapted from Rodrigues (1995).*

To enter the vacuole, sucrose has to be activated (sucrose-P). The breaking of the phosphate link generates energy for sucrose to penetrate the vacuole, where it is accumulated. As sucrose concentration is high in the inner space (vacuole), passive absorption is not processed. The mechanism of sucrose accumulation is the same, both in immature and adult tissues:

- hydrolysis of sucrose, as a prerequisite and limiter of the first step;
- formation and interconversion of hexose phosphates;
- formation of molecules similar to sucrose (maybe sucrose-P); and
- accumulation of part of the sucrose in the vacuole.

However, some differences between the accumulations in these two tissues occur, as the presence of plant regulators and the action of invertases. In immature tissues, dominated by rapid cell expansion, the accumulated sucrose is rapidly hydrolyzed by vacuolar acid invertase, moving the resulting hexoses rapidly to the cytoplasm where it is used in cell growth and development (respiration, synthesis of organic molecules, etc.). In tissues in the maturation phase, there is increased action of neutral or alkaline invertase (maximum activity at pH 7.0). There is a correlation between the level of activity of this enzyme and the concentration of hexoses. Almost no activity of vacuolar acid invertase indicates that effective accumulation of sucrose is occurring (Casagrande, 1991).

During maturation, there is a decline in acid invertase activity of intercellular spaces (apoplast), low activity of acid invertase of the cytoplasm and almost no activity of vacuolar acid invertase. In the case of growing tissues, the acid invertase of the apoplast is secreted during cell formation in the meristematic region. As the stalk cells move away from that region, they become longer due to higher concentration of sucrose, reaching the maturation process. The amount of sucrose depends on the amount of acid invertase secreted from the apoplast of the parenchymal tissue, because at this stage no more enzyme is secreted. In adult or mature cells, insoluble acid invertases are found in cell walls (apoplast). Maturation is the result of a kinetics of invertases. It is important therefore to understand the exchange of acid invertase by neutral or alkaline invertases to understand the maturation process. Fructose is a competitive inhibitor of acid invertase. High sucrose concentrations can suppress, partially or completely, the action of acid invertase. In this case, this function would be gradually carried out by alkaline invertase, which indicates that the tissue is mature and prepared for sucrose accumulation. Thus, the invertase enzymes drive carbohydrates to plant growth or to accumulate in vacuoles. In them, the increased concentration of carbohydrates will provide the ripening or maturation of stalk, which occurs when the crop has the best qualitative and quantitative yield of sugars.

2.4 Physiology of Development

2.4.1 Propagation

In commercial cultivation, propagation is asexual and made by a portion of the stalk (*setts*). The development of the root system begins immediately after planting. The buds, located at the base of the node, are lateral embryonic meristems. They remain inactive during apical dominance due to the production of auxins. Each *sett* also contains a circle of small dots above the node, which form the primordial root. Each primordium has a black center, which is the root cap, and a "halo". The bud flowers under favorable conditions and for nearly a month after sprouting. The young plant lives at the expense of reserves in the *sett* and partially uses water and nutrients supplied by the first roots. Each bud can form a main stalk from a clump (Figure 2.6A).

2.4.2 Root System

Sett roots are thin and have many branches, which support the growing plant during the first weeks after germination. Shoot roots are secondary types of roots that emerge from the base of the new shoot 5−7 days after planting. Shoot roots are thicker, are more vigorous than sett roots, and develop in the main root system (Figure 2.6A). Sett roots continue to grow for a period of 6−15 days after planting. Generally they disappear in 60−90 days. Meanwhile the shoot root system develops and takes ownership of the supply of water and nutrients.

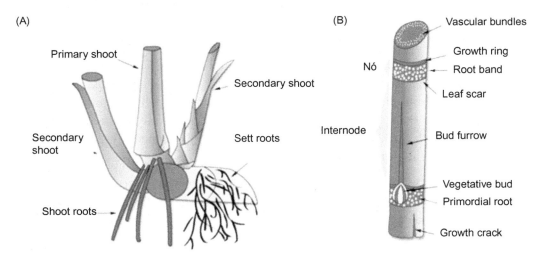

Figure 2.6
Schematic representation of the main bud and tillering (A). Internode section, showing the vascular bundles and section of the node with vegetative buds and their associated tissues (B).

Genotypes that produce many tillers usually produce many roots because each new tiller is a source of shoot roots. Similarly, crops with higher penetration of horizontal root (poor gravitropism) are more resistant to toppling than those with a stronger gravitropic root system.

2.4.3 Stalk

The stalk is a reserve organ surrounded by alternate leaves. It can be erect or decumbent and its diameter varies from about one centimeter to several centimeters. The clumps can be sparse or dense, according to the number of stalks in each one. A joint is formed of a node and an internode. The node is where the leaf is attached to the stalk and where the buds and the primordial root are found. A scar can be found in the node of the leaves when they fall (Figure 2.6B). The length and diameter of the joints diversify according to varieties and growing conditions. The stalk colors in the internodes depend on the cultivars and environmental conditions. All the colors of the stalk derive from two basic pigments: red from the anthocyanin and green from chlorophyll.

2.4.4 Leaf

The sugarcane leaf is divided into two parts: sheath and blade, separated by a blade joint. The sheath holds the leaf in the stalk, completely covering it. The sheath extends over at least one complete internode and, depending on the variety, is covered with fluff. The leaves are alternate and inserted into the nodes, forming two rows on opposite sides. A mature plant of sugarcane has top leaf surface of an average of 0.5 m^2 and the number of green leaves per stalk is around 10, depending on the variety and growing conditions.

The blade has a serrated edge, with a developed midrib that splits the blade in half, lengthwise, and several secondary veins arranged in parallel on both sides. At the junction between blade and sheath, internally, there is the ligule and externally, there is an appendage of various shapes, according to the variety. Still externally, in the junction point of the blade with the sheath, a structure called *dewlap* is formed. There are two wedge-shaped areas, the "barbs", on the blade joint.

Leaves are numbered by the Kuijper system (Casagrande, 1991) (Figure 2.7). The first leaf on the stalk from the top with clearly seen barbs is designated +1. Downwards, they receive successively the numbers +2 and +3. The leaf of the top visible barb (+3) is a diagnostic tissue often used to evaluate the nutritional status of the plant.

2.4.5 Inflorescence

When the plant reaches maturity and is under environmental stimulus, the apical meristem changes from vegetative to reproductive.

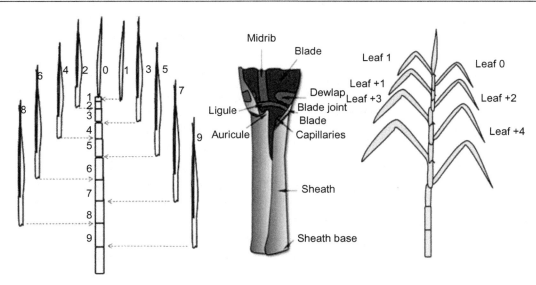

Figure 2.7
Numbering system of leaves: in the left, numbering based on morphological traits; in the right, based on the physiological development of agronomic interest. *Source: McCormick et al. (2006) and Casagrande (1991).*

The meristem stops forming the primordial leaf and begins to produce an inflorescence. The sugarcane inflorescence is an open panicle called a flag or an arrow. Each panicle consists of several thousands of flowers. Each flower is capable of producing a seed. Generally, a day length of around 12.5 h and night temperatures between 20°C and 26°C induce the beginning of flowering. Conditions for optimal growth in the vegetative phase (fertile soil, abundant supply of nitrogen and humidity) restrict the inflorescence.

2.4.6 Stages of Cultivation

Sprouting and establishment stage

This stage is comprised of the planting to the sprouting of buds. Soon after planting, the turgor process of the buds begins. Under the conditions of soil, sprouting occurs in 20–30 days after planting. Sprouting is marked by a rapid and marked increase in respiratory activity, accompanied by the beginning of active transport of substances to the growing points. After this period, the development of the roots of primary tillers, secondary tillers, and so on. Sprouting is influenced by external and internal factors. External factors are humidity, temperature and aeration of the soil, the variety, presence of sheath, the time interval between the cut of the seedling and its planting, and the presence of straw on the seedling. Internal factors are health of the bud, sett humidity, reduction of sugar content of the sett and its nutritional status. Optimum temperature for bud flowering is between 28°C and 30°C. Basal temperature for sprouting is around 12°C. Humid soil and heat ensure quick sprouting.

During sprouting, respiration increases and, therefore, good soil aeration is important. Therefore, porous, open-structure soils facilitate sprouting. In soil with good conditions, about 60% of sprouting can be ensured in a reasonable crop.

Tillering

Tillering begins around 40 days after planting and can last up to 120 days. It is a physiological process of continuous underground branching of compact node joints of the primary shoot (Figure 2.8). Tillering gives the crop the necessary number of stalks required for a good production. Several factors, such as variety, light, temperature, soil humidity (irrigation), spacing and fertilization practices influence tillering. Light is the most important factor. Adequate lighting at the base of the plant during this period results in active basal vegetative buds.

Temperature around 30°C is considered ideal for tillering. Temperatures below 20°C slow it down. Tillers formed earlier help to produce thicker and heavier stalks, while those formed later die or remain short or immature. Maximum tiller population is reached

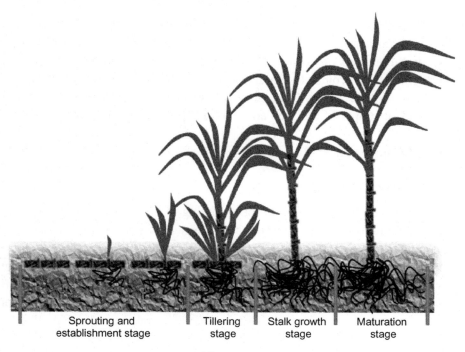

| Sprouting and establishment stage | Tillering stage | Stalk growth stage | Maturation stage |

Figure 2.8
Stages of sugarcane vegetative growth, from planting to maturation.

between 90 and 120 days after planting. At 150–180 days, at least 50% of the shoots die and a stable population is established. While six to eight tillers are produced from a bud, it has been observed that only 1.5–2.0 tillers per bud remain to form sugarcane plants.

Stalk growth stage

The stalk growth stage starts around 120 days after planting and lasts up to 270 days in a 12-month crop. This is the most important phase of the crop, because it is when there is the formation and elongation of the stalk, which results in production. Leaf production is frequent and rapid during this stage. Leaf area index reaches a value between 6 and 7. Under favorable conditions, stalks grow rapidly, almost four to five internodes per month. Irrigation, fertilization, heat, humidity and solar climatic conditions favor the elongation. Temperatures around 30°C and humidity around 80% are most suitable for this stage.

Maturation stage

In a sugarcane crop, the maturation stage lasts for approximately 6 months, starting 270–360 days after planting. The rapid synthesis and accumulation of sugar occurs during this phase, so the vegetative growth is reduced. As maturation progresses, simple sugars (monosaccharide, fructose and glucose) are converted into sugarcane (sucrose, a disaccharide). Sugarcane maturation comes from the bottom up, so the bottom contains more sugar than the upper part. Plenty of sunlight, cool nights and warm days (e.g., greater diurnal variation in temperature) and dry climate is highly conducive for maturation.

2.5 Crop Ecophysiology

Sugarcane grows in the world between latitudes 35°N and 30°S and altitudes from sea level up to 1000 m or a little more. It is regarded as essentially a tropical plant. This is a long-duration crop and therefore lives with all seasons — rainy, winter and summer — during its life cycle. The plant grows better in warm, sunny tropical areas. The "ideal" climate for maximum production of sugar from sugarcane is characterized as:

- long season, warm, with high incidence of sunlight and adequate humidity, and
- fairly dry season, sunny and cool, but with no frost for ripening and cultivation.

A total rainfall of between 1100 and 1500 mm would be appropriate if the distribution is even, abundant during the months of vegetative growth and followed by a drier season, during ripening. During the period of active growth, rainfall favors rapid growth, elongation of the stalk and formation of internodes. However, during the period of ripening, heavy rain is not desirable because it reduces the sucrose content in the juice.

The ideal temperature for sprouting is between 32°C and 38°C. Temperatures above 38°C reduce photosynthesis and increase respiration. For ripening, relatively low temperatures in the range of $12-14°C$ are desirable because they have significant influence over the reduction of vegetative growth and the increase in sucrose content. At high temperatures, an inversion of sucrose into fructose and glucose can occur. Severe cold inhibits bud sprouting and reduces plant growth. At temperatures below $0°C$, the least-protected parts freeze, such as young leaves and lateral buds.

High humidity (80−85%) favors a rapid elongation of the plant during the period of growth. A moderate rate (45−65%) along with a limited supply of water is favorable during the ripening stage.

Sugarcane grows well in areas that receive solar energy from 200 to 1200 $mol/m^2/s$. C_4 plants are capable of better converting the energy in conditions of light saturation or lack of light. High light intensity and long light exposure time promote tillering, while short, cloudy days affect the plant in the opposite way. Stalk growth increases when daylight is in a range between 10 and 14 h.

2.6 Flowering Aspects

Flowering stems from the apical bud. The flower contains a main axis or rachis, which is the uppermost internode of the shoot tip. Secondary axes stem the rachis; tertiary axes stem from secondary axes, reducing branching from the bottom up, giving the appearance of pyramidal inflorescence. The spikelets are located in tertiary branches near the base and secondary branches of the shoot tip, with one flower each.

In the process of inflorescence formation, the period in which the stimulus for the apical meristem alteration occurs, when it no longer produces leaves and stalks, but forms the inflorescence instead, must be detected. This period is difficult to define as it depends on the cultivar, on the climate, and on the changes that occur from year to year. We must determine the months of highest probability for occurrence of that stimulus. In Brazil, the stimulus for meristematic differentiation for flower formation occurs in February, March and April, giving the bloom in April, May and June. The time of stimulation, so that the meristem is altered into flower bud is 18−21 days, depending on the variety.

The process of flowering is divided into four phases: transformation of the apical meristem into flower bud, transformation of this flower bud into inflorescence, development of inflorescence and flag leaf and emission of inflorescence. Soon after induction, the rachis with the branches is developed, followed by secondary branches. At this moment, the flag leaf, which will protect the inflorescence, is formed. The flag leaf is leaf-8. Elongation of the flag leaf sheath and development of the inflorescence characterize the third stage. The leaf sheath is developed to provide room for the inflorescence, as well as to prevent it from breaking, for its tissue is still weak.

Concurrently, the spikelets are developed up to the formation of the complete structure, as well as the maximum development of flag leaf sheath. The opening of flowers and pollination follows inflorescence emission. The complete emission lasts 4−5 weeks, while the opening of flowers, fruit formation and maturation last no more than 2−3 weeks. The main damages to flowering are:

- vegetative growth of stalk is paralyzed;
- the plant comes into senescence, allowing new shoots;
- when the stalk is still in the growth phase, losses are greater;
- storage time on the field is reduced; and
- reserves of sucrose are reduced and reversed into energy, leading to pith formation.

Flowering of sugarcane is avoided either through breeding or plant growth regulators. Therefore, it is necessary to know some basic factors that control the flowering of sugarcane, from its physiology to environmental factors. It is itself quite complex, involving phytochrome, hormones, florigen, nucleic acids and various other factors. Flowering is mainly related to the following factors:

- *Photoperiod* − short days, with 11.5 h of darkness, especially in regions nearer the equator.
- *Temperature* − the higher the latitude, the more important temperature is, because flowering is favored in places with small temperature variation between day and night.
- *Latitude* − the higher the latitude, the lower the probability of flowering. Most studies indicate that above 35° latitude, plants do not flower.
- *Humidity* − under water stress, flowering is generally reduced, whereas humidity slightly above normal leads to early flowering.
- *Nutrients* − nitrogen is directly involved in flowering. High rates of nitrogen alter the carbon/nitrogen relation, which reduces flowering.

The flowering intensity depends directly on environmental conditions and genetic factors on the cultivar. Falling levels of auxin and rising levels of other phytohormones accompany flowering, mainly by the action of light on phytochrome, which responds positively to far red. Changes also occur in the distribution of water and organic and inorganic nutrients, pith-caused reduction in reserves of carbohydrates, and excretion of potassium and nitrogen by the root system. These phenomena lead to decreased crop productivity, in both weight and yield. The reduction in sucrose content reduces Brix, POL and purity, as sucrose, commonly used for the formation of seeds and the emission of lateral sprouting, is broken down into glucose and fructose. The effect is more pronounced until the fourth node and can extend to the eighth top node. The pith formation in flowered canes results in less sugarcane juice, since pith increases the fiber content, making it difficult to extract the juice. Flowering can be prevented if detected early. In advanced stages, it is not reversible, since differentiated apical tissues will already be in place. The physiological process might be reduced if the flowering-favorable conditions are changed. However, it is agronomically and economically viable to apply chemicals. Among the most used chemicals, products based on gibberellins, maleic hydrazide, monuron, diuron, diquat, paraquat and ethylene stand out. Currently, the most efficient product is etephon, which is converted into ethylene. Etephon is more efficient because it acts on the sugarcane during a large floral control period and does not act as a desiccant to the crop canopy, like other products. Additionally, this product reduces the growth of internodes developed at the time of spraying; the internodes formed later resume normal growth. Flowering can also be reduced by the use of cultivars with low potential for flowering or those with high potential for flowering only in higher, colder places. Studies using genetic transformation are promising with regard to this physiological process in sugarcane cultivation (see Chapter 12).

2.7 Aspects of Maturation

Physiologists define the maturation process as the moment of maximum accumulation of photoassimilated products in the reserve organs of the plant — in the case of sugarcane, sucrose in the stalk. The processes of elongation and thickening of the wall cells, marked increase of dry matter, gradual dehydration, increase and retention of accumulated sucrose, decreased elongation of stalk leaves and subsequent release of them, determine this phenomenon. This happens continuously, as there is production from of trioses and hexoses since the expansion of the first leaves up to the flag leaf. During this period, the accumulation of these products is directly proportional to the photosynthetic process and inversely proportional to the consumption of reserves. That is, at certain times, the leaf is no longer source but becomes sink, like other plant tissues. Reserves are accumulated from the bottom up to the shoot tip of the stalk, so that a few months after the formation of the bottom internode, it can contain sucrose concentrations similar to those found in the median internodes at harvest time. The process can be intensified by the reduction of available nutrients or water deficiency. Only immature internodes of green leaves and bottom overripe internodes (with high fiber content) do not retain appreciable amounts of sugar. Each internode accumulates its own sugar, whose values increase toward the center of the stalk and decrease towards the ends. These differences are bigger in younger internodes, probably reflecting a different distribution of invertase. The interim meristem (growth ring), in such conditions, contains much more invertase than the central tissues of the internode. Despite the efforts of breeders, cultivars still have a low content of sucrose in the last apical internodes. This can be partially solved by the use of maturation agents, plant growth regulators, cultural practices such as clipping, control of water and fertilization programs. There are many studies on the maturation processes and the determination of the point of harvest; however, studies that are more detailed are still necessary. These processes are also addressed in Chapter 10.

Bibliography

Buchanan, B.B., Gruissem, W., Jones, R.L., 2000. Biochemistry & molecular biology of plants, first ed. Courier Companies. American Society of Plants Physiologists. United States of America. 1370.

Casagrande, A.A., 1991. Tópicos de morfologia e fisiologia da cana-de-açúcar. FUNEP, Jaboticabal, 157p.

Kerbauy, G.B., 2004. Fisiologia vegetal. Guanabara Koogan, Rio de Janeiro, 452p.

Leningher, A.L., Nelson, D.L., Cox, M.M., 1998. Princípios de bioquímica, third ed. Worth Publishers, New York, 840p.

Magalhães, A.C.N., 1987. Ecofisiologia da cana-de-açúcar; aspectos do metabolismo do carbono na planta. In: Castro, P.R.C., Ferreira, S.O., Yamada, T.Y. (Eds.), Ecofisiologia da produção agrícola. Potafos, Piracicaba, pp. 113—118.

Mccormick, A.J., Cramer, M.D., Watt, D.A., 2006. Sink strength regulates photosynthesis in sugarcane. New Phytol. 171 (4), 759—770.

Rodrigues, J.D., 1995. Fisiologia da cana-de-açúcar. Botucatu.101.

Taiz, L., Zieger, E., 2004. Fisiologia vegetal. Artemed, seventh ed. Porto Alegre. 720p.

CHAPTER 3

Planting

Victor Francisco Araújo de Medeiros Barbosa
Universidade Federal de Viçosa, Viçosa, MG, Brazil

Introduction

Planting is a vital step for the proper development and good crop yield of sugarcane. Sugarcane is a semi-perennial plant and this step is responsible for the longevity of the crop. Any mistake in the planting operation, such as stand failure or spacing error, will cause problems throughout the life of the sugarcane crop, compromising production over the cuts.

3.1 Planning

Planning includes several decisions to be made that will determine the success or failure of a sugarcane crop and influence its longevity. At this stage it is necessary to approach the following issues: production environment, varieties, planting season, spacing and depth of the groove.

3.1.1 Production Environment

"Production environment" is the definition of environments producing sugarcane according to their physical, chemical and morphological characteristics, as well as weather conditions. According to Prado (2005), production environment is the sum of interactions of surface and subsurface features, considering also the degree of slope, associated with climatic conditions. The sugarcane production environment in central-southern Brazil is shown in Table 3.1.

The yield for each environment described in this text is the result of experiments performed in mills participating in the "Cana" Project, of the Agronomic Institute of Campinas.

The definition of production environments takes into consideration such factors as depth, which is directly related to water availability in soil and volume to be explored by the roots; fertility, i.e., the source of nutrients for plants; texture, related to cation exchange and quantity of organic matter; and water availability, responsible for the soluble portion present in the soil.

Sugarcane. DOI: http://dx.doi.org/10.1016/B978-0-12-802239-9.00003-7
35

Table 3.1: Sugarcane production environment in central-southern Brazil.

Environments	Productivity (TCH)	Soil Features	Soil Symbols (Embrapa, 1999; Prado, 2004)
A1	>100	ADA, e, ef, m, CTC average/high	PVAe[2], PVe[2], LVef, LVe, LVAe, CXe, NVef, NVe, MT*, MX*, GMe, GXe, GMm, GXm
A2	96−100	ADM, e, ef, CTC, average/high	PVAe[2], PVe[2], PAe[2], LVef, LVe, LVAe, CXe, NVef
B1	92−96	ADA, m, mf, CTC average/high	PVAm[2], PVm[2], Pam[2], LVmf, LVm, LVAm, LAm, CXm
		ADM, mf, m, ma, CTC average/high	NVmf, NVm, PV[1] Ama*
		ADB, ef, e, CTC average/high	LVef, LVe, LVAe, NVef, NVe, PVAe[3], PVe[3]
B2	88−92	ADM, m, mf, CTC average/low	PVAm[2], PVm[2], Pam[2], LVmf, LVm, LVAm, LAm, CXm
		ADA, ma, CTC average/low	GMma, GXma
C1	84−88	ADM, d, CTC average/high	PVA[2], PVd[2], Pad[2]
		ADM, ma, CTC average/high	LV Ama*, Lama*
		ADM, d, df, CTC average/high	LVd, LVdf, LVAd, Lad
C2	80−84	ADB, e, CTC average/low	LVe, LVAe, LAe
		ADMB, ef, CTC average/low	LVef
D1	76−80	ADB, w, wf, CTC average/high	LVw, LVwf, LVAw, LAw
		ADM, a CTC average high	PVAa[2]*, Pva[2]*, PAa[2]*
D2	72−76	ADB, ma, CTC average/high	LVma, LVAma, Lama
		ADB, e, CTC high, A chernozemico	RLe
E1	68−72	ADB, a, CTC average/low	PVAv[3], PVa[3],
		ADB, ma, CTC medium/low	PVAma[4], PVma[4], PAma[4]

(*Continued*)

Table 3.1: (Continued)

Environments	Productivity (TCH)	Soil Features	Soil Symbols (Embrapa, 1999; Prado, 2004)
E2	<68	ADMB, wf, w, a, CTC average/high	LVw, LVwf, LVAw, Law, LVa, LVAa, LAa
		ADMB, a, d, CTC medium/low	PVAa[4], PVa[4], PVAa[4], RQa, RQd
		ADMB, e, m, d, ma, a	RLe, RLm, RLd, RLma, RLa, PVAe[4]

ADA: high water availability; ADM: medium water availability; ADB: low water availability; ADMB: very low water availability; LV: Red Latosol; LVA: Red–Yellow Latosol; LA: Yellow Latosol; PVA: Red–Yellow Ultisol; PV: Red Ultisol; PA: Yellow Ultisol; NV: Red Nitosol; MT: Argiluvic Chemosol; MX: Haplic Chemosol; CX: Haplic Cambisol; RQ: Quartz–Sand Neosol; RL: Litholic Neosol; GX: Haplic Gleysol; GM: Melanic Gleysol. ef: eutroferric; e: eutrophic; mf: mesotrophic; df: dystroferric; d: dystrophic; wf: acriferric; w- acric; ma: mesoalic; a: alic.
[1] B horizon up to 20 cm depth from surface.
[2] B horizon at 20–60 cm from surface.
[3] B horizon at 60–100 cm from surface.
[4] B horizon at more than 100 cm from surface.
*Mottled or variegated B horizon.
Source: Prado (2005); Agronomic Information booklet #110 June (2005).

3.1.2 Varieties

Choosing the appropriate variety is essential for obtaining raw material with appropriate maturation during the ripening season, besides guaranteeing sprouting of stumps at different times. The choice of variety should take into account many variables, such as adaptability to mechanical harvesting, early ripening, sprouting of stumps, yield (ton/hectare, kg ATR/ha), industrial characteristics (Pol, Brix and fiber content) and suitability for production environments.

In Brazil, sugarcane breeding is carried out by the following institutions: IAC – Agronomic Institute of Campinas, which produces the IAC varieties; RIDESA – Inter-University Network for Development of the Sugarcane Sector, with RB varieties; CTC – Sugarcane Technology Center, with the CTC varieties and the SP old ones; and CANAVIALIS, with the CV varieties.

3.1.3 Planting Season

Sugarcane, in central-southern Brazil, can be planted throughout the year, but there are some restrictions related to water availability and characteristics of the variety on maturation and phenological cycle.

The main periods of planting and the names that crop receives according to this period are:

From September to early December, when sugarcane is known as "one-year sugarcane" and productivity features below 100 t/ha. This crop will be harvested as early as next season, usually at the end of it.

From January to March or April, known as "one-and-a-half-year sugarcane", with yields above 120 t/ha. In this planting season, the crop grows for a longer period, around 16–18 months, which explains its denomination. Due to the long growing season, there is no production during one season. It is the period most commonly used for planting, especially by suppliers.

From May to August — usually the dry season in the central-southern region — known as "winter sugarcane". It usually yields above 100 t/ha, but the crop requires irrigation in this period, due to low water availability in the soil. Planting in this season has advantages such as productivity and also the fact of not spending a year with no harvest, like the one-and-a-half-year sugarcane. Winter sugarcane yields approximately the same as one-and-a-half-year sugarcane due to the fact that the plant vegetates for about 14–15 months.

The one-year sugarcane presents the lowest average productivity, primarily due to the fact that the plant reaches the dry season with many stalks already formed, which leads to the formation of short internodes. The one-and-a-half-year and the winter sugarcane, on the other hand, do not suffer this influence, because the one-and-a-half-year sugarcane will reach the dry season only with leaves, or few stalks formed, and the winter sugarcane, planted in dry season, will be able to be harvested in the next dry season, reaching maturation due water stress.

3.1.4 Spacing

Spacing is critical to the plantation, since it allows better use of space, besides the optimization of planting, treatment and harvesting operations. Proper spacing contributes to increased production, because it interferes favorably in the availability of resources like light, water and temperature — determining variables for an increased production. Spacing of the planting should vary according to the production environment, terrain and characteristics of the variety to be installed.

The spacing between rows may vary from 0.9 to 1.8 m. There are simple and combined spacings. In sandy soils and poorer environments, reduced spacing between rows is recommended (0.9 or 1.2 m) in order to reduce the competition with weeds. Sugarcane covers soil more quickly in these cases. In areas where harvesting will be done with machines, it is advisable to use a spacing of 1.5 m to avoid treading on the stumps by the harvester. Areas with buried drip irrigation prefer combined spacing.

The main features of each spacing are:

Simple: Same distance between planting furrows throughout the planted area.
Combined: Alternates bands of different spacings within the same range, allowing better traffic condition of machinery in the area. In this type of spacing, used primarily in

areas with buried drip irrigation, two rows spaced 0.4 or 0.3 m, with widths of 1.4 and 1.5 m, respectively, making up a total of 1.8 m. Such spacing allows the mechanized harvest without treading on the ratoon.

Broad-based furrow: The plow makes a hole with the wide base, which allows the provision of two seedling clusters side by side instead of overlapping. It decreases competition between tillers. In this case, you can either increase the spacing between furrows (up to 1.8 m) or keep it lower (1.50 m).

3.1.5 Depth of Furrow

Depth of furrow is variable, but the ideal is around 30 cm, with an overlap of 5−8 cm of soil over the seedling at the bottom of the furrow. This variation is dependent on soil type and also time of planting.

3.2 Soil Preparation

Soil preparation is the next step after planning. It is very important for the entire crop cycle, since all subsequent operations depend on it, from planting to harvest. Due to the wide diversity of soils and management policies in sugarcane producing areas, there are many variations in soil preparation, always trying to better match the local reality.

3.2.1 Eradication of Previous Crop

According to the previous crop, eradication can be performed chemically, with the use of contact or systemic desiccants; mechanically, with the use of disk harrow; or by using these two techniques simultaneously.

The choice for one of the techniques mentioned should take into account factors such as presence of residual tree roots, stumps, rocks and so on. The best procedure is to remove these elements to take advantage of the area and avoid problems with broken equipment in future operations.

3.2.2 Infrastructure of the Ranges

Ranges are considered the basic operational unit for sugarcane crop. Their area and geometric shape vary according to soil type, local topography, regularity of the land, existing roads and property boundaries, as well as other operational characteristics. Generally, they have areas with maximum of 20 ha, although this value may vary from region to region.

3.2.3 Range Sizing

Range sizing and geometric shape depend on several variables, such as production environment, variety, topography, cultivation characteristics and harvest time.

The length of furrowing lines should be appropriate to the capacity of the equipment to be used, avoiding maneuvers in the middle of the cultivated area, especially during planting and harvesting, when traffic of heavy machinery in the area is more intense. It results in a significant operational yield.

Taking as a basis for calculating the appropriate length of the furrows and an average crop yield of 95 t/ha, the minimum distance should be 400 m and the maximum should be 700 m. That would allow the deployment of crossing dirt roads for transshipments. This is the maximum possible size for use with irrigation reels, as the hoses are 300 m long and the sprinkler coverage is 50 m.

Harvesting is the operation that will define the parameter distance between crossing dirt roads, since this is the most demanding operation on traffic of heavy vehicles in the field. Furthermore, with the increase of mechanized planting, planters available in the market have load capacity from 6 to 7 t of seedlings, which allows an autonomy similar to the transshipment of 5–6 t, used in mechanical harvesting.

3.2.4 Dirt Roads

The primary role of dust roads is for the transit of vehicles and supplies to feed the crop and to facilitate transportation to the industrial plant. Dirt roads work as points of support for major operations such as planting and harvesting. Transit of heavy vehicles is expected here, in order to avoid traffic in the cultivated area. Their width varies from 3 to 10 m, according to their importance in the logistics of the production system.

First, we must draw the path of trucks for the removal of raw materials and transportation of inputs. These paths may have bandwidth of 5 m. On roads where truck crossing occurs, we adopt a 10-m bandwidth. Dirt roads lying across furrowing lines, with light traffic, can be 3 m wide. Yet, entry and exit points of furrowing lines, near fences and other obstacles, should have dirt roads of 5 m wide, to facilitate equipment maneuvering.

3.2.5 Area Leveling and Systematization

"The different conformations of the land affect the quality of work and yield of equipments used in sugarcane crops. In addition to terrain slope, concavity and ripple should be considered. When its negative effects increase too much the cost of production, leveling or systematization techniques should take place."

Winter Sugarcane *(Storino et al., 2008).*

Area leveling and systematization can be very costly, depending on the volume of earth being moved, but the benefits for crop development and implementation of future operations can outweigh the costs. At planting and mechanical harvesting, leveling and systematization of the field are essential for better development of these technologies.

There are various equipments available on the domestic market to perform this operation. One of the simplest is the planer, which has different shapes and sizes. There is also the scraper, which acts as a blade, with a compartment like a bucket; it accumulates the surplus earth to be used in ground depressions. Despite having smaller capacity, the scraper's blades have greater agility.

3.2.6 Terracing

The construction of terraces aims to preserve the soil from the action of rainwater, facilitating its infiltration (in the case of level or infiltration terraces); or to direct rain water into a natural drain (in the case of slope terraces).

Level terraces are built on the same level (altitude) and have the function to infiltrate all the water that comes into their covering area. Their height is usually set according to the equipment used and the distance between the terraces varies with soil type, terrain slope, crop management and rainfall in the region.

Generally, for sugarcane, built-in level terraces are the most commonly used. They are constructed so that their gutter has a triangular shape, leaving the embankment that separates the gutter from the ridge almost vertical (Storino et al., 2008).

Graders and tractors equipped with blades can make the built-in terrace, apart from the terracer, which offers an operating yield of 4.6 ha/h. The header of the curves in these terraces is extremely important, so that rainwater in roads and dirt roads can be directed to the terraces, preventing erosion of the roads.

3.2.7 Conventional Tillage

Generally, conventional tillage aims at reversing and stirring a deep layer of soil; incorporating and destroying plant debris; exposing soil pests to sunshine for control; lump breaking and ground leveling.

This preparation is composed primarily of harrowing for removing the residues of previous crop. It is done still in the dry season, after subsoiling, to break up the compacted layer and could be replaced by chiseling when the compacted layer is more shallow. With a harrow or a moldboard plow, ground is turned over, burying the vegetable remains to an average depth of 15−30 cm. Together with the first plowing and harrowing, we apply fertilizers such as lime and phosphate, and pesticides. After these steps, we promote lump breaking and ground leveling with harrows. This harrowing is also used to complete the application of lime or phosphate.

3.2.8 Reduced Tillage

Reduced tillage is a conservationist technique that aims to reduce the number of operations carried out under conventional tillage and to use lighter equipment. There should be no severe impediments in the soil being prepared, neither physical impediments, such as compaction; nor chemical, such as lime and phosphate deficiency; nor even biological, such as soil pests.

The decrease in traffic aims to reduce the density and compaction of the ground and therefore do not create conditions that require conventional tillage.

3.2.9 No Tillage

Widely used in production of grains and cereals, the technique of no tillage has gained importance in sugarcane crops, especially in areas with crop rotation with soybeans and other legumes.

The furrow openers, and even the planters equipped with straw cutters at the front of the plows, will enable direct seeding even with a high amount of plant mass on the soil. For this technology, it is necessary to verify soil compaction, because there are cases where subsoiling is held to break the compacted layer and mechanical planting is done on the existing straw, without turning the soil.

3.2.10 Seedlings

Sugarcane is a semi-perennial plant — its planting is done only every 5 years or more. Therefore, the quality of the seedlings is crucial. For longilign planting, besides being appropriate to that specific environment, seedlings should be free of pests and diseases. Good quality of seedlings is the factor of production of lowest cost, but delivers higher economic returns to farmers, especially when the seedlings are produced by the farmers themselves.

For the production of seedlings, it is necessary that the basic material be of good origin, aged 9—12 months, healthy, from cane plant and first ratoon, and which has been subjected to heat treatment.

The technology used in seedling production is practically the same as that used in commercial farming, with the introduction of some phytosanitary measures, which are:

> Fumigation of the knife — when the knife used to harvest seedlings and cut them in pieces is infected, it is a major propagator of scald disease and rickets. Before and during

these operations, the knife should be disinfected with alcohol, formaldehyde, Lysol, cresol or fire. A practical, efficient and economical disinfestation is the immersion of the instrument in a 10% cresol solution (18 l of water + 2 l of cresol) for 30 min before starting the harvest of seedlings and cutting them into seedpieces. During these two operations, the knife must be plunged into this solution rapidly and several times.

Health surveillance and "roguing" — after the nursery is formed, it is essential to carry out frequent health inspections, at least once a month. The purpose of these inspections is to eradicate all clumps with pathological symptoms or different characteristics to the cultivated variety.

Crop rotation — during the renewal of the sugarcane crop, when the soil remains idle, short-cycle crops should be planted in rotation with sugarcane. Peanuts and soybeans are the most suitable.

Heat treatment — this involves subjecting the stalks to a temperature of 50.5°C for 2 h. Procedures to control rickets in the sugarcane ratoon and can be performed on seedpieces or isolated buds.

In addition to these measures, some agronomic recommendations should be taken into account to prevent disease in the sugarcane crop, such as manual straw removal of seedlings, lower density of seedlings in the furrow and the largest division of the nitrogen fertilizer.

3.3 Planting

Sugarcane planting, since the arrival of the crop in Brazil, during the colonial period, demands many workers. With the development of agriculture, equipments were made available to facilitate this operation. Initially, animal power was used to pull the first furrowing equipment, in the 18th and 19th centuries, until the mechanization of agriculture in the 20th century. Even today, sugarcane cultivation is less automated, especially at the planting stage, than the other major world crops. This operation, however, has made dramatic and rapid changes in recent years. Private companies and research centers have developed sophisticated equipment to facilitate planting, improve performance and increase demand in qualification of manpower to be employed.

3.3.1 Manual Planting

"Manual" planting is technically considered to be semi-mechanized because some of the operations are mechanized (such as opening up planting furrows (furrowing) and closing these grooves) and other mechanisms are manual (such as placing seedlings in the furrow, seedling cutting and reducing the size of seedling clusters). In some situations, for example in areas too steep or less mechanized crops, the opening and closing of furrows can be performed manually, but with low yields.

Furrowing

The operation of furrow opening is usually mechanized, according to spacing and depth, as defined in the planning. You can apply fertilizers along with this operation. Current furrowers have the option with furrow markers and stool markers to facilitate and increase operating yield.

The furrow marker helps the tractor driver, marking the spot where the tractor will be back furrowing the lines. The stool marker marks the lines to be left without furrowing to facilitate the transit of the truck with seedlings. These lines will be furrowed after the distribution of seedlings to the closure of the area.

There are numerous variations to this operation, depending on existing equipment and the number of employees participating in the front-rows. The number of lines to be furrowed for the seedlings to leave the stool can vary from six to eight lines grooved for two or three lines of seedlings in the stool (depending on the plow, if it is two lines, you may furrow six or eight lines and leave two lines of seedlings in the stool). Three-line furrowers have no stool marker option because the stool marker arm would be very large, making it difficult to work in steep areas with obstacles such as trees, fences and so on, and would also endanger the integrity of the tractor, when working near power lines. For the case of furrow openers of three lines, a small tractor should be used, to mark the grooves and the stools before the entry of the plow in the area.

Cutting, loading and transportation of seedlings

Cutting seedlings should be done close to the ground and close to the top, disregarding the leaves of the top. Loading should facilitate future distribution of seedlings in the furrow. Generally, a mechanical loader is used to put them in the truck that will transport them to the planting area.

Distribution of seedlings in the groove

The planting, i.e., the distribution and arrangement of seedlings in the furrow and cutting them into three-bud seedpieces (around 50 cm, or the size of the cutter knife), is performed in sync with the distributor truck or trailer and relies heavily on the number of people employed. Planting teams may contain between 12 and 16 people divided into feeders, cutters and planters or distributors. Feeders throw the seedlings into the furrows, planters or distributors distribute them and cutters cut them into three-bud seedpieces. The distribution of people in these three positions varies with the number of rows to be taken and the operational mode of each site.

"Covering"

After the distribution of seedlings in the furrow, they should be covered as quickly as possible, to reduce water loss from the exposed walls of the furrow and, more importantly, to prevent drying of seedpieces due to sun exposure. It is an operation that is usually mechanized. There are two or three line covers. It is important to use a cover matching the number of lines of the plow due to the parallelism of the lines.

3.3.2 Mechanized Planting

Machine planting is a fairly recent practice. The first national machinery, which were actually prototypes, were developed between 1964 and 1978 by Santal and Motocana. Initially, they were not accepted by the market due to a lack of need at the time, doubts about the germination of seedlings when planted by these machines (Stofel et al., 1984, cited by Ripoli et al., 2006), and the lower operating yield presented (3−5 ha/day) (Ripoli, 1978). The first system with a chopped raw cane harvester and seedpiece planter was presented in 1989, ushering in the new generation of planters in Brazil. Over the years there have been improvements to and assessments of the equipment tested in 1989, made by Copersucar. In 1991, the CTC (Sugarcane Technology Center) conducted feasibility studies to construct the planter in Brazil, based on a one-line Bonel planter. The CTC also evaluated other planters and made several literature reviews to guide the design of the machine in Brazil, which followed the Australian model.

Between 1993 and 1996, the CTC team tested in four cooperative mills (Santa Luiza, Usina da Pedra, São Francisco and Barra Grande) with an imported one-line Bonel planter. Aiming to create a seedpiece distributor adapted to Brazilian needs, Copersucar designed, built and tested, in Barra Grande, São João and Santa Adelia mills, between 1995 and 1996, the seedpiece distributor. With these studies, in 1997, a two-line Brazilian prototype was built and tested: the two-lines Copersucar planter used in the São João mill.

At the same time, the designer Roberto Marchini, at the request of the Itamarati mill, finalized the development of a two-line seedpiece planter: the Itamarati 1. The machine was the result of observations made on eight planters imported by Itamarati: a one-line Austoft, a two-line Austoft, three two-line Cameco and three two-line Class. The final tests of the Copersucar two-line prototype, already approved for marketing, occurred in 1998 and 1999, in São João, Santa Luiza and Bela Vista mills. During this period, it was also evaluated a planter imported from Australia, the Austoft A.

The monitoring of this machine was conducted at São Martinho mill. Between 2000 and 2001, São Martinho, in partnership with the Copersucar Center for Technology, designed, built and tested a prototype planter USM-CTC. In 2002 this planter was improved and multi-functionality for filtercake application was deployed. With the results two more units of filtercake-applicator planter were built between 2002 and 2004.

Machine planting presents a number of advantages, which result in greater control of the implementation of various phases of this operation. The opening of the furrow is concomitant with the distribution of fertilizer in depth, the placement of the seedling and its covering, thus reducing water loss by desiccation of the furrow and volatilization losses of nitrogen fertilizer. Vehicles transporting seedlings do not pass on the furrows. This reduces the demand for machines, since you do not need perform the subsequent covering, because, as stated earlier, it is held concurrently with the planting. Finally, there is a large reduction in labor for planting. This method, however, presents some restrictions, such as the fact that seedlings have to be harvested with machines, thus increasing the number of damaged buds. Moreover, the homogeneous distribution of seedlings depends a lot on the planter operator. An incorrect feeding of the planter can cause failures that will only be seen during sprouting. This explains the growing importance of operation assessment and monitoring. As a consequence, it requires more responsibility from the operators (Bernardes et al., 2001).

Mechanized planting of entire cane

In mechanized planting of entire cane, seedlings are harvested by hand and carried by mechanized loaders in wagons that are engaged in an implement that furrows, distributes fertilizers, cuts the sugarcane in clusters of approximately 50 cm, distributes them in the furrow, applies pesticides in the clusters and covers them. Less labor is needed, but the yield is less than 0.5 ha/h. Generally, it takes a tractor with minimum power output of 140 hp. This device is interesting for small and medium producers who face difficulties in getting labor for planting and have the structure of machinery (tractor, loader and truck, if necessary). Two to four people are required to feed the chopper.

Mechanized planting of chopped sugarcane

Today, the Brazilian market has six companies selling chopped sugarcane planters: DMB, SANTAL, TRACAN (Case), SERMAG, CIVEMASA and SOLLUS. The operating yield has been improved (0.8−1.2 ha/h) along with the quality of the seedlings mechanically harvested, which shows the development of this technology. In this planting method, suitable for mills and large producers, it is necessary to adapt a sugarcane harvester to cut seedlings.

In mechanical harvesting of seedlings, in order to prevent damaged buds, a number of factors must be considered. Harversters need to be adapted and seedlings should preferably be close to 9 months old, to avoid cracks in the cuttings due to their high-fiber content, ensuring the health of the sugarcanes and avoiding tipping the range to be used as seedlings.

A seedling harvester should undergo some changes so that the seedpieces are not adversely impacted, thus preventing damage to buds and making them unviable. The adaptation of the harvester for seedlings is the installation of a seedling-collector kit. These changes can be seen in Figures 3.1 and 3.2.

Figure 3.1
Removal of "lollipops" and external rubber coating at the entrance of the harvester.
Source: Donizelli (2005).

Figure 3.2
Lifting roller with rubber crack. *Source: Donizelli (2005).*

As shown in Figure 3.1, the first step in the modification of the seedling harvester is the removal of the both external divider (lollipops) and tipping roll, besides the installation of rubber flat plate at the entrance of the machine.

In Figures 3.3 and 3.4, cracks of the lifting roller are replaced rubber cracks and feeding and transport rollers are covered with rubber.

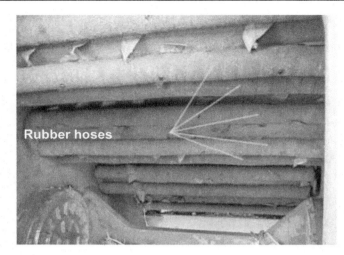

Figure 3.3
Feeding rollers with steel cracks are replaced by rubber hoses. *Source: Donizelli (2005).*

Figure 3.4
Synchronous roller only composed of a knife and the counterknife. *Source: Donizelli (2005).*

The blade set of the synchronous roller should also be replaced by a knife and counterknife, to increase the size of the cut seedpieces.

Placement of flat plates on the lift and rubber coating of the primary extractor also help in reducing injuries caused to the buds (Figures 3.5 and 3.6).

Figure 3.5
Lift floor consisting of flat plates instead of perforated ones. *Source: Donizelli (2005).*

Figure 3.6
Bulge of the primary extractor coated with rubber. *Source: Donizelli (2005).*

We highlight the need to always maintain the sharpness of the base cutter knives and chopping rollers to avoid shattering the seedpieces. The work speed is also slower than the normal harvesting speed, and might not exceed 2.5 km/h.

Figure 3.7 illustrates where the main losses occur during harvesting and the result of changes made with the seedling-collector kit suggested by the CTC.

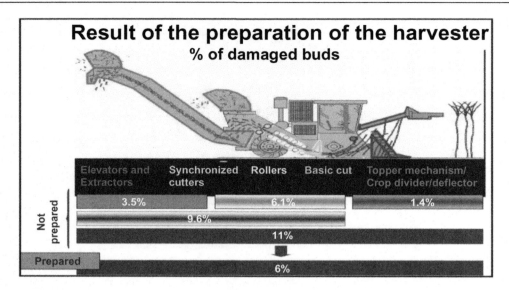

Figure 3.7
Harm reduction to buds with the modifications of the harvester. *Source: Donizelli (2005).*

This planting method requires a large structure, with a seedling-collector harvester, transfers to transport the seedlings from the harvester to the planter, a planter and also a truck-repair shop. There are two versions of planters, automotive (the SERMAG Tropicana) and tractorized (Civemasa PACC2L, DMB PCP6000, SANTAL PCP2, TRACAN PTX 7000, SERMAG SMI 10000 and SOLLUS). All tractorized planters are pulled by tractors with power exceeding 180 hp.

Mechanized planting of isolated buds

Mechanical planting of isolated buds is still a novelty in the sector and should be used in the coming seasons. Like any new technology, there are still many uncertainties about its success, especially in terms of production yield regarding bud processing. Since sugarcane is perishable, buds will have to be prepared shortly before use, which requires precise logistics. Some mills and also seedling producers already use this planting method in nurseries, manually, to the thermal treatment of seedling clusters, in order to control diseases. Yet, its large-scale use is still not usual.

The use of isolated buds can facilitate planting, as it allows better distribution of buds and reduces the number of seedlings to be used in planting. This could reduce the cost of planting, which is, along with harvesting, the most costly operation in the process.

Regardless of planting method, manual or mechanized, rigorous assessments must be carried out to diagnose failures in operation. There are several assessment methods to test

Table 3.2: Quality assessment of sugarcane planting by checking the failures in the planting.

Leaf % > 0.5	Evaluation of Planting
00—10	Excellent*
11—20	Good**
21—35	Average
36—50	Bad
>50	Very bad

*For 15 buds per meter and optimal conditions for sprouting.
**Standard — the most commonly found type.
Source: Stolf (1986).

failures. Stolf (1986) proposed a methodology that classifies every point with a distance exceeding 50 cm of the planting row without tillers as a failure. Stolf also proposed an assessment of these failures, as shown in Table 3.2.

For this analysis, failures are measured every 100 m. These distances are summed to give the percentage of failures. For example, in 100 m, we found five failures with the following measures: 50, 150, 75, 66, 274 cm, totaling 615 cm or 6.5 m of failure. According to the table proposed by Stolf, this plantation, with a 6.5% failure rate, is excellent.

Bibliography

Bernardes, M.S., Teramoto, E.R., Câmara, G.M.S., 2002. Planejamento estratégico da produção de cana-de-açúcar, Fazenda Abadia — Campos dos Goytacazes/RJ. Boletim Técnico — Escola Superior de Agricultura "Luiz de Queiroz. Universidade de São Paulo, Piracicaba, 131f.

Donizelli, J.L., 2005. Avaliação do plantio mecânico com a plantadora DMB PCP 6000. Relatório do Sistema de Mecanização do Plantio — CTC, Piracicaba-SP.

Prado, H. 2005. Encarte de Informações Agronômicas, n110.

Ripoli, T.C.C.; Cunali Ripoli, M.L.; Casagrandi, D.V.; Ide, B.Y. 2006. Plantio de cana-de-açúcar: estado da arte, first ed. Piracicaba: T.C.C. Ripoli. 216p.

Stolf, R. 1986. Metodologia de avaliação de falhas nas linhas de cana-de-açúcar. STAB 22—36, Jul/Ago.

Storino, M., Pecche, A., Kurachi, A.H., 2008. In: Denardo-Miranda, L.L., Vasconcelos, A.C.M., Landell, M.G.A (Eds.), Aspectos operacionais do preparo do solo. Cana-de-açúcar. Instituto Agronômico, Campinas, 882p.

CHAPTER 4

Nutrition and Fertilization

**Godofredo Cesar Vitti[1], Pedro Henrique de Cerqueira Luz[2] and
Wellington Sacco Altran[3]**
[1]*Universidade de São Paulo, Piracicaba — SP — ESALQ — USP* [2]*Universidade de São Paulo,
Pirassununga — SP — FZEA — USP* [3]*Raízen — Unidade Gasa, Andradina — SP, Brazil*

Introduction

The planning of activities in sugarcane crops, from planting to harvesting, is paramount in order to get high yields for compatible costs, without harming the environment. Such planning should encompass the assessment of all production components, taking into account several techniques, including the so-called yield factors, i.e., pest, disease and weed control, using varieties adapted to the edafoclimatic conditions, checking planting times, the soil's chemical, physical and biological management, and using state of the art technology regarding inputs, machinery, tools and services linked to the drafting of the physical—financial schedule. In this scenario, in order to achieve mineral nutrition of sugarcane by efficiently using fertilizers, five questions should be answered: What? How much? When? How? Is it worth it? This is the quickest, most cost-effective way to enhance the yield, and such increases are as big as the adjustments carried out regarding the productivity factors (Vitti and Mazza, 2002).

4.1 Considerations on Fertilization

Generally, fertilization may be defined by the crop's (plant) nutritional needs minus the nutrients provided by the soil times the result by a fertilizer efficiency factor (f), i.e., the portion of fertilizer that is effectively absorbed by the roots and converted into plant dry matter.

$$\text{FERTILIZATION} = (\text{PLANT} - \text{SOIL}) \times \text{f} \qquad (1)$$

To determine the scope of this equation's (plant) first parameter, basic plant nutrition knowledge is necessary, i.e., knowledge of the nutritional demands regarding a certain yield level, which can be ascertained by asking and answering the following questions:

1. What should be applied? (Nutrients to be actually applied aiming at the plant's nutrition.)
2. How much should be applied? (Amount necessary for a certain yield level.)

3. When should it be applied? (Crop's greater demand period and nutrient's dynamic once in the soil.)

4. How should it be applied? (Nutrients can be applied through the soil, in the case of cane-plant, prior to sowing the full area when using the throwing method or furrow and coverage and, in the case of potassium, while ratooning, after it has been cut. In addition to soil application, nutrients may also be applied through the foliage, through the stems or through herbicides).

In order to determine the scope of the second parameter (soil), fertility assessment (stock of nutrients in the soil) may be carried out using the following techniques:

- visual diagnosis (aspect of the shoot and root system);
- leaf diagnosis (mineral analysis of the nutrient level in the foliage);
- chemical analysis of the soil; or
- juice analysis.

Factor (f) results from the "competition" between the system (soil−plant−atmosphere) and the plant (crop) for the applied fertilizer.

Translated as losses, such competition may occur due to:

- **Erosion**: a dragging process that happens to all nutrients;
- **Lixiviation**: percolation of nutrients within the soil profile, mainly occurring on anions (NO_3^-, H_3BO_3 and $SO_4^=$) and exchangeable cations (K^+, $NH4^+$, Mg^{++} and Ca^{++});
- **Fixation**: adsorption or precipitation of anion ($H_2PO_4^-$) and of metallic cations (Zn^{++}, Cu^{++}, Fe^{++} and Mn^{++}); and
- **Volatilization**: loss of the nutrient applied through the soil to the atmosphere. It could be:
  - volatilization of the ammonia (NH_3) contained in urea, particularly when the fertilizer is applied on a surface, over sugarcane trash;
  - volatilization caused by the burning of nitrogen (N_2, N_2O), sulfur (SO_2) and boron (H_3BO_3) in the trash; or
  - biological denitrification of NO_3^-, caused by the application of ammonium nitrate over the trash when under excessive humidity conditions (reduction); resulting in volatile forms of N (N_2, NO and N_2O).

The average percentage of macronutrient usage in a conventional system is estimated according to such losses as shown in Table 4.1. When analyzing the data in Table 4.1, it

Table 4.1: Primary macronutrient plant absorption efficiency in a conventional planting system.

Nutrients	Usage (%)	Factor (f)
N	50−60	2.0
P_2O_5	20−30	3.0−5.0
K_2O	70	1.5

can be observed that efficient fertilization starts by the reduction of factor (f), i.e., by a greater usage of the nutrient through the following practices:

- direct planting, minimum tillage and raw sugarcane harvesting;
- conservationist practices;
- soil correction observing the variable rate (GPS);
- corrective practices (lime, gypsum and phosphate application);
- green fertilization in reform areas with rattle pods, soy or peanut; and
- organic fertilization using byproducts of the sugar—ethanol industry (vinasse, filtercake and ashes), animal manure (birds, pigs and cows) and food residue.

4.2 Sugarcane Mineral Nutrition

As previously mentioned, in order to determine the scope of Eq. (1) (plant) first parameter, it is necessary to ask four questions: what, how much, when and how should it be applied?

4.2.1 What Should Be Applied?

Among all nutrients, it is clear that macronutrients (N, P, K, Ca, Mg and S) must be applied. As for micronutrients, those presenting a greater probability of rendering a better yield as observed by Vale et al. (2008) are: boron (B), zinc (Zn) and copper (Cu), in descending order, and in certain cases manganese (Mn), particularly in areas located in the Brazilian Northeast, and molybdenum (Mo).

4.2.2 How Much Should Be Applied?

In order to determine the amount of nutrients to be applied, it is necessary to know how much is extracted by the crop for a certain yield level, the recycling of these nutrients by organic matter mineralization, the harvesting system (raw or burned) and the stock of nutrients in the soil.

Regarding the nutrients extracted and exported by the crop, Orlando Filho (1993) estimated the amounts of macro- (Table 4.2) and micronutrients (Table 4.3) for a 100-t stalk yield.

Table 4.2: Extraction and export of macronutrients for 100-t stalk yield.

Plant Parts	N	P	K	Ca	Mg	S
	(kg/100 t)					
Stalks	83	11	78	47	33	26
Foliage	60	8	96	40	16	18
Total	143	19	174	87	49	44

Source: Orlando Filho (1993).

Table 4.3: Extraction and export of é micronutrients for 100-t stalk yield.

Plant Parts	B	Cu	Fe	Mn	Zn
	(g/100 t)				
Stalks	149	234	1.393	1.052	369
Foliage	86	105	5.525	1.420	223
Total	**235**	**339**	**6.918**	**2.472**	**592**

Source: Orlando Filho (1993).

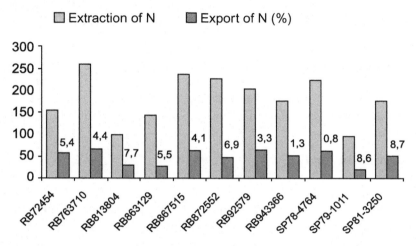

Figure 4.1
Extraction (kg/ha) and export (%) of N according to different sugarcane varieties.
Source: Oliveira (2008).

When analyzing the data in Table 4.2, it is important to observe that phosphorus and potassium, expressed by P and K, correspond to a total extraction of, respectively, 43 and 210 kg/ha of P_2O_5 and K_2O, and it should also be noted that sugarcane requires relatively larger amounts of sulfur (S) in comparison to phosphorus (P). Recently Oliveira (2008) presented the extraction of micronutrients for 11 sugarcane varieties as presented in Figures 4.1−4.3. In these figures, as quoted in Rosseto et al. (2008) extraction data may be regarded in terms of kg/ha concerning nutrients and in terms of percentage for exports. N, P and K exports varied as described below:

N: 18 and 64%;
P: 40 and 70%;
K: 38 and 83% of the total absorbed, as well as significant differences in the varieties regarding nutrient extraction.

In addition to the nutritional demands presented on Tables 4.2 and 4.3 and Figures 4.1−4.3, it is important to break down the fertilization necessary for cane-plant and sugarcane

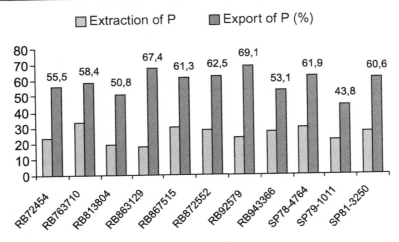

Figure 4.2
Extraction (kg/ha) and export (%) of P according to different sugarcane varieties.
Source: Oliveira (2008).

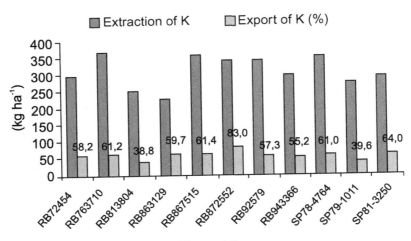

Figure 4.3
Extraction (kg/ha) and export (%) of K according to different sugarcane varieties.
Source: Oliveira (2008).

ratoon. When fertilizing cane-plant, smaller doses of nitrogen and higher quantities of phosphorus and potassium are used. The smaller quantities of N are due to the contribution given by the biological fixation of the N_2 found in the air to the stem, as well as to the mineralization of organic matter in the soil. Conversely, for sugarcane ratoon, due to the renovation of the root system, and in the absence of reserve in the stem, it is necessary to fertilize with high doses of nitrogen, as well as with high doses of potassium while not using or using phosphorus doses. As for N and K_2O, it is important to take the harvesting system into consideration, i.e., for sugarcane harvested with burning 1.0 kg/t of harvested cane should be used and between 1.3 (K soil > 1.5 mmol$_c$/dm^3) and 1.5 (K solo < 1.5 mmol$_c$/dm^3) kg/t of harvested cane of H$_2$O. In a system where the sugarcane is harvested while raw, the recommendation is to use 1.3 kg/t of harvested cane and between 0.8 and 1.0 kg of K$_2$O as will be explained below.

4.2.3 When Should it Be Applied?

For cane-plant, maintenance fertilization (N−P$_2$O$_5$−K$_2$O plus micronutrients) shall be applied in the planting furrow and on coverage prior to the closing of the sugarcane field ("quebra-lombo") particularly K$_2$O, when the recommended dosage exceeds 120 kg/ha, and mainly on sandier soils.

For sugarcane ratoon, N−K$_2$O and B should be applied right after the cutting and right on the line.

4.2.4 How Should it Be Applied?

For cane-plant, fertilization application is basically accomplished through the soil, the seasons being the only change factor.

- Full area pre-planting: lime, gypsum and phosphate application and organic fertilization.
- Planting furrow: N−P$_2$O$_5$−K$_2$O, micronutrients and organic fertilization.
- Coverage with K$_2$O: cane plant of a year and a half on sandy soils and when recommendation exceeds 120 kg/ha of K$_2$O, applied prior to the sugarcane field closing.
- Stem or foliage fertilization: micronutrients.

For sugarcane ratoon, proceed to the application of N−K$_2$O−B over the ratoon line and, when it is the case, also apply P$_2$O$_5$.

4.3 Soil Fertility Assessment

In order to assess the "stock" of nutrients in the soil, three techniques are largely used: visual diagnosis, leaf diagnosis and soil diagnosis.

4.3.1 Visual Diagnosis

Visual diagnosis is accomplished by comparing the aspect of the sample with the standard, generally comparing plant organs, in this case, the leaf. Visual indications of nutrient deficiency (morphological changes) occur after internal changes (physiological changes), initially compromising older organs in the case of the more mobile elements in the phloem

(N, P, K and Mg) and younger organs in the case of less mobile elements (Ca, S and micronutrients). By the time visual symptoms of nutritional deficiency can be visually verified, the crop yield is already in jeopardy.

4.3.2 Leaf Diagnosis

Leaf diagnosis is a nutritional assessment method for crops where certain leaves are assessed during pre-defined periods in the plant's life. Generally, foliage assessment is carried out because they are the organs that best mirror the plants nutritional condition, i.e., they show a greater response to variation in the supply of nutrients, be it by the soil, or by fertilizers. Leaf diagnosis consists, therefore, of assessing the soil by using the plant as an extracting solution.

In order to carry out an adequate foliage sampling, some aspects must be taken into consideration, that is: homogeneity of the area as to the soil, variety, age and cultural traits; kind of leaf — to gather the leaf +3, i.e., from the tip of the shoot down towards the base, it is the first leaf that has a completely open ligule clasping the stem; portion of the leaf to be used — the central 20-cm portion, discarding the central vein; time of year — gather the leaf during the greatest vegetative development stage, for cane-plant, approximately 6 months after germination of the crop and, for sugarcane ratoon, 4 months after cutting pursuant to Figure 4.4.

Table 4.4 shows ranges of adequate nutrient proportions for sugarcane according to Raij et al. (1996).

This technique is most important when interpreting micronutrient levels since, in the case of macronutrients, the greatest interference comes from dilution or concentration factors.

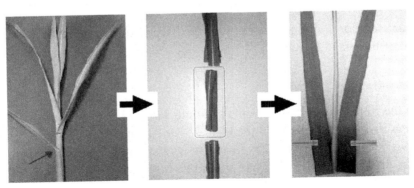

Figure 4.4
Sampling sequence and leaf preparation for analysis aiming at leaf diagnosis.

Table 4.4: Adequate nutrient proportion ranges for sugarcane.

N	P	K	Ca	Mg	S
(g/kg)					
18—25	1.5—3.0	10—16	2.0—8.0	1.0—3.0	1.5—3.0
B	Cu	Fe	Mn	Mo	Zn
(mg/kg)					
10—30	6—15	40—250	25—250	0,05—0,2	10—50

Source: Raij et al. (1996).

4.3.3 Soil Assessment

The fertilizing and correction recommendation program based on soil assessment starts with a soil sample, followed by analysis and interpretation of the results and ends with the correct utilization of inputs.

Soil sampling

For cane-plant, soil sampling must occur about three months prior to planting, allowing enough time for carrying out the analysis, quantification of necessary correctives and application of such, if there is time, in order for its reaction in the soil to start, thus bringing the desired results. To accomplish the sampling, the planting area must be consistently covered in a zigzag pattern gathering approximately 15 sub-samples of soil at depths ranging from 0 to 25 and from 25 to 50 cm. In the event, the area is a reform area in the sugarcane field where, at the time of sampling, there is still ratoon or rootstock, the sample must be gathered at 20—25 cm from the line throughout that area. Such proceedures, specific for the case of rootstocks, are necessary due to the fact that the samples gathered in the same line overestimate the levels of P and K, while samples gathered in between the lines overestimate levels of Ca, Mg, SB and V% and underestimate levels of P and K.

Fertilization and soil correction practices is adopted according to the nutrient levels and yield potential of the planted area.

Analysis interpretation

In addition to the routine fertility assessment in both sampled layers, it is paramount that a sulfur (S) assessment be carried out in the 0- to 25-cm layer for all expansion areas, particularly over pastures and in the 25- to 50-cm layer for reform areas. Regarding micronutrients, only the 0- to 25-cm layer should be assessed.

Tables 4.5—4.8 show soil nutrient level interpretations targeted at recommendations for fertilization and correction.

Table 4.5: Interpretation intervals for K and P levels in the soil.

Level	Relative Yield (%)	Exchangeable K^+ ($mmol_c/dm^3$)	P Resin (mg/dm^3)
Very low	0−70	0−0,7	0−6
Low	71−90	0.8−1.5	7−15
Average	91−100	1.6−3,0	16−40
High	>100	3.1−6,0	>40
Very high	>100	>6.0	−

Note that 10 mg/dm^3 of P in soil assessment are equal to the soil reserves of 46 kg/ha of P_2O_5 necessary for a 100 t/ha of stalks yield and that 1.0 $mmol_c/dm^3$ of K is equivalent to 96 kg/ha of K_2O.
Source: Raij et al. (1996).

Table 4.6: Interpretation of the P levels in the soil according to Mehlich 1 method.

Clay Percentage	Very Low	Low	Average	Good
61−80	<1	1.1−2	2.1−3	>3
41−60	<3	3.1−6	6.1−8	>8
21−40	<5	5.1−10	10.1−14	>14
<20	<6	6.1−12	12.1−18	>18

Source: Sousa and Lobato (1988).

Table 4.7: Interpretation intervals for Mg and S levels in the soil.

Level	Exchangeable Mg^{2+} ($mmol_c/dm^3$)	S** (mg/dm^3)
Low	0−4	<10
Average	5−8	10−15
High	>8	>15

Sources: Raij et al. (1996), Vitti (1989).

Table 4.8: Interpretation intervals for micronutrient levels in the soil.

Level	B	Cu	Fe	Mn	Zn
	Hot Water	DTPA			
	(mg/dm^3)				
Low	0−0.2	0−0.2	0−4	0−1,2	0−0.5
Average	0.21−0.6	0.3−0.8	5−12	1.3−5.0	0.6−1.2 (1.6)*
High	>0.6	>0.8	>12	>5.0	>1.2 (1,6)*

Hot water and DTPA are extractors used for determining micronutrients.
*Use this figure if the extractor used for this element is Mehlich 1.
Source: Raij et al. (1996).

Notice that for the soil assessment results, in the 0- to 20-cm layer, 1 mg/dm^3 is the equivalent of a 2.0 kg/ha reserve of the micronutrient.

4.4 Chemical Management of the Soil

The management and use of fertilizers in sugarcane crops starts with a soil fertility assessment, corrective practices (lime, gypsum and phosphate application), followed by "conservationist" practices (green and organic fertilization) ending with the application of mineral fertilizers. In other words: the following sequence is adopted in the application of management practices: Lime application*, gypsum application*, phosphate application*, green fertilization*, organic fertilization*, mineral fertilization through the soil, stem, foliage or through herbicides. The practices marked with an asterisk aim at increasing the gains from mineral fertilization (root absorption) since they foster a greater development of the root system resulting in greater water absorption and, therefore, greater nutrient absorption. Thus, the performance of corrective practices reduces the costs with mineral fertilization.

4.4.1 Lime Application

Lime application is the first practice to be adopted in the implementation and maintenance of sugarcane since, in addition to supplying calcium and magnesium as nutrients, its acidity correction features promote a greater utilization and availability of N, S, P, K and Mo, decrease the availability of Al^{+++}, Fe^{++} and Mn^{++} and increase microbial activity in the soil, thus resulting in increased mineralization of the cane trash and greater N_2 biological fixation on each plant in addition to the indirect Ca effects on soil aggregation.

Considering such effects, the following aspects should be considered when applying the lime.

The largely used recommendation criteria for sugarcane is base saturation (V%) which encompasses the chemical and physical−chemical parameters of the soil (CTC, V%), the plant (V% = 60) and of lime (PRNT). To such an end, the following equation is used for cane-plant:

$$NL = \frac{(60 - V_1{}^*)\ CEC^* + (60 - V_1{}^{**})\ CEC^{**}}{10 \times PRNT}$$

where:

NL = t/ha of lime (0−50 cm),
$V_1{}^*$ = current base saturation of the soil in the 0- to 25-cm layer,
CEC^* = Cation Exchange Capacity in the layer from 0 to 25 cm in mmol$_c$/dm^3,
$V_1{}^{**}$ = current base saturation of the soil in the 25- to 50-cm layer,
CEC^{**} = Cation Exchange Capacity in the layer from 25 to 50 cm in mmol$_c$. dm^{-3},
TNRP = Corrective Total Neutralization Relative Power.

In very sandy soil, CEC, 35 mmol$_c$/dm^3 and with very low levels of Ca^{++} plus Mg^{++} (<30 mmol$_c$/dm^3), pursuant to the base saturation criteria, lime application may not be recommended. In such cases, the Copersucar formula shall also be applied to the 0- to 25-cm layer aiming at the supplying of Ca + Mg, i.e.:

$$NL = \frac{[30 - (Ca + Mg)] \times 10}{TNRP}$$

where:

NL = t/ha of lime (0−25 cm),
Ca + Mg of soil assessment (mmol$_c$/dm^3 (0−25 cm),
TNRP = Corrective Total Neutralization Relative Power.

The recommendation is to use the criterion that presents the higher dosage.

For sugarcane ratoon a new sampling of the soil must take place right after the second cut, lime application being recommended in the event base saturation (V%) shows values smaller than 50% in the 0- to 25-cm layer. The recommendation is to use the highest allowable doses of lime i.e., 3.0 t/ha. Calculations are made using the following formula:

$$NL = \frac{(60 - V_1) \times CEC}{10 \times TNRP}$$

where:

NL = t/ha of lime (0−25 cm),
V_1 = current base saturation of the soil in the 0- to 25-cm layer,
CEC = mmol$_c$/dm^3 (0−25 cm),
TNRP = Corrective Total Neutralization Relative Power.

4.4.2 Phosphogypsum Application

Agricultural plaster (CaSO$_4 \cdot$ 2H$_2$O), a byproduct of the phosphoric acid production (H$_3$PO$_4$), is mainly comprised of CaO (26%) and S (15%).

Application of 1.0 t/ha agricultural plaster to the soil, considering a humidity of 17%, corresponds to supplying approximately 260 kg/ha of CaO, 150 kg/ha of S and 5.0 mmol$_c$/dm^3 of Ca. When solubilized in the soil solution, the plaster shows the following behavior:

H$_2$O
$$CaSO_4 \cdot 2H_2O \xrightarrow{H_2O} Ca^{++} + SO_4^- + CaSO_4^0 \downarrow$$
$$Ca^{++} + SO_4^= + CaSO_4^0$$

Approximately 50% of the product is dissociated in the form of Ca^{++} and $SO_4^=$ and about 50% remains in the ionic formula $CaSO_4^0$, being mobile within the soil's profile. As a result of these features, agricultural plaster may be used in sugarcane culture whenever the following effects are desired.

(1) Fertilizing effect: source of sulfur

When samples of the 0- to 25-cm layer (degraded pasture expansion areas) or of the 25- to 50-cm layer (already planted areas) have levels of $S < 15$ mg/dm^3, corresponding to a soil reserve of 30–40 kg/ha in the layer taken into account, it is recommended that 1.0 t/ha of plaster be applied (150 kg/ha of S), enough to maintain the high yields for 2.5–3.0 cuts, considering average extraction of 50 kg/ha of S for a 100-t/ha stalk yield.

(2) Salt level corrective

In areas with a potassium saturation (K%T > 5.0) caused by excessive application of vinasse, plaster application is recommended with ensuing irrigation or pluvial precipitation waters for the recovery of such soils through the following mechanism.

According to de Luz and Vitti (2008), the plaster dosage to be applied under such conditions may be determined pursuant to the following equation:

$$NG = (2.15 \times K/10) \times 1.7$$

where:

NG = Phosphogypsum need t/ha,
K = mmol$_c$/dm^3 (0–25 cm).

(3) Subsurface conditioning

Due to $CaSO_4^0$ mobilization to subsurface layers, agricultural plaster may be used to increase the levels of Ca^{++} and to reduce saturation by Al^{3+} (m%), through the following mechanism:

This conditioning must be used when subsurface samples (25–50 cm) show the following conditions: Ca < 5.0 mmol$_c$/dm^3 or Al > 5.0 mmol$_c$/dm^3 ou m > 30% or V% < 35.

Considering that plaster application with approximately 17% humidity increases the level of Ca^{++} in the soil to 5.0 mmol$_c$/dm^3, recommendation is made pursuant to the equation (Vitti et al., 2008):

$$NG = \frac{(50 - V_1) \times CEC}{500}$$

where:

NG = agricultural plaster need (t/ha),
V_1 = current base saturation of the soil in the 25- to 50-cm layer, and
CEC = Cation Exchange Capacity in the layer from 25 to 50 cm in mmol$_c$/dm^3.

Table 4.9: Approximate amount of plaster to be applied pursuant to the cationic exchange capacity (T) and base saturation (V) of the subsurface.

T (mmol$_c$/dm^3)	V (%)	Plaster Dosage (t/ha)
< 30	<10	2.0
	10−20	1.5
	20−35	1.0
30−60	<10	3.0
	10−20	2.0
	20−35	1.5
60−100	<10	3.5
	10−20	3.0
	20−35	2.5

Source: Demattê (1986) quoted by Demattê (2005).

Application of plaster pursuant to Table 4.9 may also be recommended.

It is emphasized that the criteria presented in Table 4.9 should be adopted in the case of soils with a maximum T of 100 mmol$_c$/dm^3 or 10 cmol$_c$/dm^3.

By recommending agricultural plaster for that end, the supply of sulfur is also satisfied ("covered"). In the event, the use of plaster is not necessary pursuant to such criteria (NG < 1.0 t/ha), the plaster should be applied at the rate of 1.0 t/ha as a source of S in the conditions already mentioned in item (1) above, although in areas with no use of sugarcane byproduct or other organic matter that may contain significant amounts of this nutrient.

4.4.3 Phosphate Application

Even though phosphate is the least extracted macronutrient by sugarcane, approximately 45 kg/100 t of stalk, it is the one that should be applied in the largest quantities in the planting furrow, approximately 150−200 kg/ha of P$_2$O$_5$, due to its high reactivity with the soil (fixation). This nutrient's efficiency within the plant may have its potential increased through corrective practices such as liming and phosphate application in the full area, a practice that aims at "stocking up" the system (fixation).

Such practice is particularly important on sandy soils with low levels of these nutrients, i.e., soils with a low CEC, consistent with clay levels (<30%), and with P resin < 15 mg/dm^3 or P-Mehlich 1 for the very low and low classifications (Table 4.6).

Once the need for phosphate application is established, the dosage should be 5.0 kg of P$_2$O$_5$ per clay percentage in the soil, generally corresponding to the tier of 100 (clay < 20%) to 150 kg/ha of P$_2$O$_5$ (clay 25−30%), being used in the full area after liming and gypsum application and before the leveling grid (superficial incorporation).

The most recommended sources of P_2O_5 for such practices are those that present average and high P_2O_5 levels soluble in citric acid (HCi) such as: natural reactive phosphate, or superphosphate (30% P_2O_5 total and 10% P_2O_5 HCi); magnesium thermophosphate (18% P_2O_5 total and 16% P_2O_5 HCi), filtercake and filtercake matter + ashes + P_2O_5 source or filtercake + ashes + animal manure, specially bird manure. When the quantity calculation for reactive phosphate considers the total P_2O_5 level.

The main practical consequences of phosphate application are: greater amounts of P in contact with the soil (greater fixation) and greater volume of soil exploited by the roots, greater water and nutrient absorption, greater tolerance to soil pests and greater benefits from the P_2O_5 applied in the furrow resulting in maintenance of high yields throughout the cutting (smaller levels of yield loss in between cuts) with a greater overall longevity of the sugarcane crop (de Luz and Vitti, 2008).

4.4.4 Green Fertilization

Upon reform of the sugarcane field, green fertilization is a compulsory practice aimed at maintaining the system's balance since it is responsible for several effects such as:

- protecting soil during the rainy season, resulting in less losses by erosion and lixiviation;
- acting as a source of nutrients, particularly N which results from symbiotic fixation;
- decreasing P fixation due to an increase of Organic Matter;
- improving soil structure;
- controling nematodes, particularly when using *Crotalaria spectabilis or Crotalaria ochroleuca*;
- faster solubilizing of the Ca, Mg, S and P contained in correctives, making them more readily available to the sugarcane; and
- eventual positive financial balance, if growing peanuts and soy.

The best crops to sow concomitantly to sugarcane, be it on rotation or meiosis, are: *Crotalaria junceae*, *Crotalaria spectabilis*, soy and peanuts depending on the sowing region.

The management of *Crotalaria junceae* consists of sowing during the October to November period (southeast region). Sowing can be accomplished through the throwing method for a full area (30 kg/ha of seeds, density of 60 seeds/m^2) or through the line method with spacing of 0.5 (25 kg/ha of seeds, density of 25 seeds per linear meter), and sowing depth between 2 and 3 cm. Upon flowering (approximately 100 days) management should occur through clearing with a knife-roll or bumper (Vitti et al., 2006a,b).

Crotalaria spectabilis has a slow initial development, thus being indicated for regions with less conservationist issues and nematode occurrence. Rattle Pods (Crotalaria) may be sowed by throwing (15 kg/ha of seeds) or furrow methods (12 kg/ha of seeds). The management may be carried out similarly to that of *C. junceae* or planting the sugarcane right in between the lines of Crotalaria, using a light bar.

For soy and peanut crops, in addition to the corrective practices, complementation with fertilizers in the planting furrow should be arranged, as well as foliage application of fitosanitary and micronutrient products.

4.4.5 Organic Fertilization

The main effects of organic matter on soil physical attributes are: increased capacity to retain soil moisture; good soil porosity (macropores); reduction in apparent density; improvement in the water infiltration rate and thermal damping, avoiding important oscillation in temperature. Furthermore, organic matter also promotes effects on the soil chemical properties such as: increasing retention of cations; supplying of macro- and micronutrients; gradual release of nutrients, and reduction in P fixation since organic radicals block the attaching sites. In addition, organic matter improves the soil biological attributes by providing better conditions for the development of living organisms (insects, annelids, etc.), especially from the soil microbiota, resulting in nutrient availability to plants.

In the sugar—ethanol industry, organic fertilization is extremely important and relevant, since this industry generates several by-products which provide technical and economic potential for application in cane sugar. The main sub-products are: filtercake, vinasse, soot or ash, decanting clay and the washing water.

From the by-products listed above, the most important are filtercake, vinasse and ashes. The filtercake is the residual material obtained after filtering the already processed juice of the sugar cane. The cake is mainly composed of remnants of bagasse and mineral material found in the raw material.

Currently, filtercake can be obtained by three types of process: rotary vacuum filter "Oliver", pressing filter and separation by diffuser. Table 4.10 shows the production of filtercake in quantitative terms, according to the generation process according to data from the industrial units.

It can be observed that the production of filtercake is significant especially for plants whose juice is extracted in mills. To give an idea, an average plant with milling capacity of 2,000,000 metric tons per season, will be able to generate approximately 60,000 metric tons of cake per season using the Oliver filter and approximately 40,000 metric tons of cake with the pressing filter. Filtercake is typically produced with about 75% of moisture if the filter is Oliver and 65% if the filter is of the pressing type, meaning that filtercake is a byproduct with high water content.

Table 4.10: Filtercake production according to different industrial processes.

Juice Extraction System	Generation Process	Production of Cake kg/t of Processed Sugar
Mill	"Oliver" filter	28–35
Mill	Pressing Filter	18–22
Diffuser	Separator	5–6

Source: de Luz and Vitti (2008).

Table 4.11: Macronutrients found in filtercake*.

Name	N Total	P$_2$O$_5$ Total	K$_2$O	Ca	Mg	S
			(%)			
Average	1.49	1.72	0.34	4.59	0.46	0.6
CV	29.9	57.2	93.7	154.8	66.4	181.4
IC-MAX	1.67	2.12	0.47	7.5	0.58	1.05
IC-MIN	1.31	1.32	0.21	1.69	0.33	0.16

*Dry base.

Table 4.12: Micronutrients found in filtercake*.

Name	Fe	Mn	Cu	Zn	Na	B
	(mg/dm^3)					
Average	22,189.2	576.9	119.1	142.9	872.2	11.3
Standard Deviation	13,884.0	271.0	68.0	87.6	699.1	8.8
CV	62.6	47.0	57.0	61.3	80.2	77.7
IC	9620.9	187.8	47.1	60.7	484.4	6.1
IC MAX	23,504.9	458.7	115.1	148.3	1183.5	14.8
IC MIN	13,821.4	224.0	10.9	26.3	618.9	69.0

*Dry base.

Table 4.13: Other parameters found in filtercake.

Name	Rel. C/N	pH CaCl$_2$	Density (g/cm^3)	UT (110°C)	M.O. Total (%)	C total (%)
Average	21.96	5.93	0.6	65.06	57.62	32.01
C.V.	16.73	22.08	27.04	19.33	32.85	32.85
IC-MAX	23.46	6.47	0.66	70.2	65.36	36.31
IC-MIN	20.46	5.4	0.53	59.92	49.89	27.72

Tables 4.11–4.13 show the different physicochemical attributes found in filtercake as revealed by a study carried out in different producing units: Table 4.11 is for macronutrients; Table 4.12 is for micronutrients and Table 4.13 for the other parameters (Light and Vitti, 2008).

Figure 4.5
Application of filtercake scattered in the total area.

In average filtercake presents: C/N ratio: 22:1; pH: 5.93; density: 600 kg/m^3 and moisture: 65%. Its main macronutrients were: P_2O_5: 1.72%; N: 1.49% and Ca: 4.59%, while micronutrients Zn and Cu responded with 143 and 120 mg/kg respectively (values in dry matter).

In terms of crop management for using filtercake, one must consider the following aspects:

a. Application

(a. 1) in total area: when reforming or expanding the sugarcane plantation and eventually also in ratoons (Figure 4.5) and

(a. 2) localized, especially at the planting furrow (Figure 4.6), being also possible to be planted on bands, in the areas of ratoons, in the line of the sugar cane.

b. Use

(b. 1) *In natura*: this is the way filtercake leaves the processing plant, with 70–75% moisture. It is used mainly in winter sugar cane *(central-southern Brazil), in the furrow, aiming at supplying nutrients together with water and to use the beneficial thermal effect of the cake. However, since its humidity is high, elevated amounts of cake are needed to achieve adequate levels of nutrients.*

(b. 2) *Conditioned* is the filtercake that went through a drying process along with the physical conditioning, to improve the application characteristics. This alternative is very interesting because the nutrients are concentrated and the flowability by applicators is improved.

(b. 3) *Enriched*: in addition to suffering the conditioning process, the filtercake undergoes an aerobic composting process, being mixed with other raw materials in order to raise the level of nutrient concentration with a product of lower humidity, extending the application areas, since the product can be transported over longer distances because of the smaller amounts needed.

Figure 4.6
Application of filtercake localized in the furrow.

Figure 4.7
Physical conditioning obtained with the help of a mechanized composter.

The composting process to enrich the cake can be done by mixing the following raw materials:

Carbon suppliers: bagasse, ash, sawdust, etc.;
Nutrient suppliers: poultry litter, poultry manure, cattle manure, pig manure, etc., and
Mineral sources: gypsum, P_2O_5 from magnesium thermophosphate and reactive natural phosphates.

For management, filtercake is recommended after physical conditioning by using a mechanized composter (Figure 4.7) at the composting area for lowering humidity and

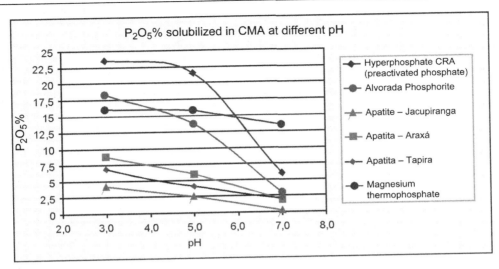

Figure 4.8

Solubility of P_2O_5 according to pH for different sources of P. *Source: From Alcarde and Ponchio (1979).*

reducing the runoff by decreasing the number of aggregates, which gives a better homogenization of the compound.

After conditioning the cake, the processing plant is qualified to the second step, i.e., to compost the cake. In this process, the cake is mixed with ash (see below) in the weight proportion of four to one. In order to enrich the compost with S and Ca, besides the possibility of reducing the loss of NH_3 by volatilization, the addition of 5% of gypsum is recommended in a weight base.

The use of filtercake in the planting of cane sugar aims at replacing fully N and partially P and K. The conditioned/enriched cake must have high content of P_2O_5 since these nutrients are normally used in the following amounts (per hectare): 30–40 kg of N; 120–150 kg of P_2O_5; and 100–120 kg of K_2O.

Thus, in the composting process, the addition of a mineral source of P_2O_5 is recommended. This mineral can be found in reactive natural phosphate (RNP) and in magnesium thermophosphate. Magnesium thermophosphate has a very interesting behavior: it is less prone to the fixation process because of its high pH, which in the composted surpasses 7.0) and it reacts with Ca, as can be seen in Figure 4.8.

Light and Vitti (2008) carried out a composting process study at the São João processing plant in Araras, SP. 90% of filtercake, 8% of bagasse and 2% of ash were mixed. The results obtained demonstrated that during the process, humidity came down from 75% to 40% (Figure 4.9); the temperature went up and subsequently came down (Figure 4.10) and the pH went from 5.50 up to 7.75 (Figure 4.11).

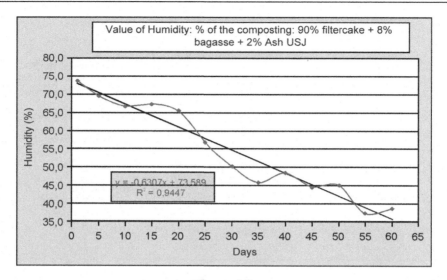

Figure 4.9
Humidity changes in compost according to time, revolving every 5 days.

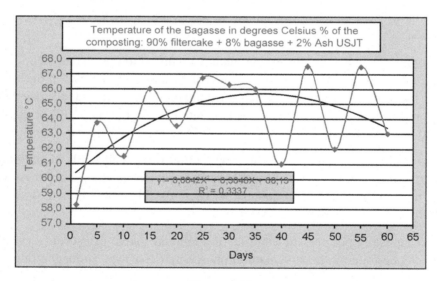

Figure 4.10
Temperature changes in compost according to time.

In using filtercake, it is well known that logistics is crucial in the distribution, with freight being an important cost factor. In this sense, a case study was carried out in the EQUIPAV Processing Plant of Promissão, SP, comparing the use of *in natura* cake against the cake enriched with 400 kg/ha of magnesium thermophosphate (Light and Vitti, 2008). Table 4.14 shows the characteristics of the three products.

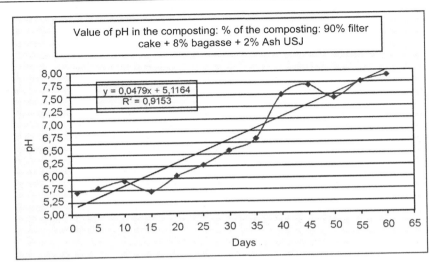

Figure 4.11
pH changes in compost according to time.

Table 4.14: Content and amount of N−P$_2$O$_5$ and K$_2$O supplied by different uses of filtercake.

Filtercake	Content (%)			U	Magnesium Thermophosphate		Source		
	N	P$_2$O$_5$	K$_2$O	(%)	Filtercake (t/ha)	(kg/ha)	N	P$_2$O$_5$	K$_2$O
In natura	0.38	0.25	0.06	75	60	—	230	152	34
Conditioned	0.70	0.47	0.1	45	33	—	230	154	35
Enriched	0.70	0.60	0.1	45	12	400	84	150	12

To determine the balance between the cost of organic fertilization (cake) in relation to mineral fertilization, the supply of N−P$_2$O$_5$ and K$_2$O by mixing granules 04:10:10 in the amount of 1000 kg/ha supplemented with 50 kg of P$_2$O$_5$ per hectare given by magnesium multiphosphate was studied.

Figures 4.12−4.14 show that the use of *in natura* cake is cheaper than the mineral up to a distance of 25.75 km of the processing plant, while the conditioned cake allows a displacement of up to 30.75 km. Moreover, enriched cake can be used at distances of 44.25 km, enabling the "socialization" of organic−mineral fertilization in wide areas.

Table 4.15 summarizes how to manage the different forms of filtercake. It is evident from this table that enriched cake can be applied to far distant sites, thus significantly increasing the area planted. In this case, 10,000 ha were fertilized with enriched cake versus 2547 ha of *in natura* cake and 3640 ha of conditioned cake.

Another byproduct from the processing plant is the boiler ash whose analytical data are presented in Table 4.16. In this case it is a "rested" ash in which there was loss of moisture

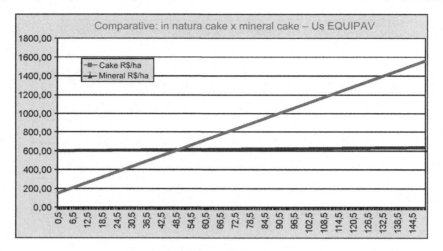

Figure 4.12
Balance point between *in natura* filtercake and mineral fertilizer.

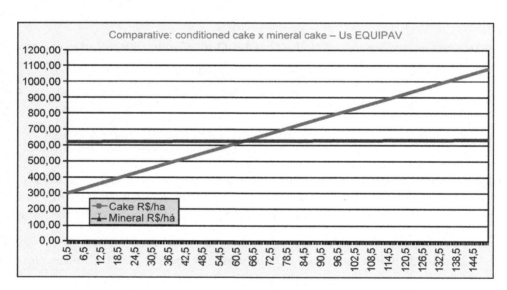

Figure 4.13
Balance point between conditioned filtercake and the mineral source.

and nutrient concentration. As mentioned before, ash can be used mixed with filtercake, thus enabling a more appropriate management.

Anyway, as ash is poor in nutrients, its isolated application requires high doses (60 metric tons per hectare) for a reasonable supply of K_2O, besides a raise in the soil pH. Its application remains thus limited to a certain distance around the processing plant. Furthermore, the mixing with filtercake results in a dilution effect. Therefore, one should not add too much ash in the mixture (maximum 20%).

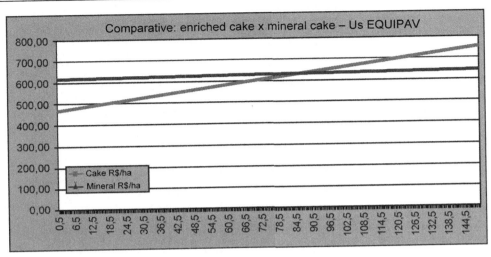

Figure 4.14
Balance point between enriched filtercake and the mineral source.

Table 4.15: Filtercake management.

Filtercake Condition	Distance (km)	é Amount (R$/ha*)	Amount (tMU/ha)	Planted Area (ha)	% Increase
In natura	25.7	624.00	60	2547	100
Conditioned	31.	626.0	33	3640	143
Enriched	44.3	639.00	12	10,000	393

*Balance point.

Table 4.16: Analysis of boiler ash in dry matter.

Type	Filtercake + ash from EQUIPAV-300407 Processing Plant					
	N	P_2O_5	Ca	S	K_2O	Moisture
Pure ash-A	0.27	0.04	3.26	0.11	0.21	19.13
Pure ash-B	0.28	0.03	3.21	0.04	0.33	31.08
Average ash	0.28	0.04	3.24	0.08	0.27	25.11

Another by product of the sugar—ethanol industry is vinasse. Its use is well established in the State of São Paulo which has specific legislation on it (CETESN Technical Norm #P-4231). Broadly speaking, this norm covers the following aspects for the use of vinasse:

Specification and characterization of areas for its application;
Physicochemical characterization of vinasse;
Sampling and soil characterization;

Storage and transport of vinasse: storage and canals;
Specification of the application system; and
Establishment of the amount to apply.

The dose of vinasse to apply is established by the following equation:

$$V = \frac{[(0.05 \times CTC - K\ soil) \times 3.744 + 185]}{K\ vinasse}$$

where:

V = volume of vinasse (m^3/ha),
CTC = capacity of cationic exchange of the soil at pH 7.0 ($cmol_c/dm^3$),
K soil = content of K in the soil ($cmol_c/dm^3$),
185 = K_2O extracted by sugar cane (kg/ha),
K vinasse = concentration of K^+ in vinasse (kg/m^3 of K_2O).

The technical principle of this equation is limit of occupancy of CEC − Cation Exchange Capacity − by K at 5%. If this limit is exceeded, the equivalent of 185 kg K_2O/ha for maintenance fertilization of the sugar cane can be applied.

In early April, all processing plants must submit their annual Vinasse Application Plans (VAP).

In the 2007/2008 season the sugar−ethanol industry used vinasse over an area of 1,100,000 to 1,200,000 ha, an estimated equivalent to 150,000 metric tons of KCl.

Vinasse must be stored covered (CETESB norm). Currently, this coverage is done with PEAD − 2.0 mm. For transportation the options are as follows:

fixed pipeline network,
mobile pipeline network,
canal, and
truck.

According to the CETESB norm, canals must be dressed up. This is presently being accomplished with PEAD − 1.0 mm or with "concrete" of stone powder, as seen in Figures 4.15 and 4.16. An aluminum-made mobile pipeline of 6′,8′ or 10′ diameter has been a well-accepted alternative in fertilization and irrigation (de Luz and Vitti, 2008).

Predominantly, application is made by mechanized spraying with coil and hydraulic cannon as shown in Figure 4.17.

Vinasse can be obtained by distilling the juice, the mixed must or the sugar cane molasses as shown in Table 4.17.

Figure 4.15
Ditch of vinasse with PEAD.

Figure 4.16
Ditch to carry vinasse.

Figure 4.17
Vinasse applier.

Table 4.17: Chemical average composition of vinasse.

Element	Vinasse		
	Sugarcane Molasses	Mixed	Juice
N (kg/m^3)	0.77	0.46	0.28
P$_2$O$_5$ (kg/m^3)	0.19	0.24	0.20
K$_2$O (kg/m^3)	6.00	3.06	1.47
CaO (kg/m^3)	2.45	1.18	0.46
MgO (kg/m^3)	1.04	0.53	0.29
SO$_4$ (kg/m^3)	3.73	2.67	1.32
Organic Matter (kg/m^3)	52.04	32.63	23.44
Fe (ppm)	80.00	78.00	69.00
Cu (ppm)	5.00	21.00	7.00
Zn (ppm)	3.00	19.00	2.00
Mn (ppm)	8.00	6.00	7.00
pH	4.40	4.10	3.70

From: de Luz and Vitti (2008).

Considering the possibilities of different vinasses dose range and interval in amounts used practiced by the processing plants, i.e., 150−300 m^3/ha when vinasse is applied as spray from reservoir or canal, and 80−150 m^3/ha when it is applied as spray from a truck, it can be stated that fertilizing ratoons with this by-product fully meets the need of K and partially

Table 4.18: Average content of nutrients found in poultry litter (%).

N	P$_2$O$_5$	K$_2$O	Ca	Mg	S	C/N	O.M.	Total Moisture
2.63	2.24	2.57	6.17	0.53	0.34	10.91	66.01	21.73

Table 4.19: Nutrients supplied by different amounts of poultry litter in ratooned sugar cane.

Amount Met (t/ha)	N	P$_2$O$_5$ Total (kg/ha)	K$_2$O	Ca	Mg	S
3.0	79	67	77	185	16	10
3.5	92	78	90	216	18	12
4.0	105	89	103	247	21	14
4.5	119	101	116	278	24	15
5.0	132	112	129	309	26	17

that of N, discounting about 30−50 kg N/ha of the planned nitrogen fertilization. S needs are also usually fully supplied by vinasse.

In addition to processing plant by-products, there are other options in organic fertilization: food industry waste (ajifer), poultry litter and manure from chickens, turkeys, cattle and swine.

Ajifer is a byproduct from the processing of sucrose or molassess to produce lysine or glutamic acid. Ajifer contains significant amounts of N and S and it is widely used for fertilizing ratoon as N source. However, ajifer must be supplemented with a source of K$_2$O.

Other raw material used as organic fertilizer which is presently attracting much interest is poultry litter. The following aspects must be pointed out in poultry litter:

Type: rice bark, peanut bark, woodchips, grass;
Number of fattening, i.e. how many poultry lots have used it; and
How the litter was taken, presence of animal waste, feathers, etc.

Thus, it is recommended to physically and chemically analyze the poultry litter to be used. Table 4.18 presents average values from analysis of poultry litter in the regions of Catanduva, Promissão, Lins, Pirassununga and Descalvado, all in the State of São Paulo. Note that the examined poultry litter had a N:P$_2$O$_5$:K$_2$O well suited for the fertilization of sugar cane ratoon, especially between N and K$_2$O. Table 4.19 shows the pattern of nutrients supplied to sugar cane ratoon by using poultry litter in sugar cane plantations with production potential of 70−110 metric tons per hectare, when the poultry litter was used exclusively on line over the ratoon.

Poultry litter can also be used mixed with filtercake + ash + gypsum in making the compost. Currently it is a very attractive option in economic terms, since its cost minus freight is around 90−110 reals per metric ton.

Table 4.20: Average content of nutrients in laying hen manure (%).

N	P_2O_5	K_2O	Ca	Mg	S	C/N	M.O.	Total Moisture
3.76	3.31	3.01	9.91	0.67	0.33	8.50	60.70	35.45

Table 4.21: Average content of nutrients in confined beef cattle.

	C/N	Moisture	C	N	P	K	Ca	Mg	S	Zn	Cu	Cd	Ni	Pb
Organic Matter		(g/kg)								(mg/kg)				
Fresh cattle manure	20	620	100	5	2.6	6	2	1	1	33	6	0	2	2
Dry cattle manure	21	340	320	15	12	21	20	6	2	217	25	0	2	1

Table 4.22: Supply of primary macronutrients by using three types of swine manure (50 m^3/ha).

	d	N	P_2O_5	K_2O
MS (%)		(kg/m^3)		
A − 1.17	1.008	1.60	1.14	1.00
B − 1.63	1.010	1.91	1.45	1.13
C − 2.09	1.012	2.21	1.75	1.25
	N		P_2O_5	K_2O
Amount (m^3/ha)		(kg/ha)		
A − 50	80		57	50
B − 50	96		73	57

Besides the poultry litter obtained from the production of broilers, another option is the chicken manure, which is the excrement of laying hens. Chicken manure is typically richer than poultry litter, because it is made of "pure" feces, but it has higher humidity than litter and it is more difficult to apply. Table 4.20 presents the average levels of chemical and physical attributes of chicken manure.

In beef cattle country, one can also use cattle manure whose values are presented in Table 4.21.

Swine manure is also an option for organic fertilization. It is used in liquid form and can be applied over the entire area or along lines. Table 4.22 presents the amount of nutrients provided by the application of 50 m^3 of swine manure.

4.4.6 Mineral $N-P_2O_5-K_2O$ Fertilization and Micronutrients

Planting fertilization

Furrow fertilization in terms of N, P_2O_5 and K_2O is a function of the areas history (for N) and of the soil assessment (for P_2O_5 and K_2O), as shown in Table 4.23.

Table 4.23: Mineral fertilization in sugarcane sowing based on soil assessment.

N (kg/ha)	P-resin (mg/dm^3)	P$_2$O$_5$ (kg/ha)	K (mmol$_c$/dm^3)	K$_2$O$^{(2)}$ (kg/ha)
40–60	0–6[1]	170	<0.7	170
	7–15[1]	150	0.8–1.5	140
	16–40	100	1.6–3.0	110
	>40	70	3.1–5.0	80
			>5.0	0

[1]In soils with a clay level <30%, use 100–150 kg of P$_2$O$_5$/ha for the full area in addition to 100 kg of P$_2$O$_5$/ha in the planting furrow.
[2]In quartz-rich sands (quartzenic neosols) and latosols, apply a maximum of 100–120 kg/ha of K$_2$O in the planting furrow and the leftover on the covering, prior to the closing of the sugarcane field in the "quebra-lombo" practice (leveling of the soil between plant rows).

Table 4.24: Mineral fertilization of burnt ratoon in relation to yield expectations.

Yield Expectations (t/ha)	N (kg/ha)	K$_2$O (kg/ha)
65–80	80	100–120
81–100	100	130–150
>100	120	160–180

In the event of nitrogen-rich fertilization, smaller doses should be adopted (40 kg/ha), in the following conditions: expanse areas and sandier soils (C, D, E production environments).

Larger doses of N are recommended for reform areas and clay rich soils (A, B production environments).

When using legumes, grown under adequate conditions, in reform areas, or when using organic waste, N fertilization during planting may be waived.

In addition to supplying primary macronutrients (N–P$_2$O$_5$–K$_2$O), the supplying of micronutrients is essential, particularly B, Zn, Cu, as will be discussed below in "Micronutrient Fertilization", and since Ca, Mg and S are supplied through corrective practices.

Ratoon fertilization: N–K$_2$O and B fertilization

Prior to defining the amounts of N and K$_2$O, it is necessary to know about the areas' administration as far as harvesting, burnt cane and raw cane.

a. Burnt-cane: N fertilization is based on the harvested yield, while K$_2$O, is based on yield and soil assessment (ratoon sample), whenever available, using approximately 1.0 kg/t of sugarcane and between 1.3 and 1.5 kg/t of sugarcane for K$_2$O, keeping the reason between N/K$_2$O in the range of 1.0 for 1.3–1.5, according to Tables 4.24 and 4.25.

Table 4.25: Potassium fertilization recommendation for burnt cane-ratoon based on soil assessments.

K (mmol$_c$/dm^3)	K$_2$O (kg/ha)
<1.5	150–180
1.6–3.0	110–140
>3.0	80

b. Raw-cane: In the raw cane harvesting system (without straw burning), a 10–15 t/ha raw matter (RM) build-up occurs. In Table 4.26, the RM mass for raw cane straw and the respective amounts of nutrients and structural carbohydrates are represented.

The data shown in Table 4.26 point out a differentiated crop management for ratoon fertilization for the following reasons:

High level relationship between C/N, C/P and C/S, pointing at low mineralization of straw organic matter, even one year after the cutting;

Paralyzing of anions (N, P and S) inside the straw;

Mineralization of the mineral cationic nutrients (K, Ca and Mg), particularly K, giving approximately 50 kg/ha of K back to the soil system; and

Clear increase in microbe activity, particularly ureases, responsible for NH$_3$ losses in urea through volatilization.

Due to these considerations regarding the supply, especially the supply of N, two factors should be paid attention to: dose and source.

Regarding dose, it should be increased by at least 30%, i.e., 3 kg of N/t of produced sugarcane. As to the source of N, the use of urea on the surface is prohibitive, since losses may go up to 70%, as observed by Lara Cabezas et al. (1997) when urea is superficially applied on corn straw. Therefore, regarding the use of urea there are only three options:

Partial burying of urea;

Usage of "protected" urea, with urease inhibitors; or

Mixing of urea with ammonium sulfate, for example formulation 32-00-00-12, resulting from the equitable mixture of those two sources.

Regarding potassium fertilization, it has been observed that the straw releases from 40 to 50 kg/ha of K, which is deducted from mineral fertilization. Thus, the K$_2$O doses in raw cane may be of the order of 0.8–1.0 kg/t of harvested cane K$_2$O, resulting in the relation N/K$_2$O of 1.0–1.3/1.0. Thus, for a production of 100 t/ha of raw harvested cane, the doses of N and K$_2$O would be, respectively, 130 and 80 to 100 kg/ha.

Table 4.26: Raw matter mass for raw sugarcane straw, amount of nutrients and structural carbohydrates, on samples collected in 1996 and in the remainder straw in 1997.

Year	RM	N	P	K	Ca	Mg	S	C	Hemicellulose	Cellulose	Lignin	Cellular Content	C/N	C/S	C/P
	(t/ha)									(kg/ha)					
1996	13.9 to	64 to	6.6 to	66 to	25 to	13 to	9 to	6.255 to	3.747 to	5.376 to	1.043 to	3.227 to	97 to	695	947
1997	10.8 b	53 to b	6.6 to	10 b to	14 b	8 b	8 to	3.642 b	943 b	6.619 to	1.053 to	2.961 b	68 b	455	552

Source: Oliveira et al. (1999).

For ratoon, the supply of the micronutrient boron is essential and, to a certain degree, so is molybdenum; these micronutrients have mass flow mechanisms (travel with water) similar to N and K_2O.

Phosphate-rich fertilization in ratoon is widely argued, mainly due to the absence or low mobility of $H_2PO_4^-$ in the soil, as well as due to the acidification of the rood zone. An adequate supply of P_2O_5 during the implementation of the crop is necessary, both in full area and in the furrow, in order to make it expandable in the ratoon. Nevertheless, for areas where P has not been adequately managed during implementation, we suggest applying about 30 kg/ha of P_2O_5, but in soil conditions with acidity duly corrected (V% > 50) and when P in the soil is less than 15 mg/dm^3 in resin.

Sugarcane fertilization with sulfur is paramount, particularly in areas with greater response probabilities, i.e.:

   Areas with no organic residue application nor gypsum;
   Areas further from the mill (without SO_2 from bagasse burning); and
   Sandier soils with low levels of organic matter.

When gypsum is applied, the 1.0 t/ha dose, for example, is enough for the supply of S for at least two cuts and its application should be renewed in ratoon after the third cut, when the levels of S in the soil for the 25- to 50-cm layers are lower than 15 mg/dm^3.

In addition to this option, sulfur may be supplied during planting, through P_2O_5, sources such as simple superphosphate (12% S), thermophosphate and magnesium multiphosphate containing S. For ratoon, particularly when raw, that has a high C/S relationship, S may be supplied through nitrogen-rich sources: ammonium sulfate (24% S), a mixture of urea with ammonium sulfate, (12% S) and, in the event of liquid fertilization, with vinasse, ajifer, or sulfuran (4% S).

4.4.7 Micronutrient Fertilization

Due their low levels in the soil, crop practices that decrease their availability and their importance to cane nutrition, adequate supply of micronutrients has become paramount for an increase in yield (Vitti et al., 2006a,b).

Micronutrients may be supplied in several ways, but application through the soil in solid formulation $N-P_2O_5-K_2O$ carries the advantage of having a greater residual effect in terms of micronutrients except on B soils, where it should be annually applied.

(a) Through the soil

   Solid fertilization: $N-P_2O_5-K_2O$ + micronutrients dosage, sources and criteria for micronutrient recommendations for sugarcane plant are shown in Table 4.27. It is

Table 4.27: Dosage and sources of micronutrients for fertilization in relation to levels of nutrients in the soil.

Soil Levels	Recommended Dose (kg/ha)*	Source
Zn (DTPA < 0.6 mg/dm^3)	3.0–5.0	Oxisulfates
Cu (DTPA < 0.3 mg/dm^3)	2.0–3.0	Oxisulfates
B (hot water < 0.2 mg/dm^3)	1.0–2.0	Ulexite
Mn (DTPA < 1.2 mg/dm^3)**	2.0–3.0	Oxisulfates

*Note: smaller doses for sandier soils and larger doses for clay rich soils.
**Northeast soils.

important to highlight that micronutrients should necessarily be aggregate to the P_2O_5 source or coating all $N- P_2O_5-K_2O$ granules. Such technologies, in addition to granting greater application homogeneity for the micronutrients, will provide greater solubility too. It should be emphasized that the current legislation for fertilizers demands that 60% of the declared micronutrient levels be soluble in amonium neutral citrate + water (CNA + H_2O) for Cu and Mn and that 60% be soluble in citric acid (HCi) in the case of the remaining micronutrients (B, Zn, Fe, etc.). Due to the Zn and Cu absorption mechanism (diffusion), i.e. their fixation in the soild with residual effect, the doses supplied during planting are sufficient for at least five cuttings, though there is a need for periodical soil assessment to confirm the presence of nutrients.

Liquid fertilization: in units that have liquid fertilization, B, Zn and Cu may be supplied with the sources of N (Aquamonia, Uran), P_2O_5 (MAP, H_3PO_4) and K_2O (chloride), in the form of boric acid or sodium octaborate and salts (Zn and Cu sulfides) or chelated products. Special notice should be given to corrosion caused by copper, thus preferably using chelated products. Dosage, upon supply of the micronutrient, through liquid fertilization, is as follows:

  B: 0.5–1.0 kg/ha
  Zn: 1.0–1.5 kg/ha
  Cu: 0.5–1.0 kg/ha

Note the smaller doses for chelated products and the larger for salt-based products.

(b) Through herbicides

Both cane-plant and ratoon accept boron application concomitantly to the herbicide, in the form of boric acid or sodium octaborate, for the dosage of 1 kg/ha of B. The guarantees for these products are shown in Table 4.28.

When preparing the solution, it is recommended to start with three quarters of water, add in the boron source, taking into consideration product solubility, i.e., the maximum possible amount to be dissolved without causing precipitation, then immediately add the herbicide to be filled up with the remainder of the water volume.

Table 4.28: Sources of boron indicted for liquid application.

Source of B	% of B	PS*
Boric Acid	17	5
Sodium Octaborate	20	10

*Product solubility (g/100 ml or kg/100 l).

(c) Through foliage

Makes the joint application with herbicides improving distribution of products when "covering" the seedlings. To carry that out, compatibility with herbicides should be checked.

Sources:

B — Boron Acid or Sodium Octaborate
Cu, Fe, Mn, Zn — salts (sulfate), chelated products, phosphates or humic and fulvic acids.

Doses

B — 300—350 g/ha of B
Cu, Fe, Mn, Zn — x f extraction
(f = 1.2—1.5 for Zn and Cu), approx. 800 g and 400 g/ha, respectively, of Zn and Cu.

(d) Through foliage

Especially upon nitrogen application through the leaves, it should be applied jointly with molybdenum, which participates in two enzyme systems directly linked to nitrogen.

N dosage should be in the range of 15—20 kg/ha and Mo in the range of 150—200 g/ha.

As suggestions of sources, there are three alternatives:

1. Application of commercial product, for example formula 26-00-00 + 0.26% Mo (d = 1,26).
2. Mixing of 350 kg urea + 320 kg of uran + 10 kg of sodium molybdenum dissolved in 1000 l of water. This gives the formula 22,6-00-00 + 0.4 Mo (d = 1.15); of which the recommended dosage is 50 l/ha or 58 kg/ha.
3. Mixing of 24 kg of urea + 9 kg of uran + amonium nitrate + 384 kg of sodium molybdenum dissolved and applied at the dosage of 50 l/ha.

Such application should be preferably carried out in crops with a high yield potential, during the closing of the field (November/December), as well as on seedlings and nurseries.

4.5 Final Considerations

To obtain high yields in the sugarcane crop, the following chemical management of the soil is recommended.

4.5.1 Cane-Plant

(i) Liming; (ii) gypsum application; (iii) phosphate application mainly on sandier soils (clay $< 30\%$ or CEC < 50 mmol$_c$/dm^3) and P resin < 15 mg/dm^3 or P − Mehlich 1 for the classes low and very low; (iv) green fertilization (*Crotalaria junceae* or *spectabilis*, soy or peanuts); (v) organic fertilization; and (vi) mineral fertilization in the planting furrow (N−P$_2$O$_5$−K$_2$O + micros, especially B and Zn).

4.5.2 Ratoon

(i) Liming (re-apply when V% < 50, in maximum dosage of 3 t/ha); (ii) gypsum application (use mainly as source of S, when sub-surface levels are lower than 15 mg/dm^3); (iii) fertilization N−K$_2$O−B of raw cane: 1.3 kg of N and 0.8−1.0 kg of K$_2$O per ton of cane yielded, and for burned cane 1.0 kg of N and 1.3−1.5 kg of K$_2$O per ton for cane yielded and about 1 kg/ha of B; and (iv) fertilization P$_2$O$_5$ − use about 30 kg/ha of P$_2$O$_5$ when P resin < 15 mg/dm^3 or P Mehlich 1 for the classes very low and low in conditions where V% > 50.

Bibliography

Alcarde, J.C., Ponchio, C.O., 1979. Ação solubilizante das soluções de citrato de amônio e de ácido cítrico sobre fertilizantes fosfatados. Revista Brasileira de Ciência do Solo. 3, 173−178.

de Luz, P.H.C., Vitti, G.C., 2008. In: Marques, M.O., et al., (Eds.), Manejo e uso de fertilizantes para cana-de-açúcar. Tecnologia na agricultura canavieira. FCAV, Jaboticabal, pp. 140−167.

Demattê, J.L.I., 1986. Solos de baixa fertilidade: estratégia de manejo. In: Sem. Agroindustrial, 5; Semana Luiz de Queiroz, 29., 1986. Anais. Piraicaba (mimeografado).

Está citado de forma correta na Paragraphina 87. Apenas retirar o último 1986, pois aparece duas vezes. Em função disso o que está errado é o rodapé da Table 4.9, ou seja, deixar escrito apenas Source Demattê 1986.

Lara Cabezas, W.A.R., Korndörfer, G.H., Motta, S.A., 1997. Volatilização de N−NH3 na cultura de milho. In: II Avaliação de fontes sólidas e fluidas em sistema de plantio direto e convencional; Congresso de Plantio Direto, vol. 1, Ponta Grossa − PR (18/03/1996).

Luz, P.H.C., Vitti, G.C, 2008. Manejo e uso de fertilizantes para cana-de-açúcar. In: Marques, M.O, et al., (Eds.), Tecnologia na agricultura canavieira. FCAV, Jaboticabal, pp. 140−167.

Oliveira, E.C.A., 2008. Dinâmica de nutrientes em cana-de-açúcar em sistema irrigado de produção. Dissertação (Mestrado Agronomia), Universidade Federal Rural de Pernambuco, Recife.

Oliveira, M.W., Trivelin, P.C.O., Penatti, C.P., Piccolo, M.C., 1999. Decomposição e liberação de nutrientes da palhada de cana-de-açúcar em campo. Pesquisa Agropecuária Brasileira. 34 (12), 2359−2362.

Orlando Filho, J., 1993. Calagem e Adubação da Cana-de-açúcar. In: Câmara, G.M.S., Oliveira, E.A.M. (Eds.), Produção de Cana-de-açúcar. FEALQ/USP, Piracicaba, pp. 133−146.

Rosseto, R., Korndörfer, G.H., Dias, F.L.F., 2008. In: Marques, M.O., et al., (Eds.), Nutrição e adubação da cana-de-açúcar. In: Tecnologia na agroindústria canavieira. FCAV, Jaboticabal, 125−139 p.

Sousa, D.M.G., Lobato, E., 1988. Adubação fosfatada. In: Simpósio sobre o Cerrado, 6., 1982, Brasília. Planaltina: Embrapa − CPAC. pp. 33−60.

Vale, F., Araújo, M.A.G., Vitti, G.C., 2008. Avaliação do estado nutricional dos micronutrientes em áreas com cana-de-açúcar. In: Fertibio 2008: Desafios para uso do solo com eficiência e qualidade ambiental. Londrina − PR. p. 80.

van Raij, B, Cantarella, H., Quaggio, J.A., Furlani, A.M.C., 1997. Recomendação de adubação e calagem para o Estado de São Paulo. second ed. Instituto Agronômico de Campinas, Campinas, 285 p. (Boletim 100).

Vitti, G.C., Mazza, J.A., 2002. Planejamento, estratégias de manejo e nutrição da cana-de-açúcar. Informações Agronômicas. 97, 16 (Encarte Técnico).

Vitti, G.C., 1988. Avaliação e interpretação do enxofre no solo e na planta. FUNEP, Jaboticabal, 37 p.

Vitti, G.C., Mazza, J.A., Quintino, T.A., Otto, R., 2006a. Nutrição e adubação. In: Ripoli, T.C.C. et al. (Eds.). Plantio de cana-de-açúcar − estado da arte. Piracicaba. pp. 102−144.

Vitti, G.C., Oliveira, D.B., Quintino, T.A., 2006b. Micronutrientes na cultura da cana-de-açúcar. In: Vanzolini, S. et al. (Eds.). Atualização em produção de cana-de-açúcar. Piracicaba. pp. 120−138.

Vitti, G.C., de Luz, P.H.C., Malavolta, E., Dia, A.S., de Serrano, C.G.E., 2008. Uso de gesso em sistemas de produção agrícola. Piracicaba − SP, GAPE. 104p.

CHAPTER 5

Management of Pests and Nematodes

Newton Macedo[1], Daniella Macedo[1], Maria Bernadete S. de Campos[2], Wilson R.T. Novaretti[3] and Luiz Carlos C.B. Ferraz[4]

[1]Imaflora, Piracicaba, SP, Brazil [2]Universidade Federal de São Carlos, UFSCAR, São João Del Rei — MG, Brazil [3]Federal University of Juiz de Fora, Juiz de Fora, MG, Brazil [4]Universidade de São Paulo, ESALQ, São Paulo, SP, Brazil

Introduction

The number of insect species that occur in sugarcane crops varies according to the plant's phenology and region; with some of them being able to develop populations that result in considerable damage to producers. The concept of "Pest" is economical, dynamic, and depends on a series of factors: ecological (population level and season occurrence); economical (economic value of the culture, control objectives and costs); social (region development and historical moment); cultural (technical level of the producer); and mainly the interaction among these factors. Thus, insect population control strategies must be rather dynamic in order to comply with the utilization of different options, according to interests of the moment.

5.1 Pests

Approximately 85 species of insects are flagged as causing damage to agro-sugarcane in Brazil. Among them, some are considered to be important pests, with either national or regional coverage.

The objective of this chapter is to summarily describe the main pests, as well as their control methods, in the sugarcane agroecosystem complex.

Additional information can be found in Pietro Guagliumi (1973), Gallo et al. (2002) and in numerous journal scientific articles and divulgation of ex-Planalsucar and Copersucar researchers.

5.1.1 Sugarcane Borer

Scientific names

Diatraea saccharalis (Fabr. 1794) and *D. flavipennella* (Box, 1931) (Lepidoptera: Crambidae).

Distribution

D. saccharalis occurs in all regions where sugarcane is cultivated in Brazil. *D. flavipennella* occurs in the Northeast, North and in some of the Southeast States, but was not flagged in the South.

Description

Moths are straw yellow, with some brownish designs, measuring around 25 mm in size (Figure 5.1). They usually lay eggs on the back side of the leaf, grouped in 5−50 eggs, looking like fish scales (Figure 5.2). After incubation, which varies from 4−8 days, larvae hatch and initially feed on the leaf parenchyma and/or sheath. Around 2 weeks later, they penetrate the soft part of the stalks, just above the node, opening galleries inside it. Their complete growth takes 40 days on average, when they reach around 30 mm. They present a milky-white coloring, with numerous back and side chestnut spots, and have a dark brown (*D. saccharalis*), or yellow (*D. flavipennella*) cephalic capsule (Figure 5.3). At the end of the larval cycle the larvae open an orifice in the sugarcane stalk, closing it with silk threads and food leftovers. The larva then transforms itself into a chrysalis, remaining in this stage for 7−14 days, when the moth is born (Figure 5.4), which exits through the previously opened orifice.

Damage and economical importance

Larvae, when attacking new sugarcane, cause "Dead Hearts" (Figure 5.5). In adult sugarcane, loss of weight, side sprouting, breakage and internode atrophy occur. Penetration of fungus (*Fusarium moniliforme* and/or *Colletotrichum falcatum*) in the galleries opened by the borer results in red rot (Figure 5.6), which determines sucrose inversion, reduction of

Figure 5.1
Diatraea saccharalis adult.

Figure 5.2
Diatraea saccharalis eggs.

Figure 5.3
Diatraea saccharalis larva.

juice pureness, and increase of gums, thus reducing sugarcane industrial productivity. There is a correlation between the infestations of the borer/rot complex — given by the formula: II% = [(damaged internodes/total internodes) × 100] — and losses noted in the field and in the industry. That correlation, in conservative numbers, is expressed by: 1% of II = 0.77% of sugarcane ton losses in the field, and 0.25% of the recoverable sugar in the industry. Lifting of II% is performed at the cutting fronts or upon arrival of the sugarcane at the industry patio, during the harvest season. At the cutting fronts (mechanized harvest), five points are

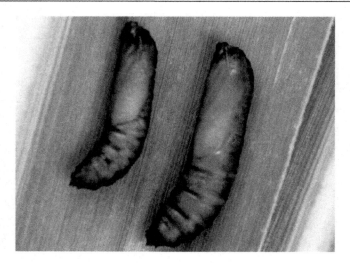

Figure 5.4
Diatraea saccharalis pupas.

Figure 5.5
Attack symptom (dead heart) of *Diatraea saccharalis*.

sampled, positioned in "X", per area of up to 50 ha, of 25 sugarcanes/point, collected: 5 sugarcanes (5 steps) + 5 sugarcanes (5 steps) + 5 sugarcanes (5 steps) + 5 sugarcanes (5 steps) + 5 sugarcanes, totaling 125 sugarcanes. In the patio, five sugarcanes must be casually retrieved per load volume that arrives at the industry. In both methods, sugarcanes

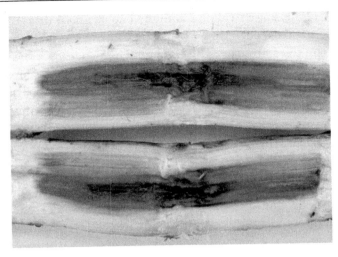

Figure 5.6
Red rot.

are opened lengthwise, counting the total number of internodes and the internodes attacked by the borer/rot complex. These data are introduced in the II% formula; the result of which is to be weighted by the area.

Control

The borer is biologically controlled, through mass releases of the *Cotesia flavipes* parasitoid, originally from India and Pakistan. In the Northeast, C. *flavipes* was introduced in 1974, by Mendonça (1974), and in the Center-South in 1978, by Macedo and Mendes (1978). It is presently mass-produced in laboratories. Release locations and timings are set according to surveys of apt larvae populations (1.5 cm or more) during plantation development. According to the composition, more susceptible varieties, first cuts, ferti-irrigated areas, and where the pest attack symptoms appear (dead heart and bored sugarcane), a larvae collection is carried out by randomly entering the plantation, searching for attacked stalks. The collection timing is set to one working hour/range, independently of the area. The ideal release timing is when the larvae are already making galleries in the stalk, with the presence of feces, and with the size greater than 1.5 cm (average and big larvae). When there are few infested areas, and availability of *Cotesia*, releases must be performed in areas that present the highest infestation indexes. When there are many infested areas, and limited *Cotesia* quantities, releases must be performed in areas where average collections are greater than 10 larvae/hour/man, covering all problem areas.

Releases must be preferentially performed during late afternoon, when more than 70% of the C. *flavipes* have already emerged in the conditioning glasses, as follows: enter the range in the sugarcane line sense, walking with the cup (with 10 masses) opened, and at every certain number of steps, according to spacing, deposit the masses into the sugarcane sheath, keeping the cup to return, cover all the problem area, range, or block, before transferring the operation to another area; and in ranges where internal walking is difficult, such as in lying sugarcane, perform the releases surrounding them and penetrating around 25 m from the tracks. It is better to release quantities from 6000 to a maximum of 9000 wasps/ha/time, repeating the operation when surveys indicate the requirement for new releases.

Maturating sugarcane fields, with 30 days or less for harvesting, must not receive those parasitoids. The parasitoids performance is also assessed through collecting biological forms 21 days after release, collecting at least 30 biological forms per release area. Collected material must be kept in a room with appropriate environmental conditions, in small boxes containing food, for 10–12 days, in order to assess the final parasitism.

Releases must be repeated in areas that present low parasitism (less than 20%) and in which larvae populations are still high. Parasitism is obtained by the following formula:

$$\%P = [(\text{total parasitized larvae}/\text{total parasitized larvae} + \text{healthy larvae}) \times 100].$$

In a natural way, numerous predators, especially several types of ants, carry out an important role in borer control by striking 70–80% of the borer eggs. *Trichogramma galloi* has not proven technically or economically feasible through mass production and release.

Chemical control (spray with insecticides compatible with biological control) can be performed in accordance with surveys at the plantation, according to the compound: more susceptible varieties, first cuts and ferti-irrigated areas. The following method must be adopted: five 25-sugarcane casual points are sampled: [5 sugarcanes (5 steps) + 5 sugarcanes (5 steps) + 5 sugarcanes (5 steps) + 5 sugarcanes (5 steps) + 5 sugarcanes], totaling 125 sugarcanes, representing the range (of any size, as long as from the same variety and same cut). In this work, the first leaf sheaths of the heart of palms region which are already slightly apart from the internode (generally the heart of palm's third or fourth sheath, counting from the top) are observed. Insecticide application is decided when 3% or more sugarcanes with the presence of live larvae are found, independently of the quantity of larvae/stalks (small second and/or third instar larvae, up to 1 cm, which have not yet penetrated the sugarcane). The insecticide used is Triflumuron (480 SC, 60 at 80 ml/ha) in air spray. Once the above index is reached, spraying is to be performed as quickly as possible, within a maximum deadline of one week. On the fifth week after spraying, the area has to be monitored once again for an eventual new spraying or release of parasitoids.

5.1.2 Giant Borer

Scientific name

Telchin licus (Drury, 1773) (Lepidoptera: Castinidae).

Distribution

Predominant in the Northeast, was recently found in other agro-sugarcane regions, such as Nova Olímpia (MT) and Iracemápolis (SP).

Description

The adults are of dark coloring, with dots and white transversal stripes on the forewing. The hindwing presents a larger transversal stripe, with white and reddish dots on the external edge (Figure 5.7). The adults (moths) have daytime habits and measure around 90 mm in size. Borers lay eggs at the base of old clumps, in quantities of 50–100 eggs. The eggs have five lengthwise borders, and their incubation period is 7–14 days. The larvae have a milky-white coloring, with brown spots next to the head. When completely developed, they measure 80–90 mm in length. At the end of the larval period, at around 110 days, the larvae build a shelter with food leftovers, staying there for 30–45 days, in the chrysalis phase, when they pop up as adults, living for 10–15 days. The complete cycle normally takes 6 months. The larvae open lengthwise galleries in the stalks, starting from the "rhizome", reaching the base internodes (Figure 5.8). When harassed, the larvae shelter

Figure 5.7
Telchin licus adult.

Figure 5.8
Telchin licus larva and damage.

themselves, moving backward in the galleries, until below ground level. Doing that, they are able to survive sugarcane cutting at harvest time, and as an additional defense resource, close up the gallery with food leftovers, in less than 1 h after the stalk is cut.

Damage and economical importance

The greatest damage occurs in ratoons, where the larva, due to its feeding habit being similar to the *Diatraea*, causes the "dead heart" in new sprouting. In adult sugarcanes, larvae cause loss of weight, a favorable situation for the development of red rot, sugarcane toppling, and consequently the partial or total destruction of the ratoons.

Control

The conventional control, through the application of insecticides and/or entomopathogenic fungus over the ratoons, has not presented satisfactory efficiency. There are indications that application of insecticides by means of equipment coupled to the sugarcane harvester, which sprinkles the insecticide syrup just after the base cutting blade action of the stalks, significantly improves the efficiency of the product. Culture treatments that perform effective destruction of ratoons and culture leftovers assure less population pressure on new plantations. Manual grooming of larvae and chrysalises, utilizing a hoe and adults captured with a net is still a recommended control method (Figure 5.9).

Figure 5.9
Telchin licus manual control.

5.1.3 Termite

Scientific names

Heterotermes tenuis (Hagen, 1858); *Procornitermes* spp.; *Neocapritermes* spp.; *Nasutitermes* spp.; *Syntermes molestus* (Burmeister, 1839); *Amitermes* spp.; *Rynchotermes* spp.; *Cornitermes* spp. (Isoptera: Termitidae).

Distribution

In all agro-sugarcane regions, with greater populations in the savanna regions.

Description

They are social insects, the laborers of which are wingless and sterile, have a white or straw yellow coloring, constituting the greatest part of the termitary population. The soldiers, also sterile, are responsible for defending the colony. The termites have a large head, of yellowish brown coloring, with well developed mandibles. Those characteristics are utilized to identify the type and species (Figure 5.10). The colonies start with the couples: King and Queen. After the bridal flight, they nestle in the ground and start the production of nymphs, which become sterile laborers and soldiers. The average lifespan of the laborers varies from 3 to 6 months, while the lifespan of the couple can last for up to 5 years. During the life of a colony, the king and/or the queen can be replaced by differentiated laborers, thus ensuring the longevity of the colony. Generally colonies are situated underground, have diffused chambers at different depths on the ground profile, and can reach to more than 4 m.

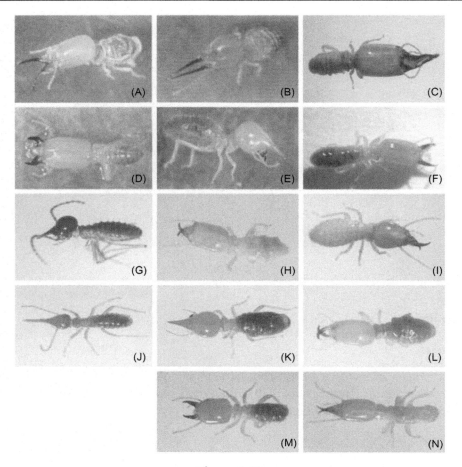

Figure 5.10
Main termite kinds that occur in the Southeast and Northeast regions of Brazil.
(A) *Ortho gnathotermes* sp., (B) *Dentispicotermes* sp., (C) *Neocapritermes opacus* and *N. parvur*,
(D) *Cornitermes cumulans e C. bequaerti*, (E) *Embiratermes* sp., (F) *Syntermes nanus*,
(G) *Nasutitemes* sp., (H) *Cylindrotermes* sp., (I) *Coptotermes* sp., (J) *Rhynchotermes* sp.,
(K) *Armitermes* sp., (L) *Amitermes* sp., (M) *Procomitermes triacifer* and P. *striatus*,
(N) *Heterotermes tenuis e* H. *longiceps*.

Damage and economical importance

They cause damage by attacking stems, damaging the buds, and consequently jeopardizing germination. In adult sugarcanes, the termites attack the roots and open galleries in the base internodes, destroying the tissues and causing stalk drying. Average losses caused by underground termites are estimated to be 10 t of sugarcane/year/ha (Novaretti, 1985).

Table 5.1: Control level, according to the percentage of points with the presence of termites or damage caused by them.

Termite Species	NC (Percentage of Points with the Presence of Termites and/or Damage)
Heterotermes, Amitermes	>05
Nasutiterme	>25
Cylindrotermes	>25
Neocapritermes	>25
Syntermes	>25
Rhynchotermes	>25
Procornitermes	>25
Neocapritermes + Procornitermes + Nasutitermes + Anaplotermes + others	>25
Cornitermes	>40

Table 5.2: Number of sampling points according to the area dimensions.

Area	Sampling Points (1 ha)	Clumps or Bait/Point
Up to 10 ha	3	20
11 to 50 ha	4	20
>50 ha	5	20

Control

Control is preventive, through spraying in the plantation, in the seedling coverage operation. Whenever possible it has to be based on sampling (in which kinds and/or species presenting damage in bait (expansion areas) or clumps (reform areas) are assessed (Table 5.1). The population note is dispensable, as it deals with social insects and is a variable datum, according to the assessment day time and soil humidity. Sampling must be of 20 sampling points/ha, with a diagram and number of points/area, according to Table 5.2 and Figure 5.11.

Whenever sampling is not possible, control must be performed based on the history of the region. The product with the best control performance is Fipronil (800WG), at the dosage of 250 g.p.c./ha.

5.1.4 Sugarcane Leaf and Root Cicada

Scientific names

Mahanarva posticata (Stal, 1855) (leaf) and *M. fimbriolata* (Stal, 1854) (root) (Homoptera: Cercopidae).

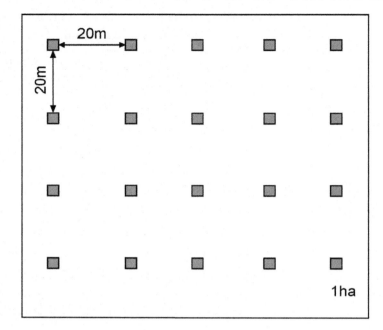

Figure 5.11
Clumps or trenches distribution diagram for sampling.

Distribution

M. posticata is predominant in the Northeast and *M. fimbriolata* in all agro-sugarcane regions.

Description

Adult leaf cicadas live in the aerial part of the plants, sucking up the juice from the leaves. The males are of brown–reddish coloring, measure around 12 mm in length, and present two spots on the apical of the forewings. The females measure around 14 mm, are darker, and without the spots presented by the males. The laying of eggs is done in the sheaths of the lower leaves, where around 100 eggs are deposited during the lifespan of the cicada. The nymphs, upon hatching, direct themselves to the sugarcane tops, lodging themselves in the leaf sheaths and protecting themselves with characteristic foam (Figure 5.12). The nymph phase lasts from 34 to 65 days, whereupon they pop up as adults. The complete evolutionary cycle is around 80 days, with the greatest swarming over occurring during the high rainfall months.

The root cicada adults are also seen in the aerial part of the plants, but prefer the clump base, sucking the juice from the roots. They are of variable coloring, with dark red predominating, and black lengthwise stripes on their wings (Figure 5.13). The females lay their eggs in the lower sheaths, next to the neck of the plant. The nymphs, soon after

Figure 5.12
Characteristic foam of the *Mahanarva posticata* nymph.

Figure 5.13
Mahanarva fimbriolata adults.

hatching, attach themselves to the superficial roots on the ground to suck the juice. They produce an exudation similar to soap foam, that serves as a shelter and protection against dissection. The complete evolutionary cycle is from 60 to 80 days, with the highest infestations occurring during hydric excedents in the soil and high temperatures.

Figure 5.14
M. fimbriolada nymph characteristic foam.

Damage and economical importance

The *M. posticata* adults, upon sucking the juice from the leaves, inject toxins that cause loss of color and later "burning", reducing the photosynthesis capacity of the plant. This results in the shortening of the internodes and loss of weight and sugar. The nymphs cause less damage, as they do not inject toxins while feeding. The *M. fimbriolata* adults, although feeding themselves on leaves similarly to their nymphs, nestle in the roots of the plant's neck (Figure 5.14), injecting toxins and causing the death of the roots. With that, the following also occur: death of the tillers, shortening of the internodes, side spring ups, stalk splitting and drying of leaves, with a substantial loss of stalk weight, sugar quantity and quality.

Control

Control must be based on population monitoring, in order to detect the appearance of the first generation, which occurs in the second semester, during Spring/Summer, after an accumulation of over 70 mm of rain, in the Center-South region. Population monitoring must be the simplest possible, in order to be agile and of low cost. One option is to sample 10 points of a meter/range, independently of the area, distributed in "X", examining the base of the clump on both sides, and counting the nymphs and adults present in the foam.

In case of data repetition, the average number of biological forms is to be calculated (adults + nymphs)/m. If there are no repetitions, it is recommended to extend the survey to 20 points/range. The survey validity is of a maximum of 15 days. Chemical control with neonicotinoid (Imidacloprid and Thiamethoxam) insecticides has resulted in a better economical return. The application of products as soon as the first nymphs appear (first generation) avoids irrecoverable agriculture losses, besides the additional benefit of the physiological effect of the products, with greater gain in productivity. *Metarhizium anisopliae* fungus is the biological control option, although, in order to obtain success, it has to be of good quality (adequate cepa, purity, and germinating capacity) and must only be utilized under the following conditions: persistent raining period and closure of the sugarcane field.

5.1.5 Migdolus

Scientific name

Migdolus fryanus (Westwood, 1863) (Coleptera: Cerambicidae).

Distribution

It is found in the States of São Paulo, Paraná, Mato Grosso do Sul, Mato Grosso, and Goiás.

Description

The males are black, with developed hind wings. The females are brown, with atrophic hind wings, and therefore do not fly. Both measure from 18 to 25 mm in length. The females, after coupling, which is performed on the soil surface or next to it, penetrate the soil, digging a new gallery or utilizing the same opened channels for their exit. A few days later, they lay eggs, which occur at different depths, along the soil profile. The larvae are milky-white, reaching 40 mm in length when completely developed (Figure 5.15). In the final larval stage they build a pupal chamber at great depth at the end of the gallery, where they are transformed into pupa. Three to four months before flying again, which occurs from December to March, the adults emerge. They initially have little mobility, remaining in the pupal chamber for a certain period of time. But they later start climbing to the surface of the soil, and are quite jittery on the occasion of their flying again (coupling). On that occasion the males perform short flights, guiding themselves by the female's sexual pheromone, for tracking and copulation purposes. Each female can copulate with several males. All development conditions can be found in depths that are greater than 2 m, and the complete evolutionary cycle lasts from 2 to 3 years.

Figure 5.15
Migdolus fryanus larvae and adults.

Damage and economical importance

The attack normally occurs in coppices, sugarcane plant, and also in subsequent cuts. The larvae initially feed on roots, in depth, moving up until the stems or "rhizome" region. The attacked "rhizomes" produce few roots, resulting in the drying of the clump at the coppices, which can be easily plucked out (Figures 5.16 and 5.17). Production losses caused by this pest are extremely great.

Control

Efficient control combines cultural practices with insecticides at the time of the sugarcane field reform.

Elimination of the ratoon is recommended when there is a greater accumulation of larvae next to the soil surface (June to September), the triple heavy leveling (using 18×36' grid), thus ensuring the elimination of the greatest possible number of *Migdolus* larvae.

Endosulfan insecticide (350CE at a dosage of 12 L/ha) must be applied before the plantation, utilizing high flow nozzle sprayers (300—400 L of juice/ha), in subsoiling or deep plowing operation, utilizing an earth board or harrow. Fipronil (800WG) is applied in the plantation for spraying the plantation furrow over the seedling, before covering, at the rate of 250 g p.c./ha.

Figure 5.16
Damages caused by *Migdolus fryanus* in formed sugarcane field.

Figure 5.17
Damages caused by *M. fryanus* in sugarcane plant.

Those operations can be restricted to the pest coppices. Endosulfan (8 L.p.c./ha) or fipronil (250 g.p.c./ha) can be applied on ratoons, directly over the plant's ratoon or on both sides of the line, in prematurely harvested sugarcane fields (April/May).

5.1.6 Elasmo Larva

Scientific name

Elasmopalpus lignosellus (Zeller, 1848) (Lepidopitera: Pyralidae).

Distribution

In all sugarcane producing regions of Brazil.

Description

The moths present a yellowish brown to dark gray coloring, and measure from 15 to 25 mm in length. The larvae have the first segment of the body and the head in a dark brown or almost black coloring. The body coloring is almost predominantly bluish green. They initially feed on leaves and then on the lower part of the bud, just below the soil level. They build a typical mixed gallery made of web and earth, connecting the shelter to the outside. The attacks occur with greater intensity in sandy soils, and during dry seasons.

Damage and economical importance

The larvae attack the buds on the neck region, opening galleries. Necrosed tissues are found around the attacked region. Damages initially cause leaf color fading and later the death of the apical yolk, characterizing it with the symptom called "dead heart". When in the sugarcane plant, the attack causes "stand" failure. Attacks in sprouting of the second cutting and harvesting of sugarcane cause severe damage, especially during long dry seasons.

Control

Cultural treatments that assure good culture germination are recommended, through appropriate soil preparation, fertilization and plantation season. Leftover host cultures must also be eliminated (rice, corn, wheat, soy, bean, etc.). As the insect occurs in a dry environment, maintaining the soil humidity helps to minimize its damages. Irrigation or utilization of vinasse reduces the problem. Chemical control is not recommended. Mechanized harvesting of the sugarcane drastically reduces elasmo attack possibility in the ratoon.

5.1.7 Hairy Borer

Scientific name

Hyponeuma taltula (Schaus, 1904) (Lepidoptera: Noctuidae).

Distribution

Distributed in the States of São Paulo, Rio de Janeiro, Espírito Santo, Alagoas, and Pernambuco.

Description

The larvae, similarly to *Diatraea saccharalis*, present a dark brown head, long bristles, and have no spots on the body (Figure 5.18). The adults are brown, presenting a circular dark spot next to the center of the forewing (Figure 5.19).

Damage and economical importance

The larvae open galleries at the base of the stalks, propitiating the invasion by fungi, which results in the death of the stalk in its young phase (tiller) or adult one (Figure 5.20).

Figure 5.18
Hyponeuma taltula larva.

Figure 5.19
Hyponeuma taltula female and male.

Figures 5.20
Damage caused by the *Hyponeuma taltula*.

Control

Application of high systemic action insecticides, such as Carbofuran and Aldicarb, in the ratoons just after harvesting, over the sugarcane line or on the sides, according to the size of the plantation.

5.1.8 Sugarcane Weevil

Scientific names

Metamasius hemipterus (L., 1765) and *Sphenophorus levis* (Varrie, 1978) (Coleoptera: Curculionidae).

Distribution

M. hemipterus is distributed in all agro-sugarcane regions of Brazil. *S. levis*, up to now, was found in the main sugarcane producing regions of São Paulo and Paraná.

Description

The *M. hemipterus* adults are similar to the *S. levis* ones, they are, however, more agile and present an oranged coloring, with very visible black dots (Figure 5.21). The females present

Figure 5.21
(A) *Metamasius hemipterus* adult and (B) *Sphenophorus levis* adult.

two dark spots on the back of the head, and the males a clear one in a half-moon form. The *S. levis* adults move slowly and pretend to be dead whenever they are touched. They are generally found in sugarcane clumps, below soil level. The males are normally smaller and more hairy than the females, mainly on the forelegs, next to their insertion in the prothorax. They have a dark-reddish coloring and measure from 10 to 15 mm in length. The *M. hemipterus* eggs are usually put in holes opened by the borer or in the stalk splittings. The larvae open irregular galleries, feeding themselves with stalk tissues, generally above soil level (Figure 5.22). In its last instar, the larva builds a shelter with food leftovers inside the stalk, transforming itself into a pupa. The *S. levis* females lay their eggs below soil level, in the rhizomes, in lesions caused by their mouthparts. Their larvae feed on tissues in that region, making irregular galleries (Figure 5.23). In the last instar they prepare a pupal chamber and transform themselves into pupas. Both species present a 3- to 8-month life cycle. The greater adult population occurs during the hottest months of the year, while the larvae prevail in colder seasons.

Damage and economical importance

The attacks occur in the sugarcane plant as well as in the ratoons. The symptoms are: plant loss of color, appearance of dead sugarcanes in the clump, new sprouting failures in sugarcane ratoons, and accumulation of sawdust in gallery openings, which are normally larger than those of the *D. saccharalis*.

Figure 5.22
Metamasius hemipterus larva.

Figure 5.23
Sphenophorus levis larva and gallery.

Control

Control must be based on larva and adult population monitoring, which consists of the plucking of two coppices/ha, casually distributed, counting the total rhizomes of the coppice; and affected rhizomes, with and without the presence of biological forms (adults/larvae/pupas). With these data, the attack percentage can be calculated. This sampling is to be performed soon after harvesting. Estimated losses, according to the tillers' infestation index, are: 5% = 4−6 t/ha; 10% = 8−12 t/ha; 15% = 12−18 t/ha; 20% = 16−24 t/ha; above 20% = 30 or more t/ha. Monitoring of the adult population can be performed by means of bait distributed in the sugarcane field and its main goal is to detect the presence of the insect in the area. The baits are 20−25 cm sugarcane stems, split, impregnated with insecticide juice (25 g of Carbaril 85% PM + 1 L water + 1 L molass), arranged in pairs in the interline and covered with straw, with the minimum of 10 baits per hectare. The adults are counted after 5−10 days. The baits must be changed after 20 days. When infestation is higher than 10% in the reform area, perform a good ratoon destruction, preferably with ratoon destruction equipment. In all infestation situations, destruction of the ratoon for the sugarcane field reform must be performed during the period of greatest larva and pupa population, which goes from May to September. It is recommended to perform the chemical control when the infestation percentage is higher than 5%, according to the calculation already presented, in ratoons that are still going to remain for more cuts. The product is to be applied over the sugarcane line in the recently harvested area, or laterally when the sugarcane is already completely sprouted. Here are some chemical products and dosage options for different situations:

    i. Fipronil 800WG + Alfa-cypermethrin 100CE or Bifenthrin 100CE (150 g + 1.5 L/ha).
    ii. Imidacloprid 480SC (1.5 L/ha).
    iii. Fipronil 800WG + Carbofuran 350SC (250 g + 6 L/ha).

Option (i) must be utilized in the ratoon from May to October, and in sugarcane plant when there are no nematodes in the area. Option (ii) is indicated for ratoons that are subject to the attack of the *Mahanarva fimbriolata*. In case of simultaneous presence of nematodes, option (iii) must be preferred in the plantation.

In order to avoid pest dissemination by the seedling, the first measure is to obtain seedlings from nurseries exempt of insects. When this is not feasible, seedling purge is required, applying an insecticide of the pyrethroid group (in 0.01% concentration, for example), spraying over the seedling, then on the cut, which must be tall. The product kills or repels the *Sphenophorus* females, assuring the transportation of the seedling without the presence of the adult insect, and/or its eggs.

5.1.9 Maggot

Scientific names

Eutheola humilis (Burm., 1847); *Ligyrus tuberculatus* (Beauv., 1808); *Sternocrates laborator* (Fabr., 1801); *Cyclocephala* spp. (Coleoptera: Scarabeidae).

Distribution

In all sugarcane producing regions of Brazil.

Description

They are brown or almost black beetles, measuring from 10 to 30 mm in length (Figure 5.24). The females lay eggs next to the sugarcane or ratoon stem. The recently hatched larvae feed on roots of new sprouts, sugarcane stems, or rhizomes in ratoons. They are of milky-white coloring, recurvated, with a brown and well-developed head. They have three pairs of well-developed thoracic legs, and a voluminous abdomen extremity (Figure 5.25). The larval period is 12–20 months. The pupas shelter themselves in the soil, inside a chamber, where they remain for 12 days, whereupon they transform themselves into adults.

Figure 5.24
Maggot adult.

Figure 5.25
Maggot larva.

Damage and economical importance

The larvae occur in sugarcane plant as well as in ratoons, feeding on underground parts. The attack is usually on cappices, and the symptoms are: plant's loss of color and appearance of dead sugarcanes in the clump. Low populations (average around three individuals/clump) during the plantation development period (Spring/Summer) is perfectly bearable by the plant and may even have a beneficial role in organic material mineralization and soil aeration.

Control

i. Reform areas — prepare the soil well, preferably during winter, with superficial destruction of the ratoon by means of grid or ratoon destruction equipment, followed by deep plowing (earth board or harrow) and subsoiling in compacted soil situations.

ii. Ratoon areas with more than five individuals/ratoon — proceed with chemical control. The best options are neonicotinoids (Imidacloprid and Thiamethoxam), in terms of control efficiency, low environmental impact, and gain in productivity. Both products can be applied directly over the sugarcane line or laterally, slightly incorporated by means of disks, or by jet spraying in already developed plantation, reaching 30% of the clump's aerial part and 70% of the base.

CHAPTER 6

Disease Management

Sizuo Matsuoka[1] and Walter Maccheroni[2]

[1]Vignis S.A., Santo Antonio de Posse, State of São Paulo, Brazil [2]AGN Bioenergy, São Paulo, State of São Paulo, Brazil

Introduction

Over a hundred diseases affect sugarcane crops worldwide (Rott and Girard, 2000). Among them, about 10 are economically relevant over a certain period of time for a specific country or region, either because of their direct impact, or because they are a future threat. It is worth noting that in sugarcane crops all diseases have their most effective control determined by the genetic resistance of the varieties, since they are created in several breeding programs around the world. Nevertheless, since most disease resistances in this kind of crop are of quantitative rather than qualitative nature, i.e., the resistance is not absolute but gradual, many varieties being cultivated may show a specific level of susceptibility to some diseases. Thus, in spite of the good level of health of sugarcane fields in general, if compared to other crops, some alternative practices of management and control may be necessary in order for the losses caused by diseases to be minimized. Unlike most other crops, sugarcane crops are not usually sprayed with agrochemicals such as fungicides and bactericides.

In this chapter, we will describe only nine most prevalent diseases that are currently considered to be the most important Brazilian sugarcane crop. Of these, the disease called orange rust is the last one arriving in Brazil. Orange rust appeared in Brazil in the second half of 2009, just when the Portuguese version of this book was in the print. Illustrations are available in Figures 6.1 and 6.2. The diseases are presented according to causal agent, either virus, bacteria, or fungus.

6.1 Diseases Caused by Viruses

6.1.1 Mosaic (Sugarcane Mosaic Virus Disease — SCMV)

History

In the beginning of the 20th century, producing countries, including Brazil, cultivated varieties of sugarcane known as noble (*Saccharum officinarum*). These varieties were,

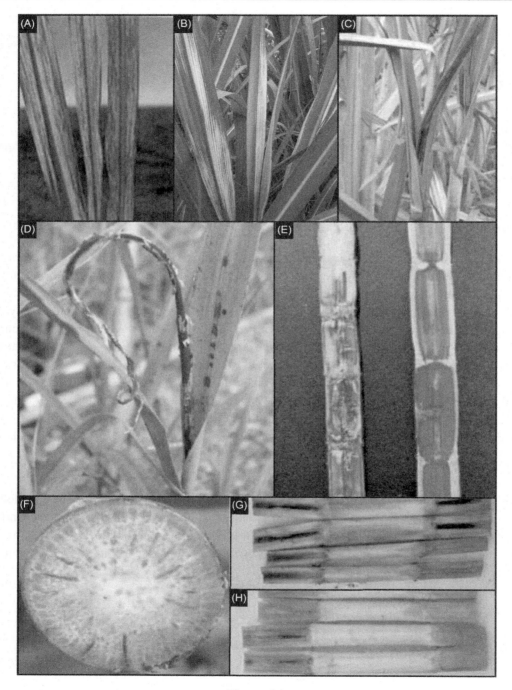

Figure 6.1
Symptoms of the main sugarcane diseases in Brazil: (A) mosaic; (B) leaf scalding; (C) red stripe;
(D) smut; (E) red rot (to the right) and fusarium rot (to the left); (F) ratoon stunting; (G)
pineapple disease (susceptible variety); and (H) pineapple disease (resistant variety).

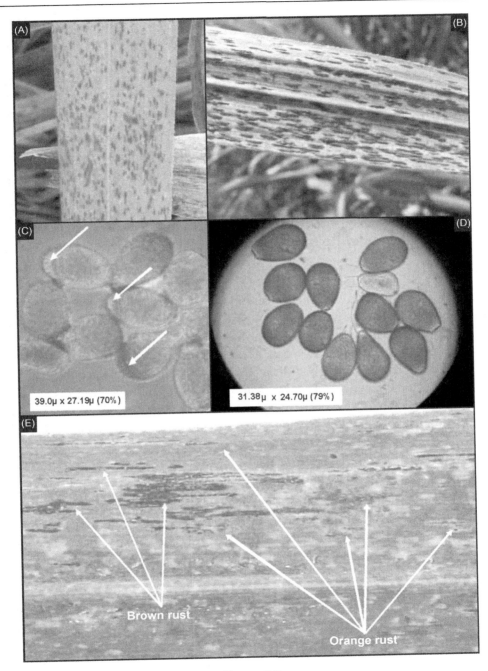

Figure 6.2

Characteristics that distinguish brown and orange rust: foliar symptom of brown rust (A); foliar symptom of orange rust (B); brown rust spores, in which the characteristic thickening of the wall is indicated by arrows and the ratio between length and width is 70% (C); orange rust spores, which have a length width ratio of 79% (D); symptoms of brown and orange rusts in the same leaf (E). *Pictures kindly released by Werner Ovalle (Cengicaña, Guatemala).*

mostly, highly susceptible to mosaic and great losses were taken. With the creation of hybrid canes by breeding programs, new varieties resistant to mosaic began to be grown and reduced the importance of the disease (Koike and Gillaspie, Jr., 1989). Nevertheless, since the virus is endemic, from time to time disease foci appear in some places.

Causal agent

Mosaic is caused by the sugarcane mosaic virus (SCMV). This virus is classified into four serum types and many strains. The diseases caused by these strains are sugarcane, sorghum, corn, and Johnsongrass mosaics (Shukla and Ward, 1994; Tosic et al., 1994). The strains differ not only in host but also in the ability to create an infection and the level of symptoms and losses. The differentiation between strains is made by specific indicator plants and distinctive symptoms. At least 14 strains, identified by letters (from A to N), have been described. In Brazil, in sugarcane, the most common form is B (Matsuoka, 1999).

Symptoms

The symptoms are basically foliar and can vary in intensity, according to the resistance of the variety in question, cultivation conditions and virus strain. The most characteristic symptom appears in the form of a foliar area with contrasting intensities of green. Frequently it shows regions of intense green circulated by a pale green background or chlorotic areas of yellowish color. The chlorotic areas are usually diffuse but in certain varieties infected with specific strands, they can be more defined. These symptoms are always more evident in new leaves. Young plants with vigorous growth are more susceptible than mature plants with slower growth. In older plants, symptoms may vanish, especially in more resistant varieties.

Dissemination and predisposing conditions

Natural dissemination of the virus happens through various species of aphids that do not have sugarcane as a host. The alate aphid, after landing on an infected plant, performs a "biting test" and upon realizing it is not an adequate plant, flies off again, now with its stylet externally contaminated by the virus and, after landing on another plant, makes another "biting test" thus transmitting the virus (Matsuoka, 1999). Weeds can be infected by virus strains and serve as alternative hosts and sources of contamination for dissemination into cane fields, but the main source is always the sugarcane plant. Another important form of dissemination is the usage of setts of infected sugarcane in the establishing of nursery or commercial cane fields. The virus is not tolerant to high temperatures, so that it is more common in temperate climates, i.e., from the central part of São Paulo towards the south.

Control

The use of resistant varieties is the most efficient method of control. The control of aphids through insecticides has no efficiency whatsoever (Matsuoka, 1999). *Roguing* to remove symptomatic plants is a very common practice when the level of infection is low, but only viable in nurseries. The conventional thermal treatment of cuttings to cure ratoon stunting has no efficiency against the mosaic virus (Benda, 1994). The usage of seedlings (plantlets) created in biofactories is the most recommended alternative, since the production process uses a combination of thermotherapy and meristem culture. Each step of the process contributes to a substantial reduction of the virus in the seedlings. Some data indicate that plants coming from thermal treatment and meristem culture are more susceptible to re-infestation with the virus. However, with a systematic roguing process, health can be maintained.

6.2 Bacterial Diseases

6.2.1 Leaf Scald Disease (Xanthomonas albilineans, Dowson)

History

This disease was identified in Australia and Java (Indonesia) in the 1920s and is now found in all producing regions of the world (Ricaud and Ryan, 1989). It was first reported on the American continent in 1943, in São Paulo (Almeida, 1994). It causes damage to production in cane fields, early reforms, and reduces the quality of the juice.

Causal agent

Leaf scald disease is caused by the gram-negative bacteria *Xanthomonas albilineans*, which belongs to the order of the Xanthomonadales. This bacterium is capable of colonizing the xylem of the plant and moving systemically in vegetable tissue. Its main host is sugarcane, but it can colonize other plants such as corn, bamboo, sorghum, and several grasses (Ricaud and Ryan, 1989). It seems there is great morphological variability of the pathogenic agent and the same strain of bacteria can have distinctive behaviors in different locations (Rott and Davis, 1994). This suggests that the expression of symptoms is dependent on local conditions, beyond the role of the different strains of bacteria.

Symptoms

When the disease manifests in susceptible varieties, it can cause up to 100% loss, with the burning of leaves and death of the stalks. Tolerant varieties can serve as a reservoir to the pathogen occurring in latency. Stresses caused by droughts, floods, and the cold can intensify symptoms. Plants attacked by the bacteria present with three possible types of infection: latent (asymptomatic), chronic, and acute. The latent form can be identified only with high sensitivity diagnostics (PCR, serum sample). Chronic infections show all foliar symptoms as thin white stripes (0.1−0.2 cm), usually throughout the leaf, parallel to the midrib and descending to the node, or larger stripes, restricted to the leaf blade. The acute form shows necrotic stripes on the leaf, usually evolving by the tip or margins, sometimes reaching full leaf necrosis as it is scalded, hence the name. In this phase, stalks can show discoloration of the xylem vessel beam in the nodal region, that is the result of necrosis of the vessel. When the tops of the stalks are severely affected by the disease, many side sprouts appear in the adult stalk, all with foliar symptoms and also with internal discoloration of the vessels. Entire clumps can wither and dry completely. The greatest effect of the disease can be seen in shoots, with the presence of faults. Acute symptoms usually manifest in susceptible or intermediate varieties, when climatic conditions stress plants.

Dissemination and predisposing conditions

The disease is transmitted mostly by cutting tools during the planting and harvesting operations and perpetuated through contaminated cuttings, Winds and rains can disseminate the disease over long distances when they spread the bacteria present in necrotic areas of the affected plants. Stress conditions, be they cold, drought, or high temperatures induce the acute phase of the disease.

Control

The most effective control comes from the resistant sugarcane varieties, allied to the production of highly healthy seedlings. For the production of healthy seedlings (stems), in some countries long thermal alternated therapy is used, with 3 h at 50.5°C preceded by 24-h immersion in water, at ambient temperature. The usual thermal treatment of 50.5°C for 2 h is not efficient, because a residual infection is still present in the treated seedlings. For production in biofactories, a combination of thermotherapy (bud sprouting at high temperature, 38°C), thermal treatment and meristem culture (Almeida, 1994). In the field, since leaf scald is transmitted by cutting tools during management of the culture, prevention by cleaning these tools is vital to avoid both the introduction of the disease into the stand as well as the dissemination from infected plants to healthy ones. The cutting tools as well as combines can be cleaned by chemical agents containing specific bactericides (quaternary ammonium is the most used product). Roguing, when carried out in two to six months old plants in the nursery, shows satisfactory results. After that period the external symptoms usually vanish and the infection remains latent in the plant until the climate conditions favor their reappearing.

6.2.2 Red Stripe (Pseudomonas rubrilineans, Stapp)

History

This disease has its origins in Asia, and today is disseminated throughout the world. In Brazil, the first report of the disease was in 1932, in the state of Rio de Janeiro (Almeida, 1994).

The disease shows restricted distribution in the country, because it needs quite specific soil and climate conditions. The greatest economic impact occurs in the states of São Paulo and Paraná and is associated with high-fertility soils (Tokeshi, 1980).

Causal agent

The red stripe is caused by the gram-negative bacteria *Pseudomonas rubrilineans*, pertaining to the order of Burkholderiales. It can be differentiated from the genus *Xanthomonas* by growing it in a culture media containing asparagine as the only source of carbon and nitrogen.

Symptoms

The disease causes the appearance of thin and long stripes ($2-3$ mm $\times$ $5-60$ cm) and rot at the top of the stalk. The first symptoms in the leaves are wet stripes that evolve, acquiring a red—brown color. With the evolution of the disease, the stripes reach the apical meristem region. In subsequent phases, the top of the plant becomes wet because of tissue death and subsequent rot. In favorable conditions to the development of the disease and in susceptible varieties, the rot that starts at the top of the plant spreads to the rest of the stalk, creating cracks through which flows the liquid resulting from the decomposition of plant tissues. This liquid has a strong and characteristic smell that can be noticed several meters away from the infected canes (Martin and Wismer, 1961). The greatest incidence of the disease is in plants from 3 to 8 months old and it normally disappears in adult plants.

Dissemination and predisposing conditions

Red stripe bacteria cannot survive in soil or in the remains of the culture for very long. Other plants, such as corn, can be alternative hosts. Favorable heat conditions (temperatures over 28°C) and high precipitation (relative humidity 90%) favor the development of the disease and its dissemination. The damage is greater in fertile soil, which promotes vigorous growth of the plants during summer months. In strong wind conditions, exudate particles (replete with bacteria) expelled by rotting tissue of the affected plant can be carried long distances and, after landing on new plants, begins a new disease focus.

Control

In regions with favorable climatic conditions, the cultivation of susceptible varieties is not advised. In the states of São Paulo and Paraná, where the disease is endemic, susceptible varieties must be cultivated only in low-fertility soils and and with no excessive fertilization. Since bacteria can survive in culture residues, establishing nurseries in affected areas must be avoided.

6.2.3 Ratoon Stunting Disease (Leifsonia xyli *subsp.* xyli)

History

The first report of ratoon stunting disease (RS) occurred in the 1940s in Australia, and today its causal agent is found in all cane fields in Brazil and in the rest of the world (Steindl, 1974; Gillaspie, Jr. and Teakle, 1989). In Brazil, the disease was first described in 1956, in the Campos Experimental Station (Campos de Goytacazes, RJ), by Frederico Veiga.

Causal agent

Ratoon stunting disease is caused by the gram-positive bacteria *Leifsonia xyli* subsp. *xyli*, which belongs to the order of the Actinomycetales. Within the *L. xyli*s species, there are two subspecies called *L. xyli* subsp. *xyli* and *L. xyli* subsp. *cynodontis*. The *cynodontis* subspecies infects grasses and causes rickets in Bermuda grass. Pathogenicity studies in several isolated samples of *L. xyli* subsp. *xyli* gathered from crops in different countries have shown low genetic variation (Gillaspie, Jr. and Teakle, 1989). The bacteria colonizes xylem vessels of plants and tends to obstruct them after causing a reaction from the plant due to its presence.

Symptoms

Although there are no external symptoms characteristic to the disease, it is possible to see, in susceptible varieties, the underdevelopment of the stalks coming from the re-sprout of the clump (hence the disease name). The result to the cane field is the retarded growth of the clump, with smaller stalks; and the more hydric stress the greater the effect is. In affected cane fields, the stunting shown by plants is not uniform and may vary from clump to clump, giving an "up and down" appearance to the stand. Thus, symptoms of starved plants, as well as their intensity, are basically determined by three factors: the variety (level of tolerance), the plant cycle (cane-plant, first ratoon or advanced ratoons, etc.) and climate conditions (drought). The intensity of symptoms can also be related to other factors affecting plant growth, such as inappropriate crop cultivation measure (compact soil, lack or fertilizer imbalance, excess competition from weeds), stresses caused by the improper usage of herbicides, and occurrence of other diseases. The most characteristic internal symptom of the disease is the development of an orange—reddish color in vascular xylem vesselss, the base of the node of the stalk. In susceptible varieties, this symptom is very evident when a cut is made at the very base of the node, transversally; then, dots, commas and concentric orange—reddish lines can be observed.

Dissemination and predisposing conditions

Bacteria transmission from one plant to another within the cane field happens via cutting tools: machete, planting and harvesting machinery, etc. These tools become impregnated with the infected plant juice and contaminate healthy plants that are cut in sequence. The bacteria has a high capability to penetrate and colonize the new stalk and, this way, it is estimated once contaminated, a knife or combine can transmit the bacteria to tens or even hundreds of new healthy clumps. The bacteria are transmitted quickly from one cycle to another by planting infected cuttings derived from contaminated cane fields. The dissemination occurs on a large scale for great extensions, and this amplification effect can result in 100% incidence of the disease in stalks and cane fields. The bacteria are more abundant in stalks than leaves, and its density grows with the maturation of the plant. The losses in productivity will depend on susceptibility of the variety and environment conditions. This way, each sugarcane producing country and even different regions of one country will have its varieties and specific environmental conditions, and thus, different losses (Matsuoka, 1984a; Gillaspie, Jr. and Teakle, 1989).

Control

The most effective control comes from the resistant sugarcane varieties to the pathogen, allied to the production of highly healthy seedlings. For producing seedlings with good health, long thermal treatment (50.5°C for 2 h) is traditionally used for cuttings to be used to establish the nurseries but it is known that this is not totally efficient (Matsuoka, 1984b; Benda, 1994). For producing seedlings in biofactories (tissue culture), a combination of thermal treatment, thermotherapy (sprouting of buds in high temperature conditions, 38°C) and meristem culture is used. In the field, since RS is transmitted by cutting tools during the management of sugarcane, prevention by cleaning these tools is vital to avoid disseminating the disease inside a healthy cane field or with a low number of infected plants, or even to avoid taking the disease from commercial cane fields to the nurseries. The cutting tools as well as combines can be cleaned by chemical agents containing specific bactericides (quaternary ammonium is the most used product).

6.3 Diseases Caused by Fungi

6.3.1 Smut (Sporisorium scitamineum)

History

Smut was firstly reported in South Africa in 1877, and appeared on the American continent in the 1940s, in Argentina. In Brazil, it was identified affecting varieties POJ36 and POJ213 in 1946 in the state of São Paulo. Despite efforts to eradicate the disease, smut spread to all states in South, Southeast and Center-West regions and recently, the Northeast. During the 1980s, smut caused great losses when it affected variety NA56-79, which occupied more than half of all the great cane field area of the Center-South (Bergamin Filho et al., 1989).

Causal agent

Smut is caused by filamentous fungus *Sporisorium scitamineum* (previously known as *Ustilago scitaminea*), pertaining to the division Basidiomycota, class Ustomycetes, order

Ustilaginales, family Ustilaginaceae. The fungus is a parasite of meristematic tissues that infects plants by teliospores (spores). The infection is established in the plant when teliospores sprout and infectious hyphae penetrate the base layers of the bud and reach meristematic tissue (Ferreira and Comstock, 1989). It is known that there are a few races of the fungus, but its relation to resistance and susceptibility with sugarcane varieties is not established yet.

Symptoms

Smut is an easy sugarcane disease to verify according to symptoms. Basically, the characteristic symptom is a structure called "whip", induced by the fungus, that develops from stalk apical meristem and can reach up to 1 meter in length. The whip is formed mostly by plant tissue and in a lesser part fungus tissue, that carries on to produce millions of spores. In the beginning of its development, the whip has an envelope made by a very thin film in silver color. Soon after, this film breaks and the whip gains a black velvet-like color, because of the fungus spores. Preceding the emission of the whip, stalks can be seen to be stretched, thinner than usual, with the leaves inserting at a more acute angle, foliar limb narrower and shorter, and often clumps with excessive tillering. Eventually certain varieties create galls, multiple buds, and an oversprouting known as witch's broom. The whips begin to emerge in plants within 2–4 months and reach maximum emission in 6–7 months. The emergence of symptoms is favored by heat and water stress. Stressed plants, even with good resistance, may eventually display symptoms. The effect to productivity of the cane field can be devastating, reaching more than 50% in highly susceptible varieties. The losses from the disease can be seen not only in productivity but also in the quality of the juice produced by affected canes.

Dissemination and predisposing conditions

The smut spores are adapted to air dispersion and can move over great distances via air currents. Whips exposed in the apex of stalks can release billions of spores a day, which are then dispersed through wind. Spores can fall directly over buds of standing canes or on bare soil. In dry air conditions and without rain, the spores can be dormant in the lateral bud until there is enough humidity, when they can then germinate and infect the meristematic cells of the bud. Following the infection, the fungus can be dormant or stimulate the bud to initiate cell division; in this case it will sprout and form a whip in the lateral bud. If, before that, the stalks which buds are contaminated are used for planting, the new plant may develop the disease and the corresponding whip. The infection of the bud after planting may also occur from spores that fell to the ground and got in contact with the bud just after planting. Since spores lose viability rapidly in high humidity and temperature conditions, high concentrations of viable spores in soil occur during cold and dry winters, so that, when planting after the beginning of rains, after the dry season, the infection is favored. There is a possibility that insects may help the dissemination when they feed from whips. Different species of smut affect different species of grasses, i.e. each species of the fungus is species-specific. For that reason, although smut may infect other plants from the *Saccharum* genus, the existence of alternative hosts seems very unlikely.

Control

The most efficient method to control smut is the usage of resistant varieties. When there are less resistant (intermediate) varieties involved, it is advisable to establish the nurseries with cuttings submitted to thermal treatment (52°C for 30 min or 50°C for 2 h), allied to a treatment with fungicide. Thereafter the nurseries should be monitored every 2 weeks until 8 months of age, roguing the symptomatic plants. Roguing in commercial fields is a high-cost and low-efficiency control method (Bergamin Filho et al., 1989).

6.3.2 Brown Rust (Puccinia melanocephala, H. and P. Syd)

History

Brown or common rust was first identified in India at the beginning of the 20th century, affecting *Erianthus ravennae*, a related species to sugarcane (*Saccharum* sp.). In the 1940s the disease was reported affecting cane fields from the majority of Asiatic and African producing countries such as India, China and South Africa. In the 1970s disease epidemics were reported in Japan, Taiwan, Australia, and some African countries. The first report on the American continent was in 1978, in the Dominican Republic (Comstock and Raid, 1994). In Brazil, rust was reported, in 1986, to be affecting cane fields in the states of São Paulo, Paraná, and Santa Catarina (Cardoso et al., 1987; Sordi et al., 1988). It is believed that the disease arrived in Brazil, as happened in the Caribbean, through wind currents coming from the African continent.

Causal agent

Rust is caused by the filamentous fungus *Puccinia melanocephala*, pertaining to the division Basidiomycota, class Teliomycetes, order Uredinales, family Puccinaceae. The fungus is an obligatory parasite that infects plants through uredospores (spores), which can have a brown, brown—orange, or brown—red color. There are reports of resistant sugarcane varieties that became susceptible through the years (Comstock and Raid, 1994). This suggests there can be variants of the fungus causing different reactions in the same plant variety.

Symptoms

Rust is a disease that affects basically the leaves of the infected plant. Initial symptoms are small chlorotic punctuations easily observed when the leaf is put against light. These punctuations evolve to elongated spots with a yellowish color and can be observed on the leaf surfaces, both over and underneath. These spots have varying sizes (2—10 cm in length, 1—3 cm width) and grow rapidly, altering its color to a redder tone, then pale-red and black in the last stages of leaf necrosis. In the center of the spots and usually on the underneath surface of the leaf, a structure named "pustule" develops. This occurs because the fungus grows inside the subepidermal part of the leaf and, when it begins to sporulate, produces a mass of uredospores, that then rupture the leaf epidermis and release abundant spores to further disseminate the disease. This epidermis rupture and the exposition of uredospores create the pustule, an embossed structure that can be detected by a characteristic harsh texture to the touch. The pustules cause loss to the foliar area, and as a consequence, reduction of the plant photosynthetic capacity, especially when they coalesce and kill part or all the leaf. Depending on the stage at which the plant was infected, the degree of susceptibility of the variety and the symptomatic level, the plant may display retarded growth, death of tillers, thin stalks and shortening of the internodes. Losses in productivity can reach up to 50% in susceptible varieties and 25% in varieties with intermediate to moderate resistance when the environmental conditions favors the disease (Comstock and Raid, 1994; Matsuoka et al., 1994; Sanguino and Cardoso, 1989).

Dissemination and predisposing conditions

Rust's main dissemination vehicle is the wind. Pustules formed in foliar lesions liberate uredospores to the environment and winds carry them on to other areas. It has been demonstrated that dissemination occurs to the downward direction of the local wind currents. The most important factors to dissemination are foliar surface humidity, where the uredospore will be deposited by the wind, and the local temperature, which will promote its sprouting. Mild temperatures (15—27°C) and high humidity favor the development of the disease. Since this fungus has a short cycle, about a week from when infection pustules are formed, and the number of spores produced is very large, an epidemic is established very rapidly when predisposing conditions are favorable. For this reason, damage caused by rust is very significant with susceptible varieties. Although humidity is important, excessive rains seem to disrupt the spread of the disease because they wash away recently exposed uredospores.

Control

The only form of controlling rust is by using resistant varieties. Since there are fungus variants, it is advised that the farmer grows simultaneously a group of resistant varieties. The usage of chemical control with fungicides is not an economically viable option.

6.3.3 Orange Rust (Puccinia kuehnii E.J. Butler)

History

This disease was described a long time ago (19th century) and its occurrence was restricted to Australia (New Guinea, Indonesia, Philippines, and Australia). Until recently it was considered a secondary disease, but from 2000 on it caused a serious epiphyty in Australia.

In a susceptible variety (Q124), used in great proportion (>50%) in the main cane field region that time, up to 40% losses were reported from the disease (Ryan and Egan, 1989). In July 2007 it was found in Florida (USA), infecting the variety CP72-2086, and in September 2007 it was detected in Guatemala. Soon after it was found in Honduras, Nicaragua, Panama, Mexico and Cuba and other countries of the Caribbean and South America, reaching Brazil in the second half of 2009.

Causal agent

Orange rust is caused by the filamentous fungus *Puccinia kuehnii*, that has taxonomic classification and a cycle similar to the causal agent of brown rust (*Puccinia melanocephala*). Its spores (uredospores) have an orange color. Because it is an old disease, it is believed that this new epidemic has been caused by a new form of the fungus that superimposed on the existing varietal resistance.

Symptoms, dissemination, and predisposing conditions

Orange rust symptoms are very similar to those of the other rust, except that the color of lesions are orangey in color, rather than brownish. This is one of the distinguishing character of both rusts, the other one being the thicker cell wall of the spores when examined under the microscope (Figure 6.2). Dissemination occurs similarly as happen in the brown rust, but the predisposing conditions differ to a certain degree, as specified below:

1. Unlike brown rust, that requires mild temperature and high humidity conditions, but not excessive rain, characteristic conditions of autumn and winter, the orange rust requires high temperature and humidity, even with heavy rains, characteristic conditions of spring and summer.
2. Also unlike Brown rust that occurs mostly in plants between 3 and 8 months of age and may show adult plant resistance, orange rust is more prevalent in adult plants.

Control

As happen in most sugarcane diseases, orange rust is also controlled by using resistant varieties and the varietal susceptibility to both rusts are distinct, but with cases of both infecting the same variety (Figure 6.2). This is the case of the variety RB72454 that is highly resistant to brown rust but susceptible to orange rust. The use of chemical control with fungicides is not an economically viable option.

6.3.4 Red Rot (Colletotrichum falcatum, Went)

History

Red rot is a disease that has been associated with sugarcane cultivation since its beginning and affects all cane fields in the world. This disease causes loss both in biomass productivity and sugar content of stalks. It is estimated that losses can amount to 70% when the fungus is associated with cane borer (*Diatrea saccharalis*). In sugarcane, there is a triple association between red rot, another rot called *Fusarium*, and cane borer (insect). Although red rot has the ability of infecting the stalk by itself, it usually has its penetration facilitated by any outer cell disruption like the hole made by the borer and, isolated or together with *Fusarium*, causes rot of the stalk. When the stalks are too affected, both the extraction of juice and the quality of sugar are prejudiced. Moreover, In too susceptible varieties this fungus can render the feedstock totally improper for milling (Singh and Singh, 1989).

Causal agent

Red rot is caused by the filamentous fungus *Colletotrichum falcatum* Went (perfect form *Glomerella tucumanensis*), pertaining to the division Ascomycota, class Pyrenomycetes, order Phyllachorales, family Phyllachoraceae. The fungus is an obligatory parasite that infects the plant through conidia (spores) and lesions in vegetable tissue. There are reported cases of resistant sugarcane varieties that become susceptible after years of cultivation. The fungus shows great genetic variability and many races. Resistance genes seem to be more present in *Saccharum spontaneum*, while *S. officinarum*, *S. sinense*, and *S. barberi* are, for the most part, more susceptible.

Symptoms

Red rot can be observed affecting many plant tissues in sugarcane. However, its worst damaging effects are to the stalks. Symptoms vary in intensity depending on factors such as susceptibility of the variety, humidity, temperature, and inoculation source. Since damage to plants is internal, the disease can go unnoticed, especially in the first stages of plant development. To best observe the symptoms, the stalk should be cut longitudinally in half. Infected tissue shows big spots of intense red color, interrupted by pale whitish transverse islands, from where the fungus can be easily isolated. These symptom, characteristic of red rot, distinguish it from other stalk rots, especially that caused by *Fusarium*. Vascular vessels with red color can be observed going out from the affected tissue and entering normal tissue. While in resistant varieties infection is restricted to the affected internode, in susceptible varieties the infection and the subsequent symptom, the reddish tissue with grayed transverse islands, can be observed through various internodes or all of the affected stalk. In leaves, symptoms can be observed in the leaf midrib as elongated lesions of red color and, in the blade, as reddish spots. The symptoms in the midrib can be mistaken with those caused by deficiency in potassium. In the breeding program the fungus can causes the complete rot of the seedling. In dry conditions, the fungus causes dry rot in roots and the base of the stalk.

Dissemination and predisposing conditions

Lesions in the midrib are considered the main source of inoculation during the growth phase of the plant. Affected stalks are also important inoculation sources. The dissemination occurs by action of winds, rains, dew, and irrigation water. Other hosts of the fungus are not considered to be representative sources of inoculation. Both dissemination and severity of symptoms are related to environmental conditions. After planting, symptoms are more severe in excessive soil humidity conditions.

Control

The fungus lives for many years in soil and leftovers of cultures in the form of clamidospores, conidia, and mycelia with thick walls. This inocula can promote primary infection of setts planted in the nursery or in the commercial plantation when environmental conditions do not favor fast sprouting. Thus, preventive control in the sprouting phase is essential to establish a good plantation and obtain high yields. Using resistant varieties is the most effective and used mode of control. Good culture practices help reducing incidence of the disease, such as eliminating culture leftovers (inoculation source), good soil draining, and good source of cuttings to ensure fast sprouting of setts.

6.3.5 Pineapple Rot (Ceratocystis paradoxa, Moreau)

History

Pineapple rot is a disease that affects a great variety of plants. In sugarcane, pineapple rot affects mainly young sprouts. In Brazil, cane fields from all producing regions are affected by the disease when the planting is made in soil and too wet and the low temperature do not favor rapid sprouting. In the Southern and Center-West states, it is a serious condition in late planting, in months between April and July.

Causal agent

Pineapple rot is caused by the filamentous fungus *Ceratocystis paradoxa* (Dade) C. Von Höhnel, belonging to the Ascomycota division, class of Ascomycetes, order of Microascales, Ceratocystidaceae family. The sexually active (perfect) stage of the fungus (*Ceratocystis*) is rarely observed in infected stems.

Symptoms

Recent planted areas that show low sprouting and death of young sprouts may indicate the presence of pineapple rot. The fungus is not able to penetrate intact tissue; so, injury is a condition necessary to establish itself. In sugarcane, the infection occurs through the cut ends of the cuttings. Once the sett is planted on contaminated soil, the fungus penetrates through the sett's ends and starts colonizing inside tissues. This colonization results in death of buds and recently emerged plants. A more precise diagnosis can be made by cutting longitudinally the sett: the tissue will show a creamy color and will exhale a pineapple smell, which gave the disease its name. Also, by exposing to air the inside part of the sett (because of the longitudinal cut), in two or three days the tissue goes black because of sporulation (clamidospores) of the fungus.

Dissemination and predisposing conditions

Pineapple rot fungus is basically a wound pathogen. Thus, the infection in a sett happens at its ends. The node constitutes a natural barrier to the growth of the fungus to the next internodes, the greater the resistance of the variety the greater the restriction. Its dissemination occurs in the form of conidia and clamidospores, which are its reproduction structures (spores). Both forms of spores survive on soil, but the most resistant form and the one which has the most influence in the dissemination of the disease is the clamidospore. Spores are disseminated by rainwater, winds and contaminated soils. Dissemination can also occur with usage of contaminated setts. Poor drained soil associated with low temperature are the most predisposing conditions favoring the disease, which in the said region occurs mainly in the autumn-winter time.

Control

The losses from pineapple disease results from the death of sprouts and the consequent gaps in the cane field. In late plantings, between April and July (Center-West, Southeast, and South regions), when low temperatures occur, buds have a slow growth, which gives opportunity for the disease to establish itself in the sett and cause damage. Plantations of exceeding depth, in bad drenched soils and under low temperatures should be avoided. Using setts coming from young parts of stalks reduce the disease incidence because they sprout faster and with more vigor. A recommended measure is using big setts, with six buds or more, or even to plant the whole cane, if the seedlings are young (Matsuoka and Gheller, 1984). Optionally, immersion of setts in fungicide before planting can be used when dealing with small plantations and valuable material, mainly in thermally treated setts. It is recommended that planting should be avoided in soils with recent history of pineapple rot. Using chemicals that enhance sprouting helps in diminishing the deleterious effect of pineapple disease as is a race of sprouting against the pathogen. It is the most damaging disease in mechanized planting because the harvesting machine cuts the setts in small pieces and further injury them, which favors infection by pineapple rot.

6.4 Final Considerations

The history of sugarcane culture records new diseases appearing from time to time, and those already known tend to periodically show recurrence. Often, disease becomes marginal in a certain region or country and, then, growers plant intermediate resistant or even susceptible varieties inadvertently, without the necessary preventive measures. Coming favorable environmental conditions, a new epidemic of the disease is often unavoidable.

Since an effective control measure for most diseases is varietal resistance, this new incidence cycle of a disease is almost always disastrous to the production sector, as resistant varieties are not always prompt available or, if they exist, it takes time to multiply them and substitute the susceptible ones. To avoid this kind of problem, or to minimize it, it is fundamental that sugarcane breeding improvement programs keep selection criteria that take into account not only the prevalent diseases but also those marginal with potential to become hazardous some day.

Bibliography

Almeida, I.M.G., 1994. Bacterial diseases of sugarcane in Brazil. In: Rao, G.P., Gillaspie, A.G., Upadhyaya, P.P., Bergamin Filho, A., Agnihotri, V.P., Chen, C.T. (Eds.), Current Trends in Sugarcane Pathology. International Books & Periodicals Supply Service, Delhi, pp. 73–84.

Benda, G.T.A.S., 1994. Serial hot-water treatments for sugarcane disease control. In: Rao, G.P., Gillaspie, A.G., Upadhyaya, P.P., Bergamin Filho, A., Agnihotri, V.P., Chen, C.T. (Eds.), Current Trends in Sugarcane Pathology. International Books & Periodicals Supply Service, Delhi, pp. 297–310.

Bergamin Filho, A., Amorim, L., Cardoso, C.O.N., Silva, W.M., Sanguino, A., Ricci Jr., A.R., et al., 1989. Epidemiology of sugar cane smut in Brazil. Sugar Cane. 1989 (2), 11–16.

Cardoso, C.O.N., Sanguino, A., Amorim, L., Bergamin Filho, A., 1987. Desenvolvimento da ferrugem da cana-de-açúcar no Brasil. Cong. sociedade dos técnicos açucareiros e alcooleiros do Brasil. Proc. 4, 266–270.

Comstock, J.C., Raid, R.N., 1994. Sugarcane common rust. In: Rao, G.P., Gillaspie, A.G., Upadhyaya, P.P., Bergamin Filho, A., Agnihotri, V.P., Chen, C.T. (Eds.), Current Trends in Sugarcane Pathology. International Books & Periodicals Supply Service, Delhi, pp. 1–10.

Ferreira, S.A., Comstock, J.C., 1989. Smut. In: Ricaud, C., Egan, B.T., Gillaspie, A.G., Hughes, C.G. (Eds.), Diseases of Sugarcane. Major Diseases. Elsevier, Amsterdam, pp. 1–10.

Gillaspie, Teakle Jr, D.s, 1989. Ratoon stunting disease. In: Ricaud, C., Egan, B.T., Gillaspie, A.G., Hughes, C.G. (Eds.), Diseases of Sugarcane. Major Diseases. Elsevier, Amsterdam, pp. 59–80.

Koike, H., Gillaspie Jr, A.G., 1989. Mosaic. In: Ricaud, C., Egan, B.T., Gillaspie, A.G., Hughes, C.G. (Eds.), Diseases of Sugarcane. Major Diseases. Elsevier, Amsterdam, pp. 301–322.

Martin, J.P., Wismer, C.A., 1961. Red stripe. In: Martin, J.P., Abbott, E.V., Hughes, C.G. (Eds.), Sugarcane Diseases of the World, vol. 1. Elsevier, Amsterdam, pp. 109–126.

Matsuoka, S., 1984a. Benefícios da prática de tratamento térmico da muda de cana-de-açúcar e eficiência de dois métodos existentes no Brasil. Cadernos Planalsucar, Piracicaba. 3 (3), 22–24.

Matsuoka, S., 1984b. Longevidade do efeito do tratamento térmico em canas infetadas pelo raquitismo-da-soqueira. Cong. sociedade dos técnicos açucareiros e alcooleiros do Brasil. Proc. 3, 244–249.

Matsuoka, S., 1999. Virus del mosaic. In: Fors, A.L. (Ed.), Manual de Enfermedades de la Caña de Azúcar. Ministerio de Agricultura y Ganadería, San Salvador, pp. 29–63.

Matsuoka, S., Gheller, A.C.A., 1984. Técnica cultural para controle da podridão abacaxi (*Ceratocystis paradoxa* [De Seynes] Moreau) em cana-de-açúcar. Summa Phytopatologica. 10 (1/2), 119–121.

Matsuoka, S., Masuda, Y., Gheller, A.C.A., Hoffmann, H.P., Vieira, M.A.S., 1994. A retrospective analyses of crop damage caused by sugarcane rust in Brazil. I.Losses in summerplanted cane. In: Rao, G.P., Gillaspie, A.G., Upadhyaya, P.P., Bergamin Filho, A., Agnihotri, V.P., Chen, C.T. (Eds.), Current Trends in Sugarcane Pathology. International Books & Periodicals Supply Service, Delhi, pp. 11–25.

Rott, P., Girard, J.-C., 2000. Sugarcane producing countries/locations and their diseases. In: Rott, P., Bailey, R.A., Comstock, J.C., Croft, B.J., Saumtally, A.S. (Eds.), A Guide to Sugarcane Diseases. CIRAD/ISSCT, Montpellier, pp. 323–339.

Sanguino, A., Cardoso, C.O.N., 1989. Influencia da ferrugem na produtividade da cana-de-açúcar. Bol. Copersucar. 47, 11–13.

Shukla, D.D., Ward, C.W., 1994. Recent developments in the identification and classification of viruses comprising the sugarcane mosaic subgroup of Potyvirus. In: Rao, G.P., Gillaspie, A.G., Upadhyaya, P.P., Bergamin Filho, A., Agnihotri, V.P., Chen, C.T. (Eds.), Current Trends in Sugarcane Pathology. Intern. Books & Periodicals Supply Service, Delhi, pp. 185–198.

Singh, K., Singh, R.P., 1989. Red rot. In: Ricaud, C., Egan, B.T., Hughes, C.G. (Eds.), Diseases of Sugarcane. Major Diseases. Elsevier, Amsterdam, pp. 169–188.

Sordi, R.A., Tokeshi, H., 1989. Cleaning leaf scald infected sugarcane plants using in vivo thermotherapy plus meristem tip culture. Summa Phytopathologica. 15, 128–132.

Sordi, R.A., Matsuoka, S., Masuda, Y., Aguillera, M.M., 1988. Sugarcane rust, a new problem in Brazil. Fitopatologia Brasileira. 13 (4), 313–316.

Steindl, D.R.L., 1974. Ratoon stunting disease. History, distribution, and control. Cong. Int. Soc. Sug. Cane Technol. Proc. 15, 210–212.

Tosic, M., Ford, R.E., Rao, G.P., 1994. Differentiation of sugarcane mosaic virus, maize dwarf mosaic virus, sorghum mosaic virus and Johnson mosaic virus on the basis of differential host reaction. In: Rao, G.P., Gillaspie, A.G., Upadhyaya, P.P., Bergamin Filho, A., Agnihotri, V.P., Chen, C.T. (Eds.), Current Trends in Sugarcane Pathology. Intern. Books & Periodicals Supply Service, Delhi, pp. 199–200.

CHAPTER 7

Weed Management

Sérgio de Oliveira Procópio[1], Antônio Alberto da Silva[2], Evander Alves Ferreira[3], Alexandre Ferreira da Silva[4] and Leandro Galon[5]

[1]*Embrapa Tabuleiros Costeiros, Aracaju, SE, Brazil* [2]*Universidade Federal de Viçosa, Viçosa, MG, Brazil* [3]*Universidade Federal dos Vales do Jequitinhonha e Mucuri, Diamantina, MG, Brazil* [4]*Embrapa Milho e Sorgo, Sete Lagoas, MG, Brazil* [5]*Universidade Federal da Fronteira Sul, Erechim, RS, Brazil*

Introduction

Sugarcane (*Saccharum* spp.) was introduced to Brazil in 1553, establishing itself in a permanent way in the Mid-South and Northeastern regions. Brazil is the greatest sugarcane producer in the world, grinding approximately 640 million tons per year, highlighting the Southeastern region with more than 60% of the total planted area (the State of São Paulo alone contributes 52% of this area), followed by the Northeastern region, with 21% of the total area. In both regions, the average productivity is 78 and 55 t/ha, respectively (IBGE, 2008).

It is important to note that the expansion of the area cultivated with sugarcane, mainly for the production of ethanol, will lead to higher demand for pesticides. In 2008, around $ 8.4 billion was spent on pesticides in Brazil, being that herbicides represented, approximately, 45% of this total (SINDAG, 2008). In weed management of the sugarcane crop, more than 10,000 tons of herbicides were used in the 2006/2007 crop, this being the second-largest crop in terms of herbicide consumption in Brazil (IBGE, 2008). The extensive use of these products in sugarcane cultivation is justified by the fact that this crop presents slow initial development, which makes its CPIP long (period that the crop needs to be free from weed interference). For this reason, it is also very common to use herbicides on sugarcane which have long residual effects on the environment which, if used without the necessary technical knowledge, can cause serious environmental impact.

We can estimate that around 1000 species of weeds inhabit the sugarcane agroecosystem in the distinct producing regions of the world (Arévalo, 1978). Despite being highly efficient in the use of the natural resources available for its growth, due to presenting a photosynthetic mechanism of the C_4 type, this crop has slow initial development, and for this reason it must be protected from the effects of competition from weeds (Procópio et al., 2003). The majority of these weeds are highly efficient in the use of the resources available in the environment (water, light, and nutrients), have fast initial growth and are present in the area at a high density (Silva et al., 2007a). Therefore, if not controlled at the beginning of the sugarcane growth, they lead to negative interference due to the better competitive skills for natural resources and the release of allelopathic substances (Silva et al., 2007a).

Besides the reduction on sugarcane tillering and on the productivity of stalks and sucrose, other negative aspects very evident due to the competition from weeds in this crop are: decrease in the sugarcane field longevity, decrease in the quality of the raw material, and difficulties in the operations of harvesting and transport (Procópio et al., 2003).

The cost of weed control in the sugarcane crop can represent up to 30% of the production cost in sugarcane ratoon and 15−25% in sugarcane plant (Lorenzi, 1996). For several reasons, such as readiness in the operation, best cost−benefit ratio, safety of the crop, and efficiency in rainy seasons, the chemical method of weed control is most commonly used in this and in other crops. However, to be efficient and safe from an environmental and technical point of view, this method needs to be supervised by a qualified professional in the weeds area, with good knowledge of the biology of these weeds and of the particular crop. Besides this, the professional must have knowledge of the chemical and physical characteristics of herbicides and their interactions with the environment.

In this chapter, basic information is available, in a summarized way, with the objective of assisting technicians, so that they can perform the management of weeds in the sugarcane crop in an adequate and safe way.

7.1 Losses Caused by Weeds in the Sugarcane Crop

Among the several losses that weed interference may cause to the sugarcane crop, the following stand out:

7.1.1 Reduction of Stalk and Sugar Productivity

Weed interference can promote losses in the crop productivity at varying levels, from 10% to 80% (Procópio et al., 2003). This differentiated effect of weed interference on the sugarcane crop can be attributed to the differentiated competitive capacity of the varieties or clones, as well as of its nutritional and sanitary state; of the cut cycle (sugarcane plant, sugarcane ratoon); of the species of weeds established in the area; of the population density and of the emergence period of the invaders; besides the availability of nutrients and water in the soil.

7.1.2 Decrease in Sugarcane Field Longevity

High weed infections, associated with deficiency in the control of these invaders, may accelerate the need to reform the sugarcane field. The natural and gradual loss of productivity in the sugarcane production areas is sharp in areas with deficient management, forcing many companies to start their ratoon destruction operation with only three cuts, when the original program was to perform at least five cuts. This occurs because the process of early deterioration of the sugarcane field is associated with lack of adequate nutrient replacement, soil compaction and the action of pests and nematodes.

7.1.3 Difficulty and Increase in Harvesting Cost

The presence of weeds in the sugarcane harvesting operation, manually or mechanically, causes operational disorders and increase of costs. When the sugarcane field is infected with weeds, the price agreed between the workers and the sugarcane harvesting company increases, due to the lower cut yield obtained by the cutters. Also, the presence of weeds may lead to higher risk to the health of the workers, being shelters to venomous animals, besides increasing the risk of accidents when handling the manual cutting instruments. In sugarcane mechanical harvesting, weeds are responsible for the operational yield loss, due to constant interruptions for cleaning and clearing of the cutting mechanisms and loading of the harvester. Besides this, premature wearing of the harvester components and difficulty in regulating the adequate cutting height can cause losses on the sprouting of the ratoon.

7.1.4 Decrease in the Industrial Quality of the Raw Material

When we harvest a sugarcane field infected with weeds, it is unavoidable that parts of the composition of these species, such as leaves, stalks, and reproductive organs, are transported together with the stalks from the crop to the industrial unit. This results in the increase of so-called vegetable impurities, which traditionally are characterized by pointers and leaves of the crop itself, but are incremented by the presence of organs and tissue pieces of the infecting weeds. These impurities make the industrial process difficult and, consequently, decrease the remuneration of the producer.

7.1.5 Shelter for Sugarcane Pests and Diseases

Several species of weeds commonly present in sugarcane crops may shelter insects and other pests or be used as hosts for determined species of fungi, bacteria, and nematodes which cause significant losses to sugarcane plants.

7.1.6 Depreciation of Land Value

The presence of determined species of weeds, such as purple nutsedge (*Cyperus rotundus*) and itchgrass (*Rottboellia exaltata*), mainly in high population densities, may depreciate the market value of the holding, or even jeopardize the dealing of partnership contracts in the infected areas. On this point, the inverse must also be practiced; when renting areas which are free from the presence of determined species of weeds, it must be required under contract that the area is returned, at the end of the contract period, with the same infection-free community. It is very common, in the sugarcane crop, to prevent the spreading of determined species of weeds, mainly the vegetative propagation ones. Special care must be taken to prevent the entry or the spreading of these species into areas without a history of occurrence.

7.2 Integrated Weed Management (Concepts and Objectives)

Integrated pest management (IPM) can be defined as the method of cultivation which considers all the factors that can provide to the plant efficient use of environmental resources and higher and better production. In this context it also includes integrated weed management (IWM) as a component of IPM. This integrated production system is gaining more ground every day in all agricultural sectors, having, in Brazil, its base enhanced in the field of entomology by pioneer works that promoted the study of the cotton problems in the Northeast of the country, which proposed a series of measures which are within the integration concept (Conceição, 2000). The premises which give foundation to the proposal of integrated management can be summarized as: quality guarantee of the harvested product, including exemption of residues of defensives in the foodstuff; environmental, economic and social sustainability of the production; keeping or increasing productivity; and guaranteeing better quality of life for the producer with regard to economic payback and higher safety in the activities which involve the use of agricultural defensives (pesticides) (Silva, 2006).

The integration is composed of the use of the several control methods available (namely cultural, mechanical, biological, or chemical) in a harmonious way, necessary for their best application, knowledge of the crop's physiology and biology and of the weeds, in order to obtain competitive advantages for sugarcane facing invaders. Management is related to decision-making, which depends a lot on monitoring. This concept is also based on "coexistence" among crops and weeds until a limit where the latter do not interfere negatively on sugarcane production. It is important to keep in mind that integrated management must be adapted to each location and situation: there is no longer a universal recommendation or "cake recipe".

To obtain success in IWM in the sugarcane crop, it is necessary to have multidisciplinary knowledge, highlighting the following areas: identification, biology and ecology of the weeds; physiology of the crop and of the weeds; mineral nutrition of plants; and sugarcane varieties available. These factors, associated with the chemical and physical characteristics of the soil and with the climate conditions, will define the degree of interference of the weeds in the crop and the need for their control. The control method to be used must take into consideration all the mentioned variables, besides the topography of the land, the costs of the operations, and mainly the environmental impact.

7.2.1 Main Infecting Weeds

Due to the Brazilian territorial dimensions and to the fact that sugarcane is present in practically all Brazilian regions, the diversity of weed species present in this crop is very large. The species which occur most frequently in the Mid-Southern sugarcane field region are highlighted in Table 7.1.

7.2.2 Description of Some Infecting Weed Species in the Sugarcane Crop

Brachiaria decumbens — *family: Poaceae (Gramineae)*

Brachiaria decumbens is a perennial plant, much tillered, upright and 30–100 cm tall. The leaves are densely pubescent and 10–20 cm long. Its reproduction occurs through seeds, rhizomes and stolons (Lorenzi, 2000). Its seeds can present viability for up to 8 years; the majority present initial dormancy and can germinate at several depths (0–8 cm).

Table 7.1: Main sugarcane infecting weed species in the Mid-Southern region.

Common Name	Scientific Name
Signal Grass	*Brachiaria decumbens*
Tanzania Grass	*Panicum maximum*
Itchgrass	*Rottboellia exaltata*
Jamaican Crabgrass	*Digitaria horizontalis*
Alexander Grass	*Brachiaria plantaginea*
Southern Sandbur	*Cenchrus echinatus*
Indian Goosegrass	*Eleusine indica*
Bermudagrass	*Cynodon dactylon*
Johnsongrass	*Sorghum halepense*
Wild Sorghum	*Sorghum arundinaceum*
Purple Nutsedge	*Cyperus rotundus*
Yellow Nutsedge	*Cyperus esculentus*
Morning Glory	*Ipomoea* sp.
Largefruit Amaranth	*Amaranthus* sp.
Puslane	*Portulaca oleraceae*
Dayflower	*Commelina* sp.
Sadamandi	*Emilia sonchifolia*
Sowthistle	*Sonchus oleraceus*
Billygoat Weed	*Ageratum conyzoides*
Milkweed	*Euphorbia heterophylla*
Wild Radish	*Raphanus raphanistrum*
Brazil Pusley	*Richardia brasiliensis*
Bristly Starbur	*Acanthospermum hispidum*
Spanish Needle	*Bidens pilosa*
Hairy Fleabane	*Conyza bonariensis*
Vente Conmigo	*Croton glandulosus*
Sida	*Sida* sp.

This species presents difficult control of its sowing, due to the irregular and prolonged germination. As a consequence of this characteristic, in sugarcane crop its control in pre-emergence requires a herbicide with long effective period of (residual) control.

Panicum maximum — *family: Poaceae (Gramineae)*

Panicum maximum is a perennial plant, robust, tillered, upright and 1−2 m tall. The stalks have dense hairiness at the nodes. The leaves are glabrous and 20−70 cm long. Its reproduction occurs through seeds and rhizomes (Lorenzi, 2000). Its seeds present low initial feasibility, increasing after a dormant period. The plants originating from seeds are very impaired and present slow initial growth; the occurrence of drought periods after emergence eliminates a good part of the young plants. After this initial phase, the growth starts to accelerate, mainly under conditions of high temperature and luminosity (metabolism C_4). The plant presents good adaptability to different types of soil and bears short drought periods, but it cannot bear long-lasting droughts or long periods in soaked soils. It presents good tolerance to shading, but low tolerance to frosts (Kissmann, 1997).

Sorghum halepense — *family: Poaceae (Gramineae)*

Sorghum halepense is a perennial plant, upright, strongly rhizomatous, cespitous, with waxy stalks and hairiness at the nodes and 1−2 m tall. It spreads through seeds and rhizomes. It is considered one of the most aggressive weeds in the world. In Brazil, its infestation area is increasing, being more frequent in São Paulo and Paraná. It is problematic to annual and perennial crops (Lorenzi, 2000).

Digitaria horizontalis — *family: Poaceae (Gramineae)*

Digitaria horizontalis is an annual plant, upright, herbaceous, much tillered and 30−80 cm tall. It has a stalk with rooting at the nodes. Its leaves measure from 6 to 12 cm long. Reproduction occurs through seeds (Lorenzi, 2000). It is very aggressive in fertile soils and is one of the first weeds to appear after preparation of the soil, from September to November, in the Mid-Southern region, being one of the most frequent annual weeds in the sugarcane fields of this region.

Brachiaria plantaginea — *family: Poaceae (Gramineae)*

Brachiaria plantaginea is an annual plant, herbaceous, tillered, upright, and 50−80 cm in height. Its stalks, in contact with the soil, may present rooting at the nodes. The leaves are glabrous and 10−25 cm long. Reproduction occurs through seeds (Lorenzi, 2000). Its seeds present low feasibility right after maturation, but the feasibility may increase for several years, if the seed can survive the low-temperature periods. In the field, only the seeds on the surface of the soil germinate; those which are positioned more deeply in the profile normally stay dormant. In well-prepared soils, after moistening, germination of most of the seeds which are close to the surface of the soil occurs. However, in low-moisture soils, only a low percentage of the seeds germinates.

Cynodon dactylon — *family: Poaceae (Gramineae)*

Cynodon dactylon is a perennial plant, herbaceous, of creeping growth habit. Its reproduction occurs mainly through rhizomes and stolons. It is a "long-living" species, even under adverse conditions. It bears acid and alkaline soils and tolerates high salinity and extreme drought conditions; however, it develops best in periods of high luminosity, temperature and moisture of the soil. This species is severely affected by frosts and practically does not develop under shade (Kissmann, 1997).

Rottboellia exaltata — *family: Poaceae (Gramineae)*

Rottboellia exaltata is an annual plant, cespitous, upright, with leaf sheaths which are densely covered by rigid bristles and 1.0—2.5 m tall. It spreads only through seeds. It is a weed of recent introduction, however, much spread already in the Mid-South of Brazil. It infests mainly annual and perennial crops, roadsides and vacant land. It is very vigorous and prolific; a single plant is able to issue up to 100 tillers and to produce more than 15,000 seeds, which stay dormant for up to 4 years (Lorenzi, 2000).

Cyperus rotundus — *family: Cyperaceae*

Cyperus rotundus is a perennial plant, upright and 10—60 cm tall. It has 5—12 basal leaves (Lorenzi, 2000). Its main reproductive mechanism includes subterranean tubers and bulbs. Its reproduction through seeds is of little significance — less than 5% of the formed seeds are viable (Kissmann, 1997). The purple nutsedge can develop in soils of different textures and within a wide range of pH. However, very saline soils are inadequate for its development.

7.3 Planting Seasons and Interference Periods

In the Mid-South region of Brazil, sugarcane is normally planted in two distinct seasons. If planted between the months of September and November, it presents its vegetative cycle with average duration of 12 months, being then denominated "cane-plant of year". When planted between the months of January and April, the sugarcane has a vegetative cycle varying from 14 to 18 months, being denominated "cane-plant of year and a half". The variations in the duration of the cycles will depend, mainly, on the planting date, climate and type of maturation of the variety used (Câmara, 1993).

All the other cuts, independently if originating from the cane-plant of year or cane-plant of year and a half, will have average duration of 12 months, being called "sugarcane ratoon".

Table 7.2: Total period of interference prevention (TPIP), period prior to interference (PPI) and critical period of interference prevention (CPIP), for the sugarcane crop, in the Mid-South.

Planting Season	TPIP (days)	PPI (days)	CPIP (days)
Cane-plant of year and a half	90−150	20−50	20−150*
Cane-plant of year	90−120	20−40	20−120
Sugarcane ratoon (sprouting May/Sep)	90−100	30−40	30−100
Sugarcane ratoon (sprouting Oct/Dec)	70−90	20−30	20−90

*Sugarcane planted in April infested by *Brachiaria decumbens* and/or *Panicum maximum*.

Several research works indicate periods of the crop cycle in which competition entails losses in the sugarcane production. However, these results cannot be extrapolated for all conditions, because these periods are influenced by several factors, such as planting season and sprouting period of the sugarcane ratoon (climatic conditions), varieties used, quality of the seedling, infecting weeds, fertilization, planting depth and spacing; in other words, factors which speed up or slow down the development of the sugarcane. In Table 7.2 average values of the periods are presented: prior to interference (PPI), total of interference prevention (TPIP) and critical of interference prevention (CPIP) for the Mid-South region (Constantin, 1993; Kuva et al., 2000, 2003, 2008).

In relation to the main species of infecting weeds by season in the Mid-South region, the trend is verified, within the months from March to September, of predominating infecting weeds with C_3 metabolism and, between the months of October and February, those with C_4 metabolism.

The PPI is approximately 20−30 days after the emergence of the bud (primary stalk) in sugarcane plant, because the maintenance of the plant in this phase depends almost exclusively on the nutritional reserves of the seedpieces, this way not entering into direct competition with the weeds in this period; in other words, the presence of these in the area does not interfere with the crop. A similar fact occurs with the sugarcane ratoon: the reserves contained in the base of the old clump keep the new sprouting for an initial period.

7.4 Weed Control Methods

7.4.1 Preventive Control

Preventive control measures aim to prevent the entry and or the spreading in the area of weed seedlings, and include:

- keeping vinasse or irrigation channels free from weeds;
- cleaning machines and implements when transferring to another range;

- using seedlings originating from nurseries with excellent weed control;
- controling the weeds in adjacent areas to the sugarcane ranges.

7.4.2 Cultural Control

Cultural control practices aim to make the sugarcane crop more competitive in relation to the weeds, and include:

- using varieties with more competitive characteristics, for example those which present high rate and high speed of tillering;
- using seedlings in good sanitational and nutritional state;
- properly fertilizing the crop in a way that favors its growth;
- reducing the spacing in areas which do not present aptitude for mechanization.

7.4.3 Mechanical Control

Mechanical control methods include: manual harvesting, manual weeding, clearing, and mechanized cultivation.

7.4.4 Biological Control

Biological control is a method which is not available or little used in Brazil for the control of weeds in the sugarcane crop, as well as crops in general.

7.4.5 Chemical Control

Chemical control is the most commonly used method in the sugarcane crop, because it is efficient, presents high yield, is low in cost in relation to other methods, and because there are several efficient herbicides registered on the market for this crop in Brazil.

In this crop, the herbicides can be applied in pre-emergence, post-emergence (initial or delayed — normally as directed spray), in the reform of the sugarcane field (for the control of sugarcane ratoon), and as a maturator in underdosing (gain of sucrose and planning of the harvest). In relation to the control spectrum, the herbicides can be classified into latifolicides (exclusive control of weeds with "broadleaves", group composed mainly of dicotyledonous); graminicides (exclusive control of weeds belonging to the grass family); herbicides of exclusive control of weeds from the sedge family ("sedge herbicides"); and broad-spectrum action herbicides (control of more than one group of weeds mentioned previously). Most of the herbicides registered for use in the sugarcane crop in Brazil belong to this last group. Table 7.3 presents more details on the action spectrum of the main herbicides registered for use in the sugarcane crop in Brazil.

Table 7.3: Main herbicides registered for use in the sugarcane crop in Brazil.

Herbicide	Modality of Application	Controlled Weeds	Important Observations
Growth Regulators			
2,4-D	Post	D[1]	Volatility
2,4-D + Picloram	Pre and post	D	Elevated persistence in soil
Photosystem II Inhibitors			
Ametryn	Pre and post	D and G[2]	Excellent control of *Brachiaria plantaginea*
Diuron	Pre and post	D and G	Normally used in mixtures with other herbicides
Metribuzin	Pre and post	D and G	High selectivity to crop
Tebuthiuron	Pre	D and G	Elevated persistence in soil
Hexazinone + diuron	Pre and post	D and G	Excellent control of *Brachiaria decumbens*
Amicarbazone	Pre and post	D and G	Possibility of use in drier periods
Cell Division Inhibitors			
S-metolachlor	Pre	G and some D and C[3]	Excellent control of *Commelina benghalensis*
ALS Inhibitors			
Imazapic	Pre	D, G and C	Possibility of use in drier periods
Imazapyr	Pre	D, G and C	Not-selective to crop
Halosulfuron	Post	C	Control almost exclusive of *Cyperus rotundus*
Trifloxysulfuron-sodium	Post	G and C	Great control of *Cyperus rotundus*
EPSPs Inhibitors			
Glyphosate	Post	D, G and C	Not-selective to crop
Inhibitors of Carotenoid Synthesis			
Clomazone	Pre	G and some D	Excellent graminicide
Isoxaflutole	Pre	G and some D	Possibility of use in drier periods
Mesotrione	Post	G and D	Efficient control of morning glory
PROTOX Inhibitors			
Sulfentrazone	Pre	D, G and C	Great control of *Cyperus rotundus*
Oxyfluorfen	Pre	D and G	Low mobility in soil
Respiratory Inhibitors			
MSMA	Post	D, G and C	Partially selective to sugarcane
Photosystem I Inhibitors			
Paraquat	Post	D, G and C	Partially selective to sugarcane when applied in low doses
Mitotic Inhibitors			
Trifluralin	Pre and PPI	G	Volatility and photodegradation

(Continued)

Table 7.3: (Continued)

Herbicide	Modality of Application	Controlled Weeds	Important Observations
Mixtures of Herbicides of Different Mechanisms of Action			
Clomazone + Ametryn	Pre and post	D and G	Can cause whitening in the sugarcane
Clomazone + Hexazinone	Pre and post	D and G	Better action spectrum in the control of grasses
Trifloxysulfuron sodium + Ametryn	Pre and post	D, G and C	Needs moisture in soil to work
MSMA + Diuron	Post	D, G and C	Application preferentially as directed spray

[1]D — dicotyledonous.
[2]G — grasses.
[3]C — sedges.

7.5 Climatic Factors Which Influence the Activities of Herbicides

With the herbicides available in the market nowadays, there is a practical solution for the chemical control of most of the weeds occurring in the sugarcane crop. In practice, the results have been a little unsatisfactory sometimes, due to lack of knowledge of application techniques and equipment and due to non-consideration of environmental conditions (temperature, moisture of the air and of the soil, wind, dew). The influence of these factors on the efficacy of chemical products is complex, because they interact with one and other.

Observations are made below in relation to the influence of these factors on the action and the properties of herbicides.

7.5.1 Solar Radiation

According to Victória Filho (1985a, b), light can increase the translocation of herbicides, because it promotes photosynthesis and, consequently, the movement of the herbicide, together with the photosynthesized products in the plant. However, in determined situations, high luminous intensity causes an increase in cuticle thickness and a higher number of trichomes, which can make the absorption of the herbicides more difficult.

7.5.2 Rainfall

The rains interfere with the action of the herbicides, depending on the time that they occur, their intensity and duration. According to Ferreira et al. (2005), the occurrence of rains a few days prior to the application of herbicides, in post-emergence, can take part of the waxes and of the alkanes from the surface of the leaves of the weeds, increasing their susceptibility to herbicides and thus improving control efficiency.

The influence of the rain on the absorption of the herbicides by the leaf also depends on the characteristics of each product, because some are absorbed quickly, while others are absorbed slowly. Generally, those formulated in oil are less affected by the rain than those conveyed in water (Victória Filho, 1985a, b). According to Silva et al. (2007b), good water content in the soil is essential for the good efficacy of herbicides used in pre-emergence.

7.5.3 Relative Air Humidity

Relative air humidity is probably the environmental factor which most influences the life cycle of the pulverization drops and the activity of the herbicides, mainly the ones which target the emerged weeds (Marochi, 1997). Victória Filho (1985a, b) states that the relative air humidity influences the absorption and translocation of herbicides applied to the leaf, because it directly affects the time of permanence of the drop on the leaf surface, as it also influences the hydration of the cuticle. Low relative humidity causes faster evaporation of the drop, making cuticular penetration more difficult and potentially causing hydrous stress in the plant.

7.5.4 Temperature

The temperature of the air influences the action of the herbicides in several ways because it can change their physical properties, such as vapor pressure and solubility, and also change the physiological processes of the plants (Beltrão and Azevêdo, 1994). Gupta and Lamba (1978) report that, normally, low temperatures (lower than $10°C$) or very high temperatures may reduce the metabolism of the plants, leading to a decrease in the toxic action of the herbicides and in the weed control. Loss of selectivity of the herbicide may also occur when it is applied in extreme temperatures. This occurs mainly when the selectivity of the crop to the herbicide is due to the differential metabolism promoted by the plant (Procópio et al., 2003).

7.5.5 Wind

According to Victória Filho (1985a, b), the wind indirectly affects the absorption of herbicides by the plants, due to the increase in the evaporation of the pulverization drop on the leaf surface. Also, plants which grow in conditions of much wind and high temperatures normally present a thicker and more pubescent cuticle, which makes the absorption of the herbicides more difficult.

When applying pesticides the wind may cause drift, which is a term used for those drops that are not deposited in the target area. Drift may cause deposition of chemical products in non-desirable areas, with serious consequences.

To reduce negative effects on environmental conditions in the application of herbicides, the following practices are recommended:

- do not apply products in adverse environmental conditions (low relative air humidity, high temperature, winds faster than 10 km/h);
- do not apply products when the weeds are in stress situations (difficult absorption and translocation of the herbicide);
- preferrably use the periods in the early morning and late afternoon to perform applications, or if the herbicide and the technological conditions allow, perform applications at night;
- incorporate herbicides sensitive to photodecomposition into the grid soil, when it is dry or with little moisture;
- if possible, use large drops in the pulverizations;
- do not exceed the recommended pressure for the pulverization tip;
- use the adjuvants recommended for each situation.

7.6 Weed Control in Green Cane

With the increase of the sugarcane areas harvested without traditional burning, due to changes in legislation as a result of environmental awareness, the current weed management in these areas represents significant changes, making more study on this new technology necessary.

The sugarcane production area destined for mechanized harvesting of green cane has increased in the last decades. The adoption of this harvesting system has resulted in important changes in cultivation techniques, such as the use of larger spacing and deposition of straw onto the soil, which directly influence weed occurrence and management (Velini and Negrissoli, 2000).

The implementation of this sugarcane harvesting system, without burning, brings some agronomically beneficial factors, such as:

- reduction of erosive processes;
- better conservation of the moisture of the soil;
- higher recycling of nutrients;
- increase of organic matter in the soil;
- increase of microbial activity in the soil;
- enhancement of the physical and chemical properties of the soil;
- prevention of the lodging of the stalks, caused by the burning;
- decrease of weed infestation;
- prevention of the loss of sugars via exudation of the stalks during and/or right after the burning.

However, some unfavorable factors can be mentioned with the adoption of this technology, as follows:

- difficulty in the sprouting of most sugarcane varieties;
- possible increase of pests which attack the sugarcane;
- likely increase in the doses of the nitrogenous fertilizers in the first years of adoption of this system;
- in colder places, it can disadvantage the sugarcane growth;
- in the "lowlands", there might be problems with sugarcane caused by the excess of moisture;
- application of herbicides only during the day, when the presence of weeds is easily detected (except for the implementation of "precision agriculture").

The preserved trash of sugarcane crop provides ground cover, which makes more difficult the emergence of weeds because it reduces the penetration of light into the soil. The release of exudates of the trash may also occur, which may present allelopathic effects on the germination of seedlings of weeds (Procópio et al., 2003).

Sugarcane harvesting without burning leaves a thick layer of straw on the soil, which can exceed 20 t/ha. This cover is very important in weed species control for influencing processes such as dormancy, germination and mortality of seeds, as well as the establishment and the reproduction of the plant (Fernandez-Quintanilla, 1988; Trezzi and Vidal, 2004). This cover also reduces erosion and evaporation, besides increasing the infiltration of water and the retention of the moisture, keeping the soil moist for a longer time (Reddy, 2003). The physical obstruction caused by the layer of straw also causes reduction of the emergence (Victória Filho, 1985a, b), which affects development of the seedlings of some weed species, causing etiolation making them susceptible to mechanical damages (Correia and Durigan, 2004). Also, the emergence of plants originating from positive photoblastic seeds and of those which require determined wavelength or thermal amplitude to germinate is reduced (Correia and Durigan, 2004). According to Almeida (1981), the dead cover can work as a weed control element, because land with a uniform and thick layer of residues presents a much lower infestation than the one which would develop if it were uncovered. The straw, associated with the technical changes necessary to implement the mechanical harvesting of the crop, created a new sugarcane production system, popularly known as green cane (Velini and Negrisoli, 2000).

Toledo et al. (2005), in a study comparing green cane to burnt sugarcane, carried out in Mexico, concluded that the first system presented lower aggressiveness of the weeds, higher biomass production (bigger and thicker stalks, besides larger quantity), juice purity, and sugar production, besides differences in the contents of organic matter, nitrogen, phosphorus, potassium, and pH of the soil. The economic analysis also showed much higher income in the green cane system. Núñez and Spaans (2008), in a similar study, comparing both systems, in Ecuador, achieved weed control costs 35% lower after the harvesting of green cane.

Velini and Negrisoli (2000) observed that sugarcane straw drastically reduced the temperature variation of the soil at 1 cm and at 5 cm of depth. According to the authors, this effect contributes in a decisive way to the reduction of germination of weeds in green cane areas, because it is known that thermal amplitude is one of most important components in the germination of seeds of several species.

It is important to highlight that the greatest efficacy of the trash in reducing the emergence of weeds depends mainly on the uniformity of its distribution on the surface of the soil, because small clearings are already enough to provide favorable conditions for the emergence of invading plants. Among the species whose population is increasing in researches carried out in green cane areas, mainly in Southeast region in Brazil, the following are highlighted: milkweed (*Euphorbia heterophylla*), morning glory (*Ipomoea* spp., *Merremia* spp.), scented shower (*Senna obtusifolia*), cissampelos pareira (*Cissampelos glaberrima*), flame vine (*Pyrostegia venusta*), bitter gourd (*Momordica charantia*), perennial soybean (*Neonotonia wightii*), and purple nutsedge (*Cyperus rotundus*). The purple nutsedge populations are being reduced by the presence of straw, however, at unsatisfactory levels. These reports show the clear trend of change of flora in the sugarcane producing areas, previously dominated by grasses in the burnt sugarcane areas and currently significant dominance of dicotyledons, mainly the ones that have large seeds and some sedges. Besides changing the composition of the infecting community, the straw resulting from the harvest without burning can change the efficiency of soil-acting herbicides. This change results mainly from the interception of the pulverization drops of the herbicide spray, preventing or making it more difficult for them to hit the soil and consequently not allowing the herbicide molecules to be positioned together in the layer of the soil where most of the seeds of weeds likely to germinate in that period are located (normally from 0 to 7 cm).

Some alternatives are being studied to improve herbicide management in areas of green cane, for example:

- Application of herbicides in post-emergence. It is advantageous to know first which species have emerged and their population density, this way the best herbicide treatment can be chosen for the situation. This strategy can reduce control costs and environmental impact. The main disadvantage is the likely need of a second application in the area. This may occur because many weed species present staggered emergence, in other words, there might be a new emergence flow in the area after the first application of herbicide, and this new flow may occur prior to the end of the critical period of interference prevention (CPIP). In this case, the need to reapply in the same area brings disorders in relation to the logistics of pulverizers, besides the increase of the production cost, increase of soil compaction, and higher probability of damages by these mechanisms to the sugarcane plants.

- Application of the herbicide treatment prior to the deposition of the straw onto the soil. This can be achieved by adapting a herbicide application system together with the mechanical harvester. In this way, the herbicide would be applied prior to the release to the soil of the harvest residues. This technology is being studied by partnerships among agricultural machine manufacturers, pesticide manufacturing companies, and research institutions, in order to be available in the near future.

- Application of herbicides which have favorable physical and chemical characteristics on the layer of straw. Research has shown satisfactory efficiency of some herbicides for weed control in pre-emergence, even when applied onto the sugarcane straw. The main advantage for applying the herbicide onto the sugarcane straw is the solubility in water. All herbicides which stand out in the experiments in green cane present high solubility in water. The occurrence of rains after application is also being pointed to as an important factor in the process of removing the molecules of herbicide which were retained on the trash of the crop. It is highlighted that the influence of the straw in the dynamics of the herbicides depends on the quantity which was deposited onto the soil. Many times, for herbicides of low solubility, small quantities of straw (less than 5 t/ha) are sufficient to impair their action.

7.7 Tolerance of Sugarcane Varieties to Herbicides

Sugarcane varieties may present differentiated responses to the herbicides used in weed control, which may cause phytotoxicity problems, and may also cause production losses. One variety may present alternate behavior, depending on the herbicide used.

In the field, some phytotoxicity symptoms of herbicide to the sugarcane are commonly observed, such as: whitening of the leaves (pigment inhibitors); leaf chlorosis followed by necrosis at the margins and tips of leaves (photosynthesis inhibitors absorbed through the leaf and respiratory inhibitors); decrease in the growth of the crop (amino acid inhibitors and photosynthesis inhibitors); stalks and roots with teratogenesis; thinner and curved internodes; thickened nodes with tumorization; curved stalks shaped like an elbow; less developed roots; and necrosis in the meristematic region near the nodes (growth regulators). Normally, these phytotoxicity symptoms disappear after 15−90 days. However, the period necessary for the recovery of the sugarcane plants depends mainly on the type of phytotoxicity symptom, on the intensity of the symptoms, and on the climate conditions after their occurrence.

Ferreira et al. (2005), when working with 11 cultivars and four clones of sugarcane, observed the differentiated sensitivity of the genotypes to the mixture of herbicides trifloxysulfuron-sodium + ametryn. Cultivar RB855113 was the most sensitive to the

Table 7.4: Effect of the pre-formulated mixture of the herbicides (ametryn + trifloxysulfuron-sodium) on the sugarcane genotypes.

Cultivars/Clones	Phytotoxicity (%)		BSPA* (%)	Sensitivity
	13 DAT	34 DAT		
RB855113	13,75 a[1]	44,40 a	33, 32 c	High
SP80-1842	7,50 b	21,16 b	50,29 b	Medium
SP80-1816	5,75 b	13,17 c	58,73 b	Medium
RB855002	6,25 b	8,33 d	94,79 a	Low
RB928064	3,75 c	5,83 e	90,51 a	Low
SP79-1011	8,50 b	16,60 c	40,35 b	Medium
SP81-3250	2,50 c	2,83 e	95,88 a	Low
RB867515	4,25 c	5,83 e	94,45 a	Low
RB957712	2,50 c	6,33 e	88,53 a	Low
RB72454	7,50 b	7,50 e	91,76 a	Low
RB845210	7,50 b	10,83	85,30 a	Low
RB947643	5,00 b	4,17 e	89,24 a	Low
RB855536	2,75 c	4,20 e	93,76 a	Low
RB835486	1,25 c	6,67 e	86,05 a	Low
RB957689	15,00 a	24,17 b	49,52 b	Medium

BSPA — relative percentage of dry mass production of the shoot in relation to the witness.
*Evaluations performed at 45 DAT.
[1]Averages followed by the same letter in the column do not differ among themselves through the Scott-Knott test.
Source: Ferreira et al. (2005).

mixture. Cultivars SP80-1842, SP80-1816, SP79-1011, and RB957689 presented medium sensitivity to the mixture of herbicides. In the other cultivars, this sensitivity was considered low (Table 7.4).

Azania et al. (2006) when working with herbicides in sugarcane reported that these products were more phytotoxic when applied at the late post-emergence phase in relation to the application at initial post-emergence phase. At initial post-emergence, the plants recover themselves completely from the intoxication effects caused by the herbicides, with lower effects on productivity. This could be justified, at least partially, by the greater number of leaves of the plants at a more advanced stage, which would maximize the interception of the herbicides. Concenço et al. (2007) state that, as the plant gets older, a set of morphoanatomical factors from it causes the herbicide to be absorbed with less efficiency. The reduction of the pore diameter of the plasmodesmata stands out as one of the factors responsible for lower absorption or translocation of herbicides in plants.

Barroso et al. (2008) when evaluating weed management in sugarcane (cultivar SP80-1816) observed that herbicide treatments caused sharp phytotoxicity to the plants (Table 7.5). The authors observed that the treatments which caused higher phytotoxicity to the plants in the first days were ametryn (300 g/L) + clomazone (200 g/L) and clomazone (500 g/L). Also in this evaluation, it was observed that, among the herbicide

Table 7.5: Phytotoxicity in sugarcane plants after the application of several herbicide treatments.

Treatments	Phytotoxicity (%)			
	7 DAA[1]	14 DAA	21 DAA	35 DAA
Clomazone (400 g/kg) + hexazinone (100 g/kg)	20,0 b[2]	11,3 b	5,8 a	0,0 a
Sulfentrazone (500 g/L)	12,5 b	9,8 b	2,8 b	0,0 a
Ametryn (300 g/L) + clomazone (200 g/L)	24,5 a	14,5 a	4,3 a	0,0 a
Clomazone (500 g/L)	23,3 a	15,5 a	5,3 a	0,0 a
Sulfentrazone (500 g/L) + clomazone (500 g/L)	17,3 c	14,3 a	4,5 a	0,0 a

[1]Days after the application of the herbicides.
[2]Averages followed by the same letter in the column do not differ among themselves by the Scott Knott test at 5% probability.
Source: Barroso et al. (2008b).

treatments, sulfentrazone (500 g/L) was the one which resulted in lower levels of damage to the crop, behavior that was maintained in the following evaluations.

7.8 Behavior of Herbicides in the Soil

The use of chemical control in weeds is an indispensable practice for sugarcane; therefore the use of herbicides is unquestionable in this crop. However, it is fundamental that these products are properly applied, so that the final quality of the harvested sugarcane is preserved, as well as the natural resources which sustain production, especially the soil and water.

When the herbicide reaches the soil, its redistribution and degradation process starts. This process can be extremely short, as in the case of some simple and non-persistent molecules, or persist for months or years when dealing with highly persistent compounds. Its period of permanence in the environment depends, among other factors, on the sorption capacity of the soil, on the dynamic of the water flow and on the transport of solutes, besides the degradation rate of the product, which is related to the microbiological activity, bioavailability and recalcitrance of the herbicide.

Although few in number, the studies involving the sorption of herbicides in Brazilian soils, in conditions of tropical climate, are also fundamental for the evaluation of weed control efficiency in Brazil, because high sorption rates can compromise the efficiency of the herbicide. With this, the importance of understanding the final destination of these molecules grows, as does the behavior in the environment where they are applied.

The study of the behavior of herbicides in the soil and in the environment has at least two main objectives: firstly, to know the environment factors, besides the herbicide itself, which directly or indirectly affect the efficiency on the control of a weed; secondly, once the herbicide is an exogenous substance to the environment, the intention is to find its interaction with the components of the soil, in a way that minimizes the possible negative effects that its presence could cause to the environment.

However, although the permanence capacity of the herbicide and its degradation in the soil are key processes in the determination of its effect on environmental quality, it is difficult to measure this capacity and the repeatability of the experiments. This is because the soil is a heterogeneous environment, under the influence of several factors, where several processes of physical, chemical, and biological order interact (Hinz, 2001).

Nowadays, the study of the behavior of herbicides in the environment has been performed by estimates of the trends which these products are subject to, according to three main processes — retention, transformation, and transportation — which interact among themselves, although they are processes described in an isolated way.

Therefore, the behavior of herbicides in the environment, mainly in the soil, depends on the sum of several involved processes, which are responsible for the final destination of these compounds. The result of the transportation, retention and transformation processes which the molecules undergo represents the capacity of contamination and persistence of these in the environment, and a detailed approach of its dynamic would be difficult encountering the different interferents related to its behavior. What is observed is that the knowledge of the characteristics of the soil, of the involved climate factors and of the herbicide—environment interaction mechanisms is fundamental to forecasting the behavior of herbicides in field conditions. This fact denotes the importance of researches, mainly on Brazilian soils, with the objective of preventing possible environmental disorders caused by these compounds. As the theme is very comprehensive, when all the involved factors interact, it is without doubt, one of the reasons for the need to implement research in this area. The implications are clear: understanding how the herbicides and other pesticides behave in the soil makes possible the use of these compounds with technical and economic efficiency. Besides this, it makes possible the prevention of environment contamination problems, besides options for recovery of affected environments.

7.9 Weed Resistance to Herbicides

Producers have preferred the use of herbicides to other weed control methods, due mainly to their high efficiency and relatively low cost. However, indiscriminate and improper use of these products has caused the development of many resistance cases to these compounds by several weed species.

The plant is sensitive to a herbicide when its growth and development are changed by the action of the product; then it could die when submitted to a determined dose of herbicide. Now tolerance is the innate capacity of some species to survive and reproduce after an herbicide treatment, even when injured. On the other hand, resistance is the acquired capacity of a plant to survive determined herbicide treatments, which in normal conditions control the other members of the population.

The repeated use of the same herbicide can select resistant biotypes of pre-existing weeds in the population, causing the increase of their number. Consequently, the population of resistant plants can increase until the point of compromising the control level, even derailing the cultivation of determined crops in the area.

The first cases of resistance to herbicides were reported in 1957, in the United States and Canada, and currently there are approximately 319 weed biotypes resistant to herbicides, belonging to 185 species, distributed in 59 countries, being 111 dicotyledons and 74 monocotyledons (Weed Science, 2008). There are weed species which present resistance to more than one herbicide mechanism with evidence in several countries. The higher number of cases refers to herbicides which inhibit the ALS and ACCase enzymes and FSH photosynthesis (triazines). It is believed that the higher number of biotypes resistant to these groups is due to the high specificities and to their efficiency, as well as the fact of being used in large areas for consecutive years.

Resistance control and management techniques try to reduce the selection pressure, control the resistant individuals before they can multiply themselves, and also expand the control alternatives likely to be adopted. This can be achieved adopting the following practices:

- use herbicides with different action mechanisms;
- perform sequential applications;
- use mixture of herbicides with different action and detoxification mechanisms;
- rotate the action mechanism;
- limit the applications of the same herbicide;
- use herbicides with lower selection pressure (residual and efficiency);
- rotate crops;
- promote rotation of control methods;
- follow changes in the flora;
- prevent suspicious plants from producing seeds;
- perform rotation of soil preparation.

7.10 Weed Tolerance to Herbicides in the Sugarcane Crop

The factors involved in the selection of species tolerant to herbicides are more complex than the ones observed in the selection of resistant biotypes of normally susceptible plants, and the changes in the population also occur more slowly (Owen, 2006). The changes in the composition of the infecting flora and the selection of species tolerant to herbicides, when they are applied repeatedly, are also influenced by environmental factors. So, when predominating within a population, the tolerant species can make it more difficult to combat the naturally tolerant species rather than reducing the frequency of plants of a resistant biotype.

The repeated application of the same herbicide or of herbicides with the same action mechanism generates selection pressure on the species which are present in the area. The two main response pathways of the weeds are specific change in the flora, through the selection of more tolerant weed species, or intraspecific selection of biotypes resistant to herbicides (Christoffoleti and Caetano, 1998).

When chemical control is the only method used in weed management, mainly when using herbicides with the same action mechanism for several consecutive years in the same place, the selection pressure for tolerant species and resistant plants is extremely high, causing a change in the weed population (Radosevich et al., 1997). According to Christoffoleti et al. (2000), any plant population which has a variable genetic base regarding the tolerance to a determined control measure, throughout time, changes its population composition as a survival mechanism, decreasing the sensitivity to this control measure. According to these authors, a good example was the use of the plow, which at the first moment eliminated practically all the weeds but, over time, new, more adapted weed species began to infect the crops. Another example was direct planting, which at first caused sharp reductions in weed incidence, but sometime after adopting this technique, there was a selection of species which adapted themselves and germinated under this new condition.

As already mentioned, the weed tolerance to herbicides is the result of the innate capacity of the species to bear applications of herbicides, at the recommended doses, without significant changes in growth and or development. Susceptibility is also an innate characteristic of a species. In this case, there are changes with significant effects on the growth and development of the plant, as a result of its incapacity to bear the action of the herbicide (Christoffoleti, 2000).

Tolerance might be related to the development stage and/or to the morphological characteristics of the species. On the other hand, resistance is the capacity acquired through biotype of a plant to survive the dose normally used (instruction dose) of a herbicide that, under normal conditions, controls the other members of the same population.

Weed tolerance to herbicides presents the same mechanisms assigned to the resistance and to the selectivity of crops, being possibly due to the development stage of the plant; to the differences in the leaf anatomy and morphology, as well as in the absorption, translocation and compartmentalization of the product; and to the metabolism of the herbicide molecule (Westwood and Weller, 1997; Vargas et al., 1999).

In the sugarcane crop, several herbicides, especially those belonging to the group of the triazines and of the substituted urea, have been frequently used in the control of Jamaican

Crabgrass. The genus *Digitaria* presents 13 morphologically similar species, being the main infecting species of sugarcane crop in the mid-west of Brazil the *Digitaria nuda*, *D. ciliaris*, *D. horizontalis* and *D. bicornis* (Dias et al., 2007). According to the authors, the Jamaican crabgrass species are being selected by the repetitive application of the herbicides used on the control of these weeds in the sugarcane crop, characterizing a process of population dynamics of specific change of weeds tolerant to herbicides.

Dias et al. (2007) developed work on Jamaican crabgrass species tolerance to the herbicides applied in the sugarcane crop, evaluating the efficacy and studying the weed's tolerance mechanisms. They concluded that species *D. nuda* was selected by the repetitive application of the herbicides used on the control of Jamaican crabgrass in the sugarcane crop, demonstrating that this species is more tolerant to the herbicides of the chemical group of imidazolinones and substituted urea, when compared to *D. ciliaris*. In this work, it was also observed, through dose—response curves, that *D. nuda* is more tolerant than *D. ciliaris* to the herbicides diuron, imazapyr, and tebuthiuron. In experiments to determine the absorption and translocation of the herbicides diuron (via leaf), imazapyr and metribuzin (via root) by species *D. ciliaris* and *D. nuda*, Dias et al. (2003) demonstrated that absorption and translocation were not the mechanisms responsible for the tolerance presented by *D. nuda* to the herbicides diuron and imazapyr.

In an assay conducted in a greenhouse, Dias et al. (2005) evaluated the efficacy of pre and post-emergence herbicides recommended for the sugarcane crop on the control of four Jamaican crabgrass species (*D. ciliaris, D. nuda, D. horizontalis* and *D. bicornis*). The pre-emergence herbicides used and their respective dose (g/ha) were: ametryn at 2.500; diuron at 2.500; tryfloxysulfuron-sodium + ametryn at 32.4 + 1.280; hexazinone + diuron at 264 + 936; tebuthiuron at 750; clomazone at 800; amicarbazone at 1.050; isoxaflutole at 112.5; and imazapic at 122.5. The post-emergence herbicides used and their respective dose (g/ha) were: mesotrione at 120; tryfloxysulfuron-sodium + ametryn at 32.4 + 1.280; ametryn at 2.000; hexazinone + diuron at 264 + 936; metribuzin at 1.440; ametryn + clomazone at 1.511.0; MSMA at 1.920; and diuron at 2.500. The authors observed that *Digitaria nuda* was the specie of more difficult control. The application of hexazinone + diuron, tebuthiuron, and imazapic, in pre-emergence, and diuron and hexazinone + diuron, in post emergence, not presented level of control satisfactory for *D. Nuda*. The other species were efficient controled by the most of the treatments, in pre-emergence as well as in post-emergence.

The glyphosate, considered to be a non-selective herbicide and recommended for desiccation in direct planting, with applications recommended to several crops, including for transgenic crops, has been used intensively in recent years, leading to the selection of tolerant species. The following are examples of species tolerant to glyphosate: tropical

Mexican clover (*Richardia brasiliensis*), morning glory (*Ipomoea* spp.), coat button (*Tridax procumbens*), and dayflower (*Commelina* spp.).

7.11 Herbicide Application Technology in Sugarcane

There are in the Brazilian market several options of herbicides for use in the sugarcane crop, for applications in pre-emergence, as well as the possibility of being used in initial, average and late post-emergence. Besides this, there are herbicides available with "systemic" or "contact" action, being some few selective to the crop. All these associated variables, as well as the difficulty for the entrance of machines in the area after a determined period, the common presence of perennial weeds of difficult control, associated to the presence of intense trash in areas of green cane harvesting, makes the weed management in this crop a complex activity, needing high quality in relation to herbicide application technology to achieve satisfactory level of weed control.

Next, some types of herbicide application in the sugarcane crop will be analyzed.

7.11.1 Application by Air

Performed by agricultural aircraft (mainly airplanes), application by air is much used in the sugarcane crop, mainly in areas of great extension. It is recommended for control in pre-emergence, and or initial post-emergence. This type of application is not recommended for weed control in average or late post-emergence, because it is not able to obtain their good coverage in more advanced stages. To obtain success in this kind of application, it is necessary to observe wind conditions, convection currents, temperature, and humidity of the air, among other factors.

7.11.2 Application by Tractor

When performed in total area, it is performed with tractored equipment, with bars which normally vary from 7 to 20 m width, working on average at speeds of 4–10 km/h, depending on the kind of machine and on the topography of the land. The applications can be performed in pre-emergence or from initial to late post-emergence.

7.11.3 Coastal Application

Coastal application is widely used in areas of irregular topography, in small sugarcane production areas, on the control of weed clumps and on "chemical scavenging", which consists of the passing of areas where some kind of control method has already been applied. The equipment for this kind of application can be hand pump or pressurized coastal pulverizers; the latter allow higher yield of the application.

The drift protection accessories in the application of non-selective herbicides, such as paraquat, glyphosate, and MSMA are efficient, because significant reductions were observed in the intensity of the intoxication symptoms of the crop plants with paraquat (Rodrigues and Almeida, 2005).

7.12 Weed Management in Green Cane

The harvesting process preceded by the burning of the sugarcane field is being substituted by the harvesting of green cane, due to changes in legislation and environmental awareness in Brazilian society. In this way, weed management after green cane harvesting and the physical, chemical, and biological properties of the soil have been changed, making it necessary to obtain more scientific information about this new technology. The adoption of this harvesting system has resulted in changes in cultivation techniques, such as the use of larger spacing between lines and the deposition of straw onto the soil, which influence directly weed occurrence and management (Velini and Negrisoli, 2000).

Some agronomic benefits, such as reduction of erosive processes, better conservation of the soil moisture, better nutrient recycling, increase of organic matter, and microbial activity of the soil, improvement of the physical and chemical properties of the soil, decrease of weed infection, besides the decrease of sugar losses via exudation of stalks during and/or right after the burning, are advantages observed in the green cane harvesting system (Velini and Negrisoli, 2000). However, some unfavorable factors to the adoption of this technology can be mentioned, such as: difficulty in sprouting for most of the varieties of sugarcane underneath the trash, increase of soil pests which attack the crop, increases in the doses of nitrogenous fertilizers, damages to the growth of the sugarcane in cold places, maintenance of moisture excess in lower areas, and limitation of the herbicide application for the daytime period when the presence of weeds is easily detected − except for the implementation of "precision agriculture" (Correia and Rezende, 2002).

The preserved sugarcane trash provides coverage of the soil, which makes the emergence of weeds more difficult, because it reduces the penetration of light into the soil. The release of exudates by the straw, which presents allelopathic effects on the germination of weed seedlings, may also occur (Procópio et al., 2003). According to Pitelli and Durigan (2001), the physical effect is not restricted only to the blocking of quantitative and qualitative solar radiation passing, but it also includes the softening of the amplitude of thermal variation, and of the moisture variation in the superficial layer of the soil. According to the same authors, another important effect of the trash is biological, with the maintenance of a microbial community, which can act on the survival of seeds and seedlings of the weed species.

The green cane cultivation system represents an important change in weed management. In this system, harvesting with the deposition of straw onto the soil causes changes in weed populations, favoring species with higher capacity of germination underneath a thick layer of straw which retains some herbicides, decreasing their efficacy. However, the deposition of trash onto the soil prevents germination and the establishment of species with little reserves stored in the seed due to the physical and/or allelopathic effects, acting on the integrated management of weed species reducing the dependence on herbicides and improving the quality of the soil.

Bibliography

Almeida, F.S., 1981. Plantio direto no Estado do Paraná. IAPAR, Londrina, 244p. (Circular, 23).

Arévalo, R.A., 1978. Matoecologia da cana-de-açúcar. Ciba-Geigy, São Paulo, 16p.

Azania, C.A.M., et al., 2006. Seletividade de herbicidas: III — aplicação de herbicidas em pós emergência inicial e tardia da cana-de-açúcar na época da estiagem. Planta Daninha. 24 (3), 489—495.

Barroso, A.L.L., et al., 2008. Manejo de plantas daninhas na cultura da cana-de-açúcar (cana-soca em pós-emergência), em região de cerrado. In: XXVI Congresso Brasileiro da Ciência das Plantas Daninhas e XVIII Congreso De La Asociación Latinoamericana De Malezas, 2008b, Ouro Preto-MG. Anais. Sete Lagoas-MG: SBCPD, CD-ROM.

Beltrão, N.E.M., Azevêdo, D.M.P., 1994. Controle de plantas daninhas na cultura do algodoeiro. Embrapa — CNPA, Campina Grande, 154p.

Câmara, G.M.S., 1993. Ecofisiologia da cultura da cana-de-açúcar. In: Câmara, G.M.S., Oliveira, E.A.M. (Eds.), Produção de cana-de-açúcar. FEALQ, Piracicaba, pp. 31—64.

Christoffoleti, P.J., Caetano, R.S.X., 1998. Soil seed banks. Sci. Agríc. 55, 74—78.

Christoffoleti, P.J., et al., 2000. Plantas daninhas na cultura da soja: controle químico e resistência a herbicidas. In: Câmara, G.M. (Ed.), Soja: tecnologia da produção. ESALQ, Piracicaba, pp. 179—202.

Conceição, M.Z., 2000. Segurança na aplicação de herbicidas. In: Congresso Brasileiro da Ciência das Plantas Daninhas, 22., 2000, Foz do Iguaçu. Palestras. Foz do Iguaçu: Sociedade Brasileira da Ciência das Plantas Daninhas, pp. 46—91.

Concenço, G., et al., 2007. Sensibilidade de plantas de arroz ao herbicida bispyribac-sodium em função de doses e locais de aplicação. Planta Daninha. 25 (3), 629—637.

Constantin, J., 1993. Efeitos de diferentes períodos de controle e convivência da Brachiaria decumbens Stapf. com a cana-de-açúcar (Saccharum spp.). Dissertação (Mestrado), Universidade Estadual Paulista, Botucatu, 98f.

Correia, N.M., Durigan, J.C., 2004. Emergência de plantas daninhas em solo coberto com palha de cana-de-açúcar. Planta Daninha. 2 (1), 11—17.

Correia, N.M., Rezende, P.M., 2002. Manejo integrado de plantas daninhas na cultura da soja. Editora UFLA, Lavras, 55p. (Boletim Agropecuário, 51).

Dias, A.C.R. et al. 2006. Eficácia agronômica de herbicidas pré e pós-emergência no controle de capim-colchão (D. Ciliaris, D. Nuda, D. Horizontalis e D. Bicornis) na cultura de cana-de-açúcar. In: Congresso Brasileiro D.A. Ciência D.A.S. Plantas Daninhas, 25., Brasília, 2006. Resumos... Brasília: SBCPD/UnB/Embrapa Cerrados, p. 328.

Dias, A.C.R., et al., 2007. Problemática da ocorrência de diferentes espécies de capim-colchão (Digitaria spp.) na cultura da cana-de-açúcar. Planta Daninha. 25 (2), 489—499.

Dias, T.C.S., Alves, P.L.C.A., Lemes, L.N., 2005. Períodos de interferência de Commelina benghalensis na cultura do café recém-plantada. Planta Daninha. 23 (3), 397—404.

Fernandez-Quintanilla, C., 1988. Studyng the population dynamics of weeds. Weed Res. 28, 443—447.

Ferreira, E.A., et al., 2005. Composição química da cera epicuticular e caracterização da superfície foliar em genótipos de cana-de-açúcar. Planta Daninha. 23 (4), 611—619.

Gupta, O.P., Lamba, P.S., 1978. Modern Weed Science. Today and Tomorrow's Printers and Publishers, New Delhi, India, 421p.

Hinz, C., 2001. Description of sorption data with isotherm equations. Geoderma. 99, 225−243.

IBGE − Instituto Brasileiro de Geografia e Estatística. Disponível em: < www.ibge.gov.br/home/estatistica/ indicadores/agropecuaria > Acesso em: 21 de novembro de 2008.

Kissmann, K.G., 1997. In: Tomo, I. (Ed.), Plantas infestantes e nocivas, second ed. BASF, São Paulo, SP, p. 825.

Kuva, M.A., et al., 2000. Períodos de interferência das plantas daninhas na cultura da cana-de-açúcar. I − Tiririca. Planta Daninha. 18 (2), 241−251.

Kuva, M.A., et al., 2003. Períodos de interferência das plantas daninhas na cultura da cana-de-açúcar. III − capim-brachiaria (*Brachiaria decumbens*) e capim-colonião (*Panicum maximum*). Planta Daninha. 21 (1), 37−44.

Kuva, M.A., et al., 2008. Padrões de infestação de comunidades de plantas daninhas no agroecossistema de cana-crua. Planta Daninha. 26 (3), 549−557.

Lorenzi, H., 1996. Tiririca-uma séria ameaça aos canaviais. Boletim Técnico Copersucar. 35, 3−10.

Lorenzi, H., 2000. Plantas daninhas do Brasil: terrestres, aquáticas, parasitas, tóxicas e medicinais, third ed. Plantarum, Nova Odessa, 608p.

Marochi, A.I., 1997. Pontos chaves para o sucesso de aplicações noturnas de herbicidas. In: Congresso Brasileiro de Plantas Daninhas, 21, 1997, Caxambu, MG. Palestras e Mesas Redondas. Caxambu, MG: SBCPD, pp. 147−154.

Núñez, O., Spaans, E., 2008. Evaluation of green-cane harvesting and crop management with a trash-blanket. Sugar Tech, Springer India. 10 (1).

Owen, M.D.K., 2006. Update on Glyphosate-Resistant Weeds and Weed Population Shifts. Academic Press, Iowa, 03p.

Pitelli, R.A., Durigan, J.C., 2001. Ecologia das plantas daninhas no sistema plantio direto. In: Rossello, R.D. (Ed.), Siembra directa en el cono sur. PROCISUR, Montevideo, pp. 203−210.

Procópio, S.O., Silva, A.A., Vargas, L., Ferreira, F.A., 2003. Manejo de plantas daninhas na cultura da cana-de-açúcar. Editora UFV, Viçosa, 150p.

Radosevich, S., Holt, J., Ghersa, C., 1997. Associations of weeds and crops, Weed Ecology − Implications for Management. second ed. Wiley & Sons, New York, pp. 163−214.

Reddy, K.N., 2003. Impact of rye cover crop and herbicides on weeds, yield, and net return in narrow-row transgenic and conventional soybean (*Glycine max*). Weed Technol. 17, 28−35.

Rodrigues, B.N., Almeida, F.S., 2005. Guia de herbicidas, fifth ed. Londrina, 592p.

Silva, A.A., 2006. Manejo integrado de plantas daninhas. MAPA. II Coferência Internacional sobre Rastreabilidade de Produtos Agropecuários, Brasília, DF, pp. 269−284.

Silva, A.A., et al., 2007a. Herbicidas: classificação e mecanismo de ação. In: Silva, A.A., Silva, J.F. (Eds.), Tópicos em manejo de plantas daninhas. Universidade Federal de Viçosa, Viçosa, pp. 83−148.

Silva, A.A., et al., 2007b. Biologia de plantas daninhas. In: Silva, A.A., Silva, J.F. (Eds.), Tópicos em manejo de plantas daninhas. Universidade Federal de Viçosa, Viçosa, pp. 18−61.

Sindicato Nacional da Indústria de Produtos para Defesa Agrícola − SINDAG. Disponível em: <http://www. sindag.com.br/upload/compimp0105.xls> Acesso em: 10 de fev. de 2008.

Toledo, E.T., et al., 2005. Green sugarcane versus burned sugarcane − results of six years in the Soconusco region of Chiapas, Mexico. International Media Ltd, Great Britain. v. 23, n. 1, pp. 20−23.

Trezzi, M.M., Vidal, R.A., 2004. Potencial de utilização de cobertura vegetal de sorgo e milheto na supressão de plantas daninhas em condição de campo: II − Efeitos da cobertura morta. Planta Daninha. 22, 1−10.

Vargas, L., et al., 1999. Resistência de plantas daninhas a herbicidas. JARD Prod. Gráficas, Viçosa-MG, 131p.

Velini, E.D., Negrisoli, E., 2000. Controle de plantas daninhas em cana crua. In: Congresso Brasileiro da Ciência das Plantas Daninhas, 22., 2000, Foz do Iguaçu. Anais. Foz do Iguaçu: Sociedade Brasileira da Ciência das Plantas Daninhas, pp. 148−164.

Victória Filho, R., 1985a. Fatores que influenciam a absorção foliar dos herbicidas. Informe Agropecuário. 11 (129), 31−37.

Victória Filho, R., 1985b. Potencial de ocorrência de plantas daninhas em plantio direto. In: Fancelli, A.L., Vidal Torrado, P., Machado, J. (Eds.), Atualização em plantio direto. Fundação Cargill, Campinas, pp. 31−48.

Weed Science, 2008. Disponível em: <http://www.weedscience.org/in.asp>. Acesso em: 17 de agosto de 2008.

Westwood, J.H., Weller, S.C., 1997. Absorption and translocation of glyphosate in tolerant and susceptible biotypes of field bindweed (*Convolvulus arvensis*). Weed Science. 45, 658−663.

CHAPTER 8

Irrigation Management

Rubens Alves de Oliveira, Márcio Mota Ramos and Leonardo Angelo de Aquino

Federal University of Viçosa, Viçosa, MG, Brazil

Introduction

Irrigation consists of efficient water application onto soil, in an adequate amount, and at the right moment, with the purpose of keeping moisture at adequate levels which favor the full development of the crop.

In Brazil, the irrigated sugarcane area is still in the minority, being smaller than 5% of the cultivated total. This is mainly because of the high resistance of the crop to water deficits and to the geographical location of the sugarcane cultivations, where the rainy season coincides with vegetative growth and the maturation phase coincides with the dry period.

The daily water consumption of the sugarcane crop varies with its development stage, the climate, the type of soil, the population of plants and the variety, generally between 2.0 and 7.0 mm. According to Doorenbos and Kassam (1979), the water requirement of sugarcane is from 1.500 to 2.500 mm per vegetative cycle.

The water deficit in the sugarcane crop is not limited to the arid and semi-arid regions, because the irregularity of the rainfall in humid regions can cause water deficiency in the plants, reducing potential productivity.

Brazilian sugarcane production in the 2007/2008 harvest was 493 million tons. The State of São Paulo is the greatest producer, with 60% of national production, where almost all the sugarcane produced is cultivated without the use of irrigation techniques. However, the plant responds positively to irrigation in situations where the rains are not enough to fulfill its water needs.

Sugarcane irrigation brings several benefits, such as increase of stalk productivity and of sucrose content, precocity in the harvesting, longevity of the sugarcane field, low tipping rate, making the mechanized harvesting easier, and providing higher resistance to pests and diseases. There are also social—economical benefits, such as the increase in the number of jobs and in regional income.

Sugarcane productivity is not determined only by good water availability in the soil, as in the case of irrigated crops, but also by other determinant factors of production, for example climate, soil fertility, variety with great productive potential, management of the crop, including phytosanitary control, period of cutting, etc.

In Brazilian tropical and subtropical regions, sugarcane productivity in rainfed conditions can reach 100 t/ha; however, in most of the producing regions, the yield is well below this value. In the irrigated cultivation, yields higher than 200 t/ha can be reached, in favorable climate conditions, soil fertility, and adequate conduction of the crop, and irrigation management. However, the lack of knowledge of the benefits originating from irrigation restricts the use of this technique. Guazzelli and Paes, mentioned by Santos (2005), studied the variety SP80-1842, cultivated with drip irrigation in the region of Ribeirão Preto, SP. They obtained 173 t/ha with total irrigation and 144 t/ha without irrigation, with an increase of 29 t/ha, corresponding to an addition of 20%.

In sugarcane irrigation management it is important to characterize the development phases of the crop, in order to allow adequate application of water throughout the cycle. Therefore, for this crop, four development stages are defined: Stage I — Germination and emergence, with approximated duration of 1 month; Stage II — Tillering and establishment of the crop, from 2 to 3 months; Stage III — Vegetative development, from 6 to 7 months; and Stage IV — Maturation, with duration of 2 months, approximately.

Sugarcane presents good resistance to water deficit. This comes from its physiological adjustments, as well as from the capacity of its root system to adapt to new conditions of water stress in the soil. Water deficit in the soil is most critical in the first two development stages, possibly causing a reduction in the population of plants. In the vegetative development phase, this deficit does not affect productivity so much as in the two previous phases. But in the maturation phase, the sugarcane responds well to water deficit in the soil, with increase of sugar content in the plant.

There are several irrigation management techniques which are applied to the sugarcane crop: saving irrigation, regulated deficit irrigation, supplementary irrigation, and total irrigation.

In saving irrigation, two or three applications of water or vinasse are performed after planting, in the case of cane-plant, or after cutting, in the case of ratoon, trying to guarantee the germination and the initial development of the seedlings. In this modality of water and vinasse application, there is no rigidity in fulfilling the real water need for the sugarcane crop. This type of water or vinasse application is recommended in places with clay soil and humid climate, being normally applied 40—80 mm in 1 month. In the case of sandy soil, the number of irrigations can be increased reducing the amount of water applied each time.

In regulated deficit irrigation, irrigations are performed applying water only to fulfill part of the water needs of the plants. This management can be applied throughout the whole cycle or in the development and maturation phases.

In supplementary irrigation, the application of water is performed when the water need of a crop in a period is larger than the amount of water available in the soil, due to lack or irregularity of rains.

In total irrigation, sufficient irrigations to completely fulfill the water need of the crop are performed. This type of management is the least common due to economic reasons and should only be applied in arid regions, where the contribution of rainwater is insignificant.

Irrigation is an important practice for success for most crops, including sugarcane, but first we should evaluate whether there is the need and feasibility to irrigate, depending on the local climatic conditions, on the type of soil, on the availability of water, and on economic aspects. The decision to irrigate or not will depend, therefore, on the quantity and distribution of rains in the region, on the response of the sugarcane to irrigation, on the financial capacity of the producer, and on the economic feasibility for investment. If the technical and economic conditions justify the cultivation of irrigated sugarcane, the most appropriate irrigation system must be selected.

The costs of sugarcane irrigation vary with the method used, with the land relief of the area, the type of soil, the climatic conditions, and with the management technique of the irrigation used. Soares et al. (2003) estimated the average costs, by hectare, of the following irrigation systems for the sugarcane crop, in the region of Juazeiro, BA: furrows with channel, R$450.00; furrows with windowed tube: R$1513.00; center pivot: R$5870.00; linear system: R$6562.00; and drip: R$6243.00.

8.1 Most Used Irrigation Methods in the Sugarcane Crop

The most used irrigation methods in the sugarcane crop are sprinkler, surface, and microirrigation.

8.1.1 Sprinkler Irrigation Method

In the sprinkler irrigation method, the water is conducted in tubes, under pressure, and launched into the atmosphere in the form of a jet, which fragments into drops, being distributed onto the crop and the soil of the parcel to be irrigated, being similar to a rain.

This method adapts to different types of soil and crops, being possible to use where the terrain presents a little irregular topography. On the other hand, the sprinkler irrigation method presents limitations, such as low efficiency, when used in places subject to wind

Figure 8.1
Sugarcane irrigated by center pivot.

with average speed above 4 m/s, and high interference in phytosanitary treatments, due to washing the chemical products which were pulverized in the shoot of the plants.

The most used sprinkler irrigation systems in the sugarcane crop are described next.

Center pivot

Center pivot equipment is composed of suspended steel tubing, containing sprinklers and sustained by metal frames with pneumatic wheels, denominated towers, and by a command system (Figure 8.1).

The tubing of the center pivot is made of galvanized steel, with diameter of 168 mm (6⅝"), 219 mm (8⅝"), and 254 mm (10"). The most common have tubing of 168 mm of diameter. The center pivot manufacturers also make available equipment with tubes containing internal coating, for protection against corrosion, in the case of performing ferti-irrigation with the application of vinasse in the sugarcane crop.

The center tower is a pyramid-shaped metal structure, fixed on a concrete base built in the center of the irrigated area. The suspended tubing turns around the center tower, through the pivoflex, irrigating circular areas. In the center tower, the control panel and the elevation tube of the center pivot are installed.

In the control panel of the pivot, there are the following components: main key, selecting key of rotation direction, percentage relay, voltmeter, light indicating that the system is on, lights indicating defects, hourmeter, and timer against excess of water, among others. There are options for analogical (conventional) and digital control panels.

The main key controls the energization of the panel and towers for activating the center pivot. This key has two positions, on and off, and operates in rated voltage of 480 V, as the other components of the pivot.

The selecting key of the rotation direction allows turning the pivot in clockwise or counterclockwise direction, depending on the area to be irrigated.

The percentage relay allows the control of the rotational speed of the pivot, in percentage terms, being able to be operated even with the equipment running. The indicated values express the motor energization time of the last tower of the center pivot. For example, if the percentage relay is set at "0%", the last tower will not move and, consequently, the equipment will remain stopped. If it is set at "50%", the last tower will move during a determined period of time, for example 30 seconds, it will remain stopped for the same time, this process being repeated continuously, while the equipment is on. If it is set at "100%", the last tower will move continuously at maximum speed, which results in a shorter time for a complete turn of the center pivot.

The tubes of the parts are interconnected through a spherical pin, with its own housing, and through special hoses. This allows movement of the suspended tubing for adaptation of the equipment to irregularities in the terrain surface.

The length of each part varies between 33.2 and 62.5 m, depending on the manufacturer of the center pivot. The longest distance between towers is the most economical, with the part mounted with nine tubes, and must be used only in conditions of smooth and uniform topography and in soils which present good stability when wet, avoiding atoll problems of the center pivot.

The free height in relation to the soil is 2.70 m, 3.60 m, 4.60 m, or 5.60 m, however these values can vary, depending on the manufacturer of the center pivot. In the case of sugarcane irrigation, the free height of the pivot must be of 4.60 or 5.60 m.

Each tower has a gearmotor with ¾", 1, or 1.5 hp, mounted on the base-beam, which activates, through elastic couplings, the final reducers of the wheels.

The wheels of the tower are made of steel, with tires of the type used in the rear axis of agricultural tractors. Both wheels are mounted in the ends of the base-beam of the tower, in a way that the tires have opposite gripe, so they can cross themselves in the terrain, once the equipment can turn in two directions.

The displacement speed of the last tower is regulated by the control panel of the center pivot. The speed and alignment of the other towers are controlled by an individual box with electromechanical components and by control metal bars existing in the upper part of each of them.

Inside the individual control boxes, there is a motion microswitch, responsible for starting and turning off the tower, and a safety microswitch, which shuts the whole system down when there is excessive misalignment of the side line, as in the case of atoll of the center pivot.

The last part of the tubing stays swinging, with the purpose of increasing the irrigated area at a lower cost. Some center pivots have a sector sprinkler of the rain gun type installed at the end of the swinging tubing, with the same purpose mentioned previously; however, this practice is in disuse due to the high intensity of water application, to the formation of rains with large drops, which may cause damage to the soil and to the crops, and to the higher susceptibility to water drift by the wind, reducing the efficiency of the water application. To compensate the reduction of irrigated area caused by the removal of the rain gun, some manufacturers have increased the length of the swinging tubing.

Spool winder

This irrigation system is composed of a suction tube, a pump set, an adducing tube and a main tube with hydrants, where the spool winder is connected.

This equipment is composed of a metal platform with wheels, a steel reel, a polyethylene hose of medium density, and a sprinkler installed on a cart. Depending on the spool winder model, the hose can be from 200 to 550 m long, with diameter varying from 50 to 140 mm.

The reel has a propelling system, generally a hydraulic turbine. The polyethylene hose is connected to the reel and to the ascension tube of the sprinkler.

In the operation of the equipment, the spool winder is placed in the middle of the carrier, next to a hydrant in the main line, referent to the strip to be irrigated. The cart with the sprinkler is displaced with the use of a tractor to the opposite end of the strip, promoting the unwinding of the hose. The entry nozzle of the spool winder is connected to the hydrant of the main line through another hose, approximately 5 m long.

After the actuation of the pump set, the hydrant is opened, putting the system into operation. While the reel turns, the winding of the hose occurs; and the cart, with the sprinkler operating, starts to displace towards the metal platform. At the end of the route, the stopping device is actuated automatically, interrupting the displacement of the cart.

The procedure is repeated for irrigating the strip located on the other side of the main line, after the reel turns 180° on the metal platform. After the irrigation of two contiguous strips, the whole set is displaced and positioned next to another hydrant of the main line, repeating the previously described operations until the whole area is irrigated.

There are manufacturers of spool winders in the market which make available several models for the irrigation of small and large areas, generally varying between 15 and 60 ha.

Direct mounting

This irrigation system is composed of a suction tube with hose and foot valve, a pump set with device for primer, a crane, and a sector sprinkler of the rain gun type. The set is mounted onto a frame with four pneumatic wheels.

The crane is used as holder for the hose, allowing the maintenance of the foot valve with sieve at an adequate height inside the channel, besides assisting in the transportation of the system.

The direct mounting system is widely used for application of effluents of industries and distilleries (vinasse) onto the soil, mainly in sugarcane areas, and for the application of water in the irrigation of several crops. The components of the system are manufactured with appropriate materials to resist corrosion.

The equipment is normally mounted next to a channel, where it will operate during the necessary time to apply the desired depth. The operating time in each position will depend on the intensity of the application of the sprinkler and on the water depth or on the effluent to be applied.

The distance between the equipment mounting points depends on the operating range of the sprinkler. The spacing between the channels and between places of installation of the equipment generally is 100 m. The spacing between channels can be increased using extensions composed of tubes with line valves spaced at 100 m.

After operating in one position, the system is shut down and transferred to the following position with the use of an agricultural tractor. In the new position, the foot valve with sieve is immersed in the channel and the set is put into operation. This procedure is repeated until the whole area is irrigated.

Portable rain gun

This irrigation system is generally composed by a pump set, galvanized steel tubes with quick coupling, accessory parts and one or more large sprinklers of the rain gun type.

In this system, the side line has one hydrant in each sprinkler connection position, operating with only one rain gun during the sufficient time to apply the necessary water depth. When finishing the irrigation in one position, the rain gun sprinkler is displaced to the following position on the same side line, and this way successively, until completing the other sprinkling positions. When finishing the irrigation of the parcel corresponding to the determined position of side line, it is disassembled, transported and mounted once again in another position in the main line, repeating the procedure until the whole area is irrigated.

8.1.2 Subsurface Drip Irrigation

This irrigation method is characterized by the application of water at low intensity directly underneath the surface of the soil, using buried tubing, with high frequency, to keep the moisture of the soil next to the field capacity.

The subsurface drip irrigation system is composed of pump set, control units, PVC tubing and buried drip tubes or drip tapes.

In the control unit there are filters, chemical product injector, valves, among others. In the case of automated systems, there are several components, such as digital controller, for programming the irrigation in the different sectors, solenoid valves, and hydraulic or electric command valves.

In the field, at the beginning of each sector, there is a structure called an easel, where control valve, manometer, cupping glass, and anti-vacuum valve are installed. For dealing with systems with buried emitters, the use of screen or disc filters on the easel are recommended.

The tubing can be made of different materials. The pumping line and the main line are normally buried, PVC being the most commonly used material. The derivation lines, usually buried, are made of PVC or polyethylene. The side lines, composed of drip tubes, or drip tapes, are generally made of polyethylene.

The side lines are buried mechanically during the planting of the sugarcane, generally 30 cm deep and with spacing of 1.40 m.

In the drip irrigation system, one of the bigger problems is the occurrence of obstruction of the emitters. The causes can be physical, chemical, and biological nature of the irrigation water and, in the case of subsurface drip, the obstruction can also occur by root intrusion into the drips. In sugarcane, the intrusion problem is aggravated due to its fascicule root system, composed of thin roots, which grow and penetrate into the drips. Another factor which also causes the obstruction of the emitters is the occurrence of vacuum in the buried side line, which can suck particles of the soil to the interior of the drips.

Apart from these problems, the subsurface drip has been used in the sugarcane crop for increasing productivity with optimization of water and energy, besides not interfering in cultural practices.

8.1.3 Furrow Irrigation

In this irrigation method, furrows are opened parallel to the sugarcane planting rows, in which water is applied (Figure 8.2).

After the application of determined flow in the beginning of the furrows, the water starts to drain and, while it advances towards the lower part of the area, the infiltration occurs. In an irrigation conducted adequately, the water must run off at the end of the furrows in enough time for the application of the necessary water depth. This way, part of the infiltrated water is retained in the soil profile which is explored by the roots of the plants, providing the water storage for use during the period between two consecutive irrigations.

Figure 8.2
Windowed tube applying water to the furrows in the sugarcane irrigation.

The spacing between furrows depends on the type of soil and, mainly, on the distance between rows of sugarcane plants to be irrigated. In general, in the sugarcane crop, one furrow is used for each row of plants.

The length of the furrows depends on the format and on the size of the area, on the type of soil and on the applied flow, generally varying between 50 and 400 m.

In most cases, the furrows are built with decline varying between 0.1% and 2.0%, the first value being considered as a desirable minimum to avoid overflows, making the drainage of the water easier, and the second value as a maximum to avoid serious problems of soil erosion.

The use of the furrow irrigation system demands a lot of labor per unit of area, used mainly in the control of water in the channels and in the application of the flow into the furrows.

8.2 Irrigation Management

Irrigation management refers to a set of technical decisions involving the characteristics of the crop, climate, water, soil, and irrigation system. The adequate management of irrigation, associated with the other cultivation techniques, allows the producer to reach elevated productivity levels, saving water and energy, besides contributing to the preservation of the environment.

Well-conducted management consists of defining the adequate moment to start irrigation and of determining the necessary amount of water for the crop, utilizing knowledge for the time of water application or the displacement speed of the irrigation equipment.

To irrigate sugarcane efficiently, it is necessary to know some parameters related to the soil, climate, water, plant, and irrigation system which will allow quantification of the water which must be applied.

8.2.1 Important Parameters in Irrigation Management

The soil is a natural reservoir of water for plants. The water stored in the soil and available to the plants is comprehended between the field capacity and the permanent wilting point.

Field capacity

Field capacity (Cc) corresponds to the superior limit of available water and represents the moisture of the soil after drainage of the water contained in the macropores by gravity action. This moisture condition favors higher absorption of water and nutrients by the plants.

Usually, field capacity is determined in the laboratory, by the retention curve method. In this method, the value of the field capacity moisture is represented by the balance moisture with tension of 6−33 kPa, depending on the texture, structure and content of organic matter in the soil.

Permanent wilting point

The permanent wilting point (Pm) corresponds to the inferior limit of available water. This moisture condition severely restricts the absorption of water by the plants, which will die if there is no replacement of the water in the soil.

In general, the permanent wilting point is also determined in the laboratory, by the retention curve method. In this method, the moisture value of the wilting point is represented by the balance moisture with tension of 1.500 kPa (Figure 8.3).

To draw the retention curve, soil moisture values are obtained after submitting samples to different tensions in the Richards Extractor.

Despite its dynamic character, for practical purposes of irrigation, field capacity is usually obtained with tension value of 10 kPa (0.10 atm) in sandy soils and 33 kPa (0.33 atm) in clay soil. On the other hand, the moisture corresponding to the permanent wilting point is obtained with the tension of 1.500 kPa (15 atm).

Soil density

Soil density is the relation between the mass and the volume of a dry soil sample. In its determination, an Uhland Sampler can be used, whose cylinder is inserted in the soil, in the medium depth of the soil layer which is explored by the roots of the plants. After the removal of the cylinder, the sample is prepared and taken to the oven to dry for 24 h, at an

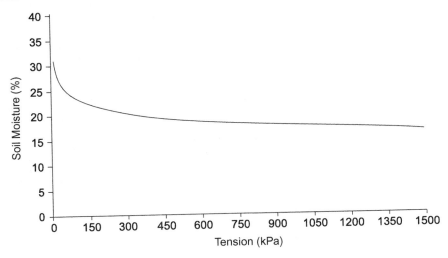

Figure 8.3
Retention curve of water in soil.

Figure 8.4
Moistening of the soil (A) and insertion of the PVC tube for the removal of sample (B).

approximate temperature of 105°C, to determine its mass. The volume is determined with the use of the diameter and height values of the soil sample.

In the sampling for determination of the soil density, the method recommended by Oliveira and Ramos (2008), denominated PVC Tube Method, can also be used.

The PVC Tube Method is performed as follows: level the soil surface beforehand, moisten it (Figure 8.4A), and then insert a tube into the soil with a nominal diameter of 50 mm and length of 15 cm (Figure 8.4B), with one of the ends beveled, until the upper edge of the

Figure 8.5
Edge of the PVC tube close to the soil surface (A) and removal of the tube after opening of trench (B).

Figure 8.6
Removal of the soil excess at the base of the tube (A) and closing of its ends with adhesive tape (B).

PVC tube is at the soil surface (Figure 8.5A). After that, the soil around the PVC is dug, to facilitate access to its lower end. With a knife, cut the soil at the base of the tube (Figure 8.5B), remove the set (tube with soil) and trim the base of the sample, to eliminate the excess of soil (Figure 8.6A). Then, clean the tube and seal the ends of the soil sample with adhesive plastic tape (Figure 8.6B). For irrigation purposes, at least three repetitions in the irrigated area are recommended.

The PVC tubes with the samples must be sent to a laboratory, requesting the determination of the soil density. After the determination of the density, a compound sample can be used to obtain the field capacity and permanent wilting point values.

In the laboratory, the soil contained in the PVC tube must be removed and put into the oven at a temperature of 105°C, for 24 h. After this time, the dry soil is weighed and its density is calculated with the application of the following equation:

$$D_s = \frac{m_s}{V_s} \tag{1}$$

where D_s: soil density, g/cm^3; m_s: dry soil mass, g; and V_s: soil sample volume, cm^3.

While inserting the PVC tube, generally compaction occurs, with degradation of the soil surface inside the tube, as can be observed in Figure 8.5A. The error due to compaction is eliminated in the PVC Tube Method when considering, in the soil density calculation, the internal volume of the PVC tube, and not the volume of the compacted soil sample.

The internal volume of the PVC tube (V_s) is calculated multiplying the transversal section area of the tube by its length.

$$V_s = \frac{31416 \, D^2}{4} C \tag{2}$$

where: D = internal diameter of the PVC tube, cm; and C = length on the PVC tube, cm.

So, when considering the soil sample volume equal to the internal volume of the PVC tube, the error coming from soil compaction is eliminated, which normally occurs during sampling.

Effective depth of the root system

The effective depth of the root system (Z) is the one where at least 80% of the roots of the plants concentrate. Its value varies according to the crop and its development stage, the type of soil, and its management, besides the management of the irrigation itself. For the irrigated sugarcane crop, the following values of effective depth of roots can be used: stage I — 15 cm, stage II — 27 cm, and stage III — 40 cm.

Factor of water availability in soil

If moisture is in the field capacity, with the occurrence of evapotranspiration, the level of water in the soil decreases, making each time more difficult for the absorption of water and nutrients by the plants. In irrigation, the content of water in the soil must not be allowed to reach the permanent wilting point. In this way, in adequate irrigation management, a moisture limit must be considered, the value of which is between the field capacity and the permanent wilting point.

The factor of water availability in soil (f) is important in the calculation of the moisture limit and of the water depth required for the crop. The value of f represents the fraction of the total of water stored in soil, between the field capacity and the permanent wilting point,

Table 8.1: Values of Kc for the different phenological stages of the sugarcane crop.

Age (months)		Development of the Sugarcane		
Cane-Plant	Cane-Ratoon	Development Phases	Development Stages	Kc
0−2	0−1	Planting up to 25% of closure	I	0.40
2−4	1−2.5	25−75% of closure	II	0.65−0.95
4−14	2.5−10	75−100% of closure	III	1.10−1.25
14−18	10−12	Maturation	IV	0.75

which can be used by the crop, in such a way that the plants do not suffer water restriction at a level that can compromise development and reduce productivity. For the sugarcane crop, a maximum value of $f = 0.65$ can be used.

Evapotranspiration of the crop

The process which associates the transferring of water from the soil and from the plants to the atmosphere, in the form of steam, is called evapotranspiration (ET). It represents, in practice, the water consumption of a crop, generally expressed in millimeters per day (mm/day). One millimeter represents the height of the depth formed by the application of one liter of water in an area of $1\ m^2$ ($1\ mm = 1\ L/m^2$).

Evapotranspiration varies according to the type of crop, due to the individual characteristics of the plant species. This way, there is a need to define the evapotranspiration for a reference crop (ET_o) and, from there, estimate the evapotranspiration of the crop of interest (ET_c).

In irrigation management, several methodologies can be used to determine ET_c, among them the Class A tank, equations of evapotranspiration estimation, and the Irrigation Water Meter.

In the case of using a Class A tank, ET_c is determined by multiplying the value of the evaporated depth by the coefficients of the tank and of the crop. The coefficient value of the tank (K_t) depends on the conditions of its installation in the field and on the climate. On the other hand, the coefficient value of the crop (K_c) (Table 8.1) depends on the type of crop, on their phenological stages, on the cultural practices adopted, on the climate and on the irrigation frequency.

In the equations of evapotranspiration estimation, meteorological elements data are used, generally obtained from automatic meteorological stations. The reference evapotranspiration reference (ET_o) is calculated and its value is multiplied by K_c to obtain the evapotranspiration of the crop (ET_c). In this case, irrigation management must be conducted with the use of tables or computer programs.

Figure 8.7
Irrigation Water Meter used in irrigation management in center pivot.

The Irrigation Water Meter is a device which introduces great simplicity to water management in irrigated areas by estimating ET_c directly, besides answering the three basic questions of irrigation management: the moment to irrigate, the required water depth for the crop and the irrigation time (Figure 8.7). In the cases of mechanical drive systems, such as the center pivot and the linear system, the Irrigation Water Meter provides the displacement speed of the equipment, instead of the irrigation time. The Irrigation Water Meter also quantifies the rainfall in the cultivated area, allowing the optimization of rainwater use and, consequently, reducing energy consumption.

The Irrigation Water Meter must be pre-adjusted for the conditions of the soil, the crop and the irrigation equipment which exist on the agricultural property. With this done, the decision regarding water management comes from the irrigation equipment operator, who does not need to have specialized technical information. The operation of the Irrigation Water Meter is performed easily, without the need for calculations or the use of computer programs. The operation of the device consists simply of opening and closing valves, following a predefined sequence.

The Irrigation Water Meter presents the following advantages:

- it is simple, easy to install and use, and of relatively low cost;
- measures the water depth, allowing easy inclusion of rain in the irrigation management;
- directly provides the estimation of the crop evapotranspiration value; and
- indicates directly to the person who is irrigating when to irrigate the crop and the time of operation of the irrigation equipment, or its displacement speed in percentage terms, without the need to perform calculations.

Efficiency for water application of the irrigation system

In irrigation, only part of the water applied is effectively used by the crop, due to losses by evaporation, entrainment by the wind, surface drainage, percolation and leaks in the tubing.

Generally, when the irrigation system is well dimensioned and managed adequately, an efficiency of water application (Ea) of around 90% for subsurface drip can be considered; 85% for center pivot and linear system; 75% for the spool winder systems, direct mounting, self-propelled, and portable rain gun; and 60% for furrow irrigation systems.

8.3 Irrigation Management Strategies

8.3.1 Irrigation Without Water Deficit

This case covers management with total irrigation or with supplementary irrigation.

Irrigation management with fixed watering shift

The watering shift (TR) is the interval, in days, between consecutive irrigations in a particular area. Management with fixed watering shift consists of performing irrigations at intervals of defined duration, in other words, daily, every 2 days or every 3 days, and so on.

Once the watering shift is established, it is necessary to quantify the total water depth to be applied, allowing calculation of the displacement speed, in the cases of the center pivot systems, linear system, spool winder or self-propelled; the time of water application by position, in the cases of the direct mounting systems and portable rain gun; and the irrigation time in each sector, in the cases of the subsurface drip and furrow systems.

The determination of the water quantity to be applied per irrigation is generally performed through evaluation of soil moisture or evapotranspiration estimation of the crop.

Irrigation management based on the soil moisture

After defining the watering shift, the soil moisture must be evaluated prior to each irrigation event. This allows calculation of the water depth to be applied by the irrigation system, in order to return the soil moisture to the field capacity. After the irrigation, a parcel of water stored in the soil will be used by the crop during the next period corresponding to the watering shift. In a similar way, a new evaluation of the moisture must be performed and new calculation of the depth to be applied, with this procedure being repeated in all irrigation events of the crop.

The total irrigation depth is calculated applying the following equation:

$$LI = \frac{(Cc - Ua)}{10\,Ea} Ds\,Z \tag{3}$$

where: LI = total irrigation depth, mm; Cc = field capacity, % in weight; Ua = soil moisture prior to the irrigation, % in weight; Ds = soil density, g/cm^3; Z = effective depth of the root system, cm; and Ea = efficiency of water application, decimal.

In the case of sugarcane irrigated with center pivot or linear system, the displacement speed must be calculated by:

$$V = \frac{100 \, L_p}{LI} \tag{4}$$

where: V = displacement speed of the center pivot or linear system, %; and L_p = Project depth of the pivot for the speed of 100%, mm.

In the case of spool winder and self-propelled, after calculating the total irrigation depth, the catalog of the equipment manufacturer should be consulted to obtain the speed of corresponding displacement.

In the case of the direct mounting, portable rain gun and subsurface drip systems, the time of operation of the equipment in each position or sector must be calculated by:

$$t = \frac{LI}{I_a} \tag{5}$$

where: t = irrigation time in each position or sector, h; LI = total irrigation depth, mm; and I_a = water application intensity of the irrigation system, mm/h.

Irrigation management based on the estimation of the crop evapotranspiration

In the case of use of Class A tank or of automatic meteorological station, the crop evapotranspiration is calculated from the ET_o. In the first case, a pluviometer must be used to quantify the rain. The values of ET_c are accumulated daily throughout the period corresponding to the watering shift, providing the water depth to be replaced to the soil by the irrigation system.

The total irrigation depth is calculated applying the following equation:

$$LI = \frac{\sum ET_c}{Ea} \tag{6}$$

where: LI = total irrigation depth, mm; ET_c = sum of the values of ET_c which occurred during the period corresponding to the watering shift, mm; and Ea = efficiency of water application, decimal.

After calculating the total irrigation depth, proceed in a similar way to the previous case to calculate the displacement speed of the equipment (Equation 4) or to calculate the time per position or per sector (Equation 5).

If there is rain in the period, verify whether the precipitated depth was enough to replace the water deficit in the soil existing until the moment when the rain occurred. If this has happened, zero the sum of the values of ET_c existing, once the deficient depth was replaced to the soil by the rain. During the remaining days of the watering shift, accumulate again the daily values of the ET_c. The total irrigation depth is calculated applying Equation 6 and the displacement speed of the equipment or the time per position or per sector, which are obtained as described previously.

If the precipitated depth is lower than the corresponding depth to the sum of the ET_c, the difference between them will provide the current deficient depth after the rain occurs. In this case, this difference will be added to the new daily values of the ET_c verified during the remaining days of the watering shift. The total irrigation depth and the displacement speed of the equipment or the irrigation time are also obtained as described previously.

In the case of use of the Irrigation Water Meter, prior to beginning management, the equipment should be prepared according to the development stage which the sugarcane is at in the field. Two situations can occur: (1) beginning of management with the planting of the cane-plant; and (2) beginning of management with the crop already implemented.

In situation (1), the Irrigation Water Meter must be prepared with Side 1 of the management ruler facing forward and with the mark of the sliding rod in the direction of the value indicated for stage I of the crop, according to the recommendation of the manufacturer.

In situation (2), it is necessary to certify which development stage the sugarcane is in. If the crop is in stage I, the Irrigation Water Meter must be prepared with side 1 of the management ruler facing forward and with the mark of the sliding rod in the direction of value 2.5 cm, indicated for stage I of the crop (Figure 8.8). If the crop is in stage II, the Irrigation Water Meter must be prepared with side 2 of the management ruler facing forward and with the mark of the sliding rod in the direction of value 3.5 cm, indicated for stage II (Figure 8.9). If it is in stage III, side 3 of the management ruler must be facing forward with the mark of the sliding rod in the direction of value 4.5 cm, indicated for stage III of the crop (Figure 8.10). In case it is necessary to irrigate the sugarcane in stage IV, it is only necessary to position the mark of the sliding rod in the direction of the indicated value for stage IV of the crop, keeping side 3 of the management ruler.

Management must be started right after the occurrence of an irrigation. In this moment, the Irrigation Water Meter must already be prepared.

Once the watering shift is defined, the quantity of water necessary for the crop is directly indicated in the feeding tube of the Irrigation Water Meter. According to the characteristics of water application of the center pivot or of the linear system, the Irrigation Water Meter

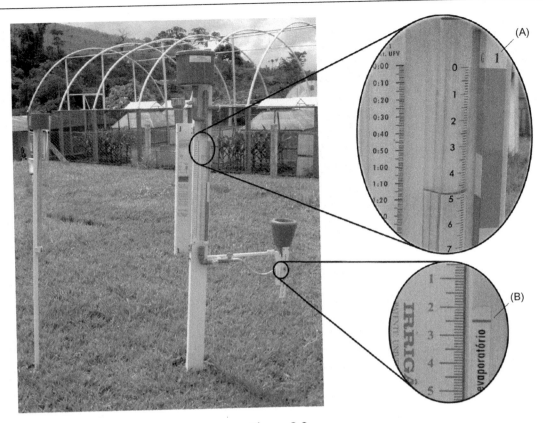

Figure 8.8

Crop in development stage I: side 1 of the management ruler is facing forward (A) and the mark of the sliding rod stays in the direction of value 2.5 cm on the level ruler (B).

comes equipped with a percentage ruler, which will directly define the displacement speed of the equipment. In the cases of the direct mounting, portable rain gun and subsurface drip systems, the Irrigation Water Meter comes equipped with a temporal ruler, whose model corresponds to the liquid intensity of application.

The Irrigation Water Meter also allows whoever is irrigating to easily consider whether the amount of rain water fulfills the needs of the plants, making it possible to reduce water and energy consumption. In the case of having rain, the operator of the Irrigation Water Meter must measure the precipitated depth and verify, after that, whether it was enough or not to fulfill the water deficit which existed in the soil prior to the occurrence of the rain. If the water level in the feeding tube of the Irrigation Water Meter is in the direction of the blue strip of the management ruler, there is no need to irrigate, postponing the irrigation to the next event established by the watering shift.

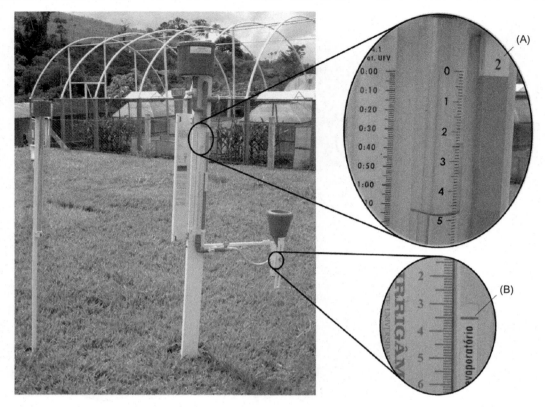

Figure 8.9
Crop in development stage II: side 2 of the management ruler is facing forward (A) and the mark of the sliding rod stays in the direction of value 3.5 cm on the level ruler (B).

Irrigation management with variable watering shift

Management with variable watering shift consists of performing the irrigation when the moisture of the soil reaches the moisture limit value, or when the crop has consumed the real water depth, in the case of the evapotranspiration estimation.

Irrigation management based on the soil moisture

The irrigation must be performed when the limit moisture of the soil is reached. It is calculated applying Equation 7:

$$U_L = Cc - f(Cc - Pm) \tag{7}$$

where: U_L = limit moisture defined by factor f, % in weight; C_c = field capacity, % in weight; f = factor of water availability in soil, dimensionless; and P_m = permanent wilting point, % in weight.

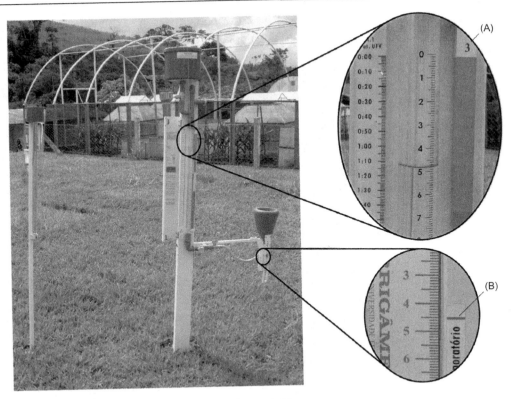

Figure 8.10
Crop in development stage III: side 3 of the management ruler is facing forward (A) and the mark of the sliding rod stays in the direction of value 4.5 cm on the level ruler (B).

The total irrigation depth is calculated applying Equation 3, substituting the variable U_a by U_L, and the displacement speed must be obtained by Equation 4. In the systems where management is performed based on the irrigation time, Equation 5 must be used.

Irrigation management based on the evapotranspiration estimation of the crop

In the case of variable watering shift, each irrigation event must occur when the sum of the evapotranspired water depths is equivalent to the real water depth, calculated by:

$$LR = \frac{(Cc - Pm)}{10} Ds\ f\ Z \tag{8}$$

where LR is the real water depth, in mm.

In the case of use of Class A tank or of automatic meteorological station, the evapotranspiration of the crop is calculated daily from the ET_o. In the case of use of the Class A tank, a pluviometer must be used to quantify the rain.

The values of ET_c must be accumulated daily, until the sum is approximately equal to the value of the LR. The total irrigation depth is calculated by applying Equation 6 and the displacement speed of the center pivot or of the linear system must be calculated with the use of Equation 4. In the cases of spool winder and self-propelled, after calculating the total irrigation depth, it is necessary to consult the catalog of the manufacturer of the equipment to obtain the corresponding displacement speed. In the cases of direct mounting, portable rain gun and subsurface drip systems, the operation time of the equipment in each position or sector is calculated with Equation 5.

In the case of use of the Irrigation Water Meter, management must be started right after the occurrence of an irrigation. At this moment, the Irrigation Water Meter must already be prepared.

The adequate moment to irrigate the crop is easily indicated in the management ruler, it being only necessary to observe the water level in the feeding tube.

When the water level is in the direction of the blue strip, the indication is not to irrigate, because there is high availability of water in the soil. Irrigation in this condition causes soaking of the soil and the loss of nutrients to the deeper layers.

If the water level is in the direction of the green strip, it is indicative of good water availability in the soil and also, in this case, there is no need to irrigate the crop.

When the water level drops to the point of reaching the beginning of the yellow strip, it is the moment to irrigate. The length of the yellow strip establishes a safety margin indicative of the time to irrigate. In this case, the decision to irrigate or not is of whoever is irrigating. If there is a safety margin or evidence of probability of rain, whoever is irrigating can wait for the following day.

If the water level drops to the point of the red strip, the Irrigation Water Meter indicates low water availability in the soil, showing to the producer that the moment of irrigation has already passed. So, there are serious risks of significant reductions in the productivity of the crop, which increase the lower the level of the water.

Once the decision to irrigate is taken, it is necessary to observe the quantity of water necessary for the sugarcane crop, which is directly indicated in the feeding tube of the Irrigation Water Meter. According to the characteristics of water application of the center pivot or of the linear system, the Irrigation Water Meter will indicate directly, in the percentage ruler, the displacement speed of the equipment. If the equipment is a spool winder or self-propelled, the Irrigation Water Meter will also indicate directly its displacement speed. In the cases of the direct mounting, portable rain gun and subsurface systems, the Irrigation Water Meter will indicate directly, in the temporal ruler, the time of irrigation in each position or sector.

8.3.2 Irrigation with Water Deficit

This case covers sugarcane irrigation management with regulated water deficit. This type of management allows a better use of rain water and stimulates deepening of the root system, increasing the volume of the soil which is explored by the roots of the plants. The applied water depth in each irrigation is lower than the water quantity necessary for the crop, but its value must be enough not to significantly affect the development and productivity of the sugarcane.

The control of the quantity of water to be applied in each irrigation is performed adopting crop coefficient values which are lower than the ones recommended in Table 8.1.

In the case of irrigation management with an Irrigation Water Meter, the regulated water deficit is established simply with the change of the water level in the interior of the evaporation reservoir, positioning the mark of the sliding rod in a lower value in the level ruler than the one recommended by the manufacturer of the equipment for each development phase of the crop (Figures 8.8—8.10). Therefore, the decrease of the water level in the evaporation reservoir reduces the evaporating surface area, resulting in lower evapotranspiration estimation of the sugarcane.

Bibliography

Doorenbos, J., Kassam, A.H., 1979. Yield response to water. FAO, Rome, 193p. (Irrigation and Drainage Paper, 33).

Oliveira, R.A., Ramos, M.M., 2008. Manual do Irrigâmetro. Viçosa.144p.

Santos, M.A.L. Irrigação suplementar da cana-de-açúcar (*Saccharum* spp.): Um modelo de análise de decisão para o Estado de Alagoas. 2005. Tese (Doutorado), Escola Superior de Agricultura Luiz de Queiroz, Universidade de São Paulo, Piracicaba. 100p.

Soares, J.M., Vieira, V.J.S., Junior, W.F.G., Filho, A.A.A., 2003. Agrovale, uma experiência de 25 anos em irrigação da cana-de-açúcar na região do Submédio São Francisco. ITEM. 60, 55—62.

CHAPTER 9

Precision Agriculture and Remote Sensing

Carlos Alberto Alves Varella[1], José Marinaldo Gleriani[2]
and Ronaldo Medeiros dos Santos[3]

[1]*Federal Rural University of Rio de Janeiro, Rio de Janeiro, RJ, Brazil* [2]*Universidade Federal de Viçosa, MG, Brazil* [3]*University of Brasilia, Brasília, DF, Vicosa, MG, Brazil*

Introduction

Precision agriculture (PA), sometimes called "prescription farming" or "variable rate technology," is a set of techniques that can be used in several areas of agricultural science. PA can be defined as a new management technology based on georeferenced information for the control of agricultural systems. It is based on the detailing of georeferenced information through the application of monitoring processes and integration of characteristics of soil, plant, and climate (Stewart and McBratney, 2000; Plant, 2001). Stafford (2000) emphasizes the need to also develop new techniques, particularly in the area of systems for remote sensing and mapping of spatial variability, prior to precision agriculture being widely practiced. According to Molin (2008), it is estimated that the area cultivated with sugarcane which uses PA techniques represents around 10% of the total area cultivated with this crop in Brazil, and that the simplest way to adopt PA in sugarcane crops is to apply inputs at variable rates based on soil characteristics.

9.1 Data Acquisition in Precision Farming

Data in precision farming can be acquired directly in the field or with the aid of ground or satellite sensors. In precision agriculture, all information is georeferenced and linked to a square-shaped area, called a cell, pixel, or grid of the map (Figure 9.1). Sampling has been a limiting factor to the use of precision agriculture, due to its being time-consuming, to the cost of laboratory analyses and to the lack of intelligent computational systems capable of transforming data into information. In general, the data can be viewed on a regular grid or directed grid.

9.1.1 Sampling in a Regular Grid

Sampling in a regular grid does not require prior knowledge of the area. However, maps generated from regular grids may not capture the spatial variability of the attribute. In this

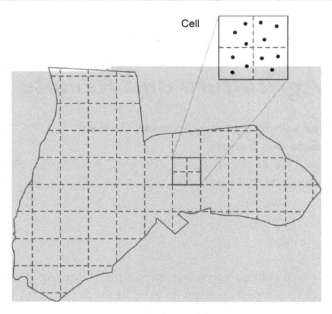

Figure 9.1
Cell, pixel, or grid of the map.

type of sampling, systematic errors may occur, besides the possibility of subsampling and oversampling of attributes. Several observations are collected to form a composite sample of attribute values. The value of the attribute for each cell will be the average of the composite sample. There is no methodology for defining the dimensions of the sample cells, nor is there consensus on what the appropriate size of those should be. According to Pocknee et al. (1996), soil-composed samples should provide between 15 and 20 observations, and the sampling radius must be equal to 25% of the cell size. Figure 9.2 illustrates a sample in a regular grid.

9.1.2 Sampling in Directed Grids

Sampling in directed grids is used when there is prior knowledge of the spatial variability of the attribute. The sample's resolution varies with the spatial variability of the attribute. You can increase or decrease the amount of observations in different regions sampled, that is, vary the sample resolution, in accordance with *a priori* knowledge of the spatial variability of the attribute (Figure 9.3). This type of sampling consists of dividing the total area into regions. The shape and size of sampling regions will depend on the spatial variability of the area and the type of management required. *A priori* information about the spatial variability of the area can be obtained from yield maps, topographic maps, satellite imagery, or aerial photographs, among other means.

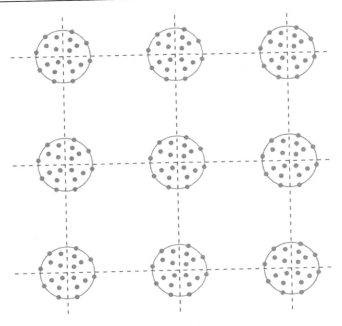

Figure 9.2
Sampling on a regular grid.

Figure 9.3
Sampling in a directed grid.

9.2 *Applications of Remote Sensing Imagery in Precision Agriculture*

Imagery that comes from remote sensors makes it possible to obtain a great deal of information about specific "targets" on the surface. Such information includes qualitative and quantitative aspects, such as shape, geographical context, temporal behavior, size, physical properties, intrinsic aspects, and/or state-specific factors. Some of these properties can be obtained with reasonable accuracy. However, this approach does not apply, or sometimes does not present the same levels of performance, when dealing with quantitative operations for which the required level of discrimination between targets is the pixel. For these cases, among others, a varied number of techniques is available, which, assisted or implemented in computational environments, allow for the handling, processing, and extraction of information from digital imagery originating from remote sensing. Digital processing techniques for remote sensing imagery can be used both in the "preparation" and adaptation of the images for further analysis, as well as in the generation of a final product. For the former, geometric correction and contrast enhancement operations can be made, for example, while the latter comprises sorting and channels/bands ratio operations, among others. The main techniques applied in the process of analysis and information extraction from imagery are shown in Table 9.1, which also presents the purpose, principle, and the types of approaches associated with each technique. The application of one or more techniques depends on the type of product purchased and the purpose for which the outcome of the study is intended.

When an incident flow (Φ) of electromagnetic radiation (EMR) interacts with the target, three basic phenomena occur: absorbance (α), transmittance (τ), and reflectance (ρ). The latter is the component of interest, obtained with the aid of ground, airborne, or satellite sensors. The graphical representation of the value of reflectance as a function of wavelength (l) is what is known as the spectral curve or spectral reflectance of the target. The entire potential for application of remote sensing imagery in precision agriculture is based on the use of the principle of interaction between electromagnetic radiation (EMR) and the different targets in the surface (Table 9.2).

In general, the spectral response exhibited by a given material is unique. The validity of this assertion, however, depends on factors related to characteristics of the product (imagery) to be used, which in turn depends on the structural and operational characteristics of the sensor, such as spatial resolution, radiometric resolution, and spectral resolution. Considering that the imagery is able to adequately distinguish between the different spectral behaviors presented by the material contained in a scene, the use of such information is one of the most immediate applications of remote sensing in precision agriculture.

9.2.1 *Spectral Behavior of Vegetation*

The spectral reflectance of a green leaf is low in the visible region; high in the near infrared region; and medium in the mid-infrared region. In the visible region (0.4−0.7 mm), the

Table 9.1: Main techniques for processing of imagery from remote sensing.

Technique	Purpose	Principle	Approaches
Geometric Correction	Mitigate the effects of geometric distortions arising from imagery processing and assign coordinates to the pixels of an image corresponding to their actual geographic locations in space	Mathematical operations such as rotation and translation, performed by means of polynomials involving the coordinates of the image and control points whose real coordinates are known beforehand	Rectification, orthorectification, record
Radiometric Correction/ Calibration	Convert the digital numbers (gray levels) of image pixels to physical data regarding spectral radiance and reflectance; mitigate the effect of the atmosphere in the spectral response of the targets	Application of mathematical relationships between spectral reflectance/radiance and gray levels from a 'gross' image; application of mathematical functions for correction or mitigation of atmospheric effects for the spectral response of targets on the surface	Radiometric calibration; atmospheric correction, cloud filtering, etc.
Filtering	Remove isolated pixels (*noise*) or pixels of discrepant values in relation to their surroundings; highlight the differences in spectral response between the different targets in the image; extract edges and features provided with preferential orientation; mitigate the effect of texture	The central pixel is replaced by the result of a mathematical operation performed on its neighbors (*window*)	Medium-value filter; filter of the most frequent value; central-value filter, filter for edge detection (*Sobel*), etc.
Contrast enhancement	Highlight the differences in spectral responses between the targets; highlight a specific target, corresponding to a particular range of values for gray levels; improve the visual aspect of an image	A mathematical function is applied to the histogram of the original image, in order to perform a better distribution (spreading) of the frequency of gray levels	Functions, linear or nonlinear, applied across the entire histogram or only in the part (gray levels) where you want to increase contrast levels
Classification	Discriminate between different materials and states present in a satellite imagery "scene", having as a basis their spectral response	Each *pixel* / set of *pixels* in an image is associated with a particular pre-determined class of material or target (soil, water, natural vegetation, irrigated crop, etc.)	Non-supervised *pixel by pixel*; supervised *pixel by pixel*; supervised by *region growth*. A classifier algorithm can be applied to each of the approaches, for example, *minimum distance, parallelepiped, maximum likelihood, or neural network*
Channels/ bands ratio	Discriminate the dominant reflectance (indirect measure of the dominant material in the "scene"); mitigate the effect of topographic shading; generate vegetation indices (highlight photosynthetically active vegetation or indirect measurements of biomass).	Arithmetic operations of division between bands.	Vegetation indices using different mathematical formulations involving the red and near infrared channels, etc.

Source: Crósta (1992), Curran (1985) and USACE (2003).

Table 9.2: Facts of the interaction between the EMR and the various targets on the surface.

Facts	Applications
The spectral response for a particular material is unique	Distinction between vegetation, water and soil
The spectral response for a particular type of vegetation is unique	Distinction between types of vegetation
The spectral response for a given configuration of the canopy is unique	Distinction between crop stages and amount of biomass (harvest prediction)
Once the state of the plant changes, the pattern of the plant's spectral response will also change.	Distinguish between crop stages, detection of phenotypic changes due to water or nutrition stress, pests, etc.

predominance is of absorbance by leaf pigments, such as chlorophyll, carotenoids, and anthocyanins. The more frequently used wavelengths are in the blue and red regions, with a slight increase in the green region. In the near infrared region (0.7–1.3 mm), the responsible factor is the leaf structure – upon crossing the leaf tissues (palisade and spongy parenchyma), the electromagnetic wave undergoes multiple reflections and refractions caused by variation of the refractive index n (n = c/v) which happens in its path. The variation of n is due to hydrated cells (n_{water} = 1.425) and intercellular spaces (n_{air} = 1.0). Obviously, the more compact the parenchyma is and the less variation exists in the path of n, the lower the reflectance will be (Gausman, 1974). The opposite effect is obtained with parenchyma with spaced cells or with layers of superposed leafs, described by the LAI parameter (Leaf Area Index).

9.2.2 Water Spectral Behavior

The spectral behavior of water is limited to the visible region (0.4–0.7 mm), since in the infrared region reflectance is low. The spectral behavior of water is linked to three elements found in it: inorganic particulate matter, particulate organic matter, and phytoplankton. The particulate organic matter scatters incident radiation, and this scattering is a function of the size and color of the particles; regarding the inverse relationship with reflectance, dissolved organic matter greatly absorbs radiation, mainly from 0.4 to 0.6 mm. The phytoplankton and, more precisely, its proliferation, is associated with the presence of organic matter and eutrophication caused by carrying of agricultural nutrients, which tints the water with a green shade.

9.2.3 Spectral Behavior of Soils

Among the factors that influence the spectral response of soils we can include: iron oxides, humidity, organic matter, granulometry, and clay mineralogy. In tropical soils, iron oxides

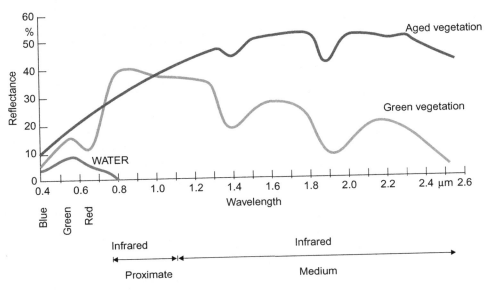

Figure 9.4
Typical spectral responses for vegetation, soil, and water.

and ferrous cations play an important role, because they present in high concentrations in this type of soil. Thus, the quantity and quality of the oxides influence the color of the soil.

Figure 9.4 illustrates the typical spectral response for soil, vegetation and water, while Figure 9.5 illustrates the application of this information in the identification of the type of cover or land use, analyzing only the reflectance of materials in the spectral region corresponding to red.

The differences in the spectral behavior of water, vegetation and exposed soil found in the region of the electromagnetic spectrum corresponding to red allow for a good level of distinction between these attributes (Figure 9.5). For this spectral region, fully exposed soils have higher reflectance values and, therefore, appear in light shades of gray in the image. Subsequently, the pixels in darker shades of gray include areas occupied by different types of photosynthetically active vegetation and with well-developed canopies. Water bodies, being of small width, were identified by shape (sinuosity and interconnectivity) and context (riparian forests).

9.2.4 Interaction between Radiation and Plant Canopy

The plant canopy is composed of all components of the plant which are above the ground; the union of these components forms a three-dimensional body composed of trunks, branches, flowers, leaves, and fruits (Hurcom et al., 1996). Despite the fact that all the

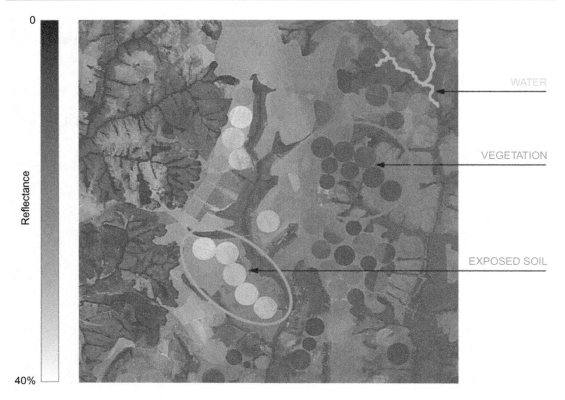

Figure 9.5
Identification of the type of cover or land using the red band of the spectrum.

components interact with electromagnetic radiation, the leaves have special attention, as they are present during most of the cycle of the plant and have a high surfaced area compared to the other parts of the plant canopy. Several factors influence the reflectance of a plant canopy, among which we mention: acquisition and canopy geometry, soil-beds, canopy architecture and shading. Quantifying the energy reflected by a surface is no easy task, and in a plant canopy, this difficulty is even greater.

9.2.5 Acquisition and Canopy Geometry

The plant canopy is one of the most anisotropic agricultural land targets. During development, the canopy changes from a geometry of rows (incomplete canopy) to a homogeneous geometry, which covers 100% of the soil (full canopy). The major changes to the value of energy reflected, mainly in the visible region, are observed during the incomplete canopy stage. Canopy reflectance is a sum of components: sun-lit vegetation, shaded vegetation, shaded soil, and sun-lit soil. The proportion of each component depends

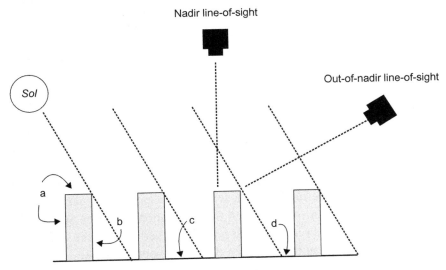

Figure 9.6

Components of canopy reflectance: (a) sun-lit vegetation, (b) shaded vegetation, (c) shaded soil, and (d) sun-lit soil. *Source: Adapted from Kimes and Kirchner (1983).*

on the spacing between rows, the developmental stage of the crop, and the viewing geometry of the sensor (Figure 9.6).

9.2.6 Soil-Beds

A major work on the interference of soil in the spectral response of the plant canopy is found in Huete (1988). Using a radiometer, Huete made measurements in a canopy of cotton plants throughout the cycle, and, among the rows, switched trays with different soil types and humidity levels. For example, when the canopy covered 40% of the soil, the energy reflected increased approximately 200% after the substitution of *Cloverspring-loam* dry soil with *Superstition sand* dry soil. For the near infrared region, the increase was of about 87% for the same replacement. The variation of spectral response with the changing of the soil is more intense in the early stages of the cycle; however, even with the canopy at 100% coverage, there is spectral influence of the soils due to high transmittance levels in the near infrared region (Huete, 1988).

9.2.7 Architecture of the Plant Canopy

The architecture of the plant canopy is characterized according to the spatial arrangement of the leaves. Bunnik (1978) described six vegetation canopy architectures, according to the cumulative distribution of the angles regarding the horizontal axis. These architectures are: (1) planophile – leaves oriented horizontally are more frequent; (2) erectophile – vertically oriented leaves are more frequent; (3) plagiophile — high frequency of leaves with oblique angles; (4) extremophiles — low frequency of leaves with oblique angles; (5) spherical — leaves distributed as if they were on the surface of a sphere; and (6) uniform — leaves with an equally frequent distribution of angles.

9.2.8 Distinction between Types of Vegetation

Spectral differences between different types of vegetation are the result of the interaction of *EMR* with different leaf and canopy structures, particular to each species. In the case of leaves, the structure, the pigments, and the typical amounts of water are the factors that predominate in defining the spectral behavior, while in the case of the canopy the prevailing factors are the number of layers and the orientation of leaves and branches. In Figure 9.7 the application of such information on the distinction between natural vegetation and crops is illustrated, as well as that between different types of crops.

The image in Figure 9.7 is a color composite R(3)-G(2)-B(1) of the scene of an agricultural area cultivated with green-leaf vegetables. The spectral pattern exhibited by natural vegetation ("1") is largely the result of the effect caused by the canopy, while the behavior of the other vegetable-cover areas (crops) can be explained by differences in typical pigmentation and leaf structure. In the case of "6", high reflectance is the result of increased exposure of the soil and lower water content in the leaves of dry vegetation. The main use of the distinction between types of vegetation is the possibility of systematic mapping of crops, which can be used both for monitoring the temporal evolution of the

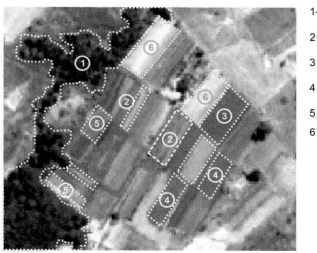

1– Natural vegetation

2 – Lettuce

3 – Parsley/cilantro

4 – Butter collard

5 – Cabbage

6 – Dry undergrowth
(fallow land)

Figure 9.7
Application of satellite imagery on the distinction between types of vegetation.

planted area and for its quantification as to what regards harvest predictions. Due to the high likelihood of mixing between states/types of vegetation and other attributes, for example, the soil exposed, a simple visual analysis of the spectral behavior is not enough to perform this separation, the adoption of imagery processing techniques, combined with field incursions for pattern recognition, then being required.

9.2.9 *Distinction between Crop Stages and Amount of Biomass (Harvest Prediction)*

The stage of development of a crop necessarily reflects the amount of biomass produced, which stabilizes when the plant reaches maturity. Once associated with the size and vigor of the canopy, the amount of biomass has a high degree of correlation with the spectral behavior exhibited by a productive system, a relationship that can be used in the process of remotely estimating or predicting harvests. The prediction of harvests through remote sensing has a classic paradigm, based on the quantification of the planted area and on yield estimates. The area occupied by a crop can be determined with reasonable accuracy, applying visual interpretation and imagery processing techniques. As for yield estimation, various approaches have been proposed, and it is worth highlighting those which use direct relationships between vegetation or reflectance indices in the near infrared region and yield measurements performed in the field.

In Figure 9.8, several models proposed by Machado (2003) are presented to estimate the crop yield of sugarcane. Spot-specific values of yield, determined based on samples taken in the field, were compared with the real spectral reflectance (corrected for atmospheric effects) on the channel/band 4 (near infrared region) of the *Landsat 7 − ETM +* satellite.

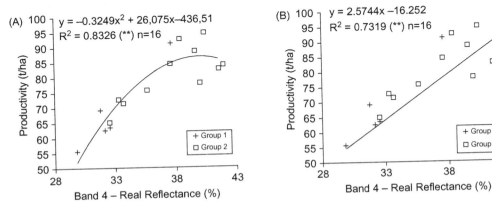

Figure 9.8
Relationship between productivity and reflectance in channel 4 of Landsat 7-ETM+.
Source: Machado (2003).

In Figure 9.8A, the coefficient of determination (R_2) was higher than in Figure 9.8B. Although the adjusted models present a coefficient of determination higher than 0.7, the equations have a strictly local character, with further studies needed for the development of models that are appropriate to the particular characteristics inherent to each place of interest, such as the varieties planted and the handling of the cutting subsystem, among others.

9.2.10 Detection of Phenotypic Changes (Stress or Change in Crop Stage)

Considering the leaf, the plant, or the crop, plant systems phenotypically express the consequences of stressful situations, such as water or nutritional deficits, incidence of pests (weeds, fungi, insects, etc.), and changes in crop stage. Thus, based on knowledge of the spectral behavior of a "healthy" system, one can associate changes in the behavior considered to be standard with one or more possible causes of that change. Figure 9.9 illustrates the influence of the cultural stage in the spectral response of vegetation.

For the case of humidity levels found in the leaves, there is a widespread increase in reflectance if water levels decrease, as illustrated in Figure 9.10, which presents an example for the corn crop.

Using this principle, other factors that cause changes in the normal physiological state of the plant can also be identified by remote sensing satellite imagery, for example infestations by fungi or nematodes in banana crops, which cause progressive desiccation and death of the leaves, resulting in loss of efficiency in plant respiration and a "break" in productivity.

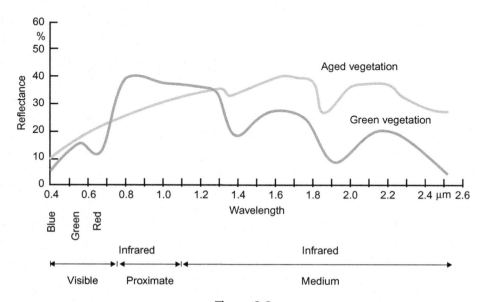

Figure 9.9
Influence of the crop stage in the spectral response of vegetation. *Source: Moreira (2001).*

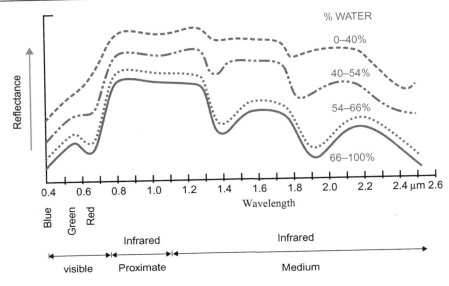

Figure 9.10

Influence of water content in the spectral reflectance of corn leaves. *Source: Adapted from Ponzoni and Shimabukuro (2007).*

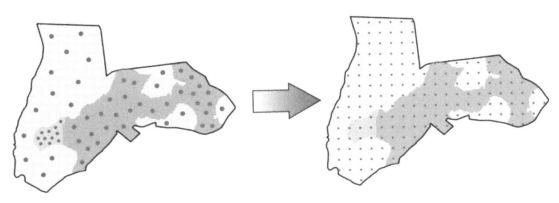

Figure 9.11

Interpolation of raw data to generate a regular grid.

9.3 Mapping Spatial Variability in Precision Agriculture

For the mapping of spatial variability of a given attribute, one must have a database that presents the value and location of the attribute. This database is usually obtained from a non-regular directed grid sampling of points within the area evaluated. These data, often called raw data, are then converted into a regular grid by interpolation (Figure 9.11).

Moore (1998) tested several interpolation methods for mapping the productivity of maize using a Massey Ferguson combine with a mapping system to generate an average yield value for planting cells of $20-25$ m^2 in area. The researcher concluded that interpolation by the moving average method can be used for the mapping of productivity with harvesters. According to this same author, the choice of the method of interpolation of raw data to generate regular grids depends on the level of detail desired, that is, one should take into account the purpose of the map. The most commonly used interpolation methods in precision agriculture are: nearest neighbor, moving average, inverse distance, and kriging.

9.3.1 Nearest Neighbor

Nearest neighbor interpolation is the simplest of the interpolation methods. It simply uses the attribute value nearest to the node of the grid to estimate the interpolated value (Figure 9.12). The observed values are not modified, being only redistributed in a regular grid. This method is used when one wants to turn raw data into a regular grid without modification of the observed values.

9.3.2 Moving Average

The interpolator estimates the values of attributes of regular grid points simply averaging values of attributes from selected points around each node. The points are selected according to the number of neighbors or the search radius. In the example illustrated in Figure 9.13, the parameters for interpolation are: search radius = 10 m or number of neighbors = 8.

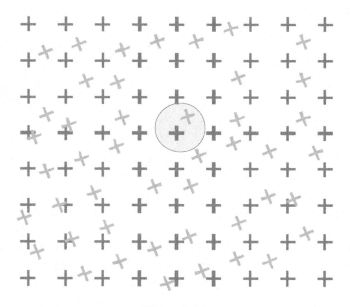

Figure 9.12
Interpolation by the nearest neighbor method.

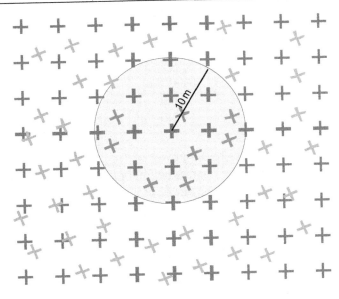

Figure 9.13
Interpolation by the moving average method.

9.3.3 Inverse Distance

This interpolator uses the statistical model called "inverse distance". The model is based on spatial dependence, that is, it is assumed that the closer a point is to the other, the greater the correlation between their values must be. Thus, greater weight is attributed to observations which are closer than to more distant ones. Therefore, the model consists of multiplying the values of the observations by the inverse of their respective distances to the reference point, and dividing that sum by the sum of the inverse of the distances to the reference point (Equation 1).

$$z = \frac{\sum_{i=1}^{n} \frac{1}{d_1} z_i}{\sum_{i=1}^{n} \frac{1}{d_i}} \tag{1}$$

Where:

z = estimated value for point z,
n = number of observations,
z_i = observed values and
d_i = distances between observed and estimated values (z_i and z).

Equation 1 can be adjusted to include a power of "p" to the inverse of the distances "d" (Equation 2).

$$z = \frac{\sum_{i=1}^{n} \frac{1}{d_i^p} z_i}{\sum_{i=1}^{n} \frac{1}{d_i^p}} \tag{2}$$

Thus, one can assign different values to the power of "p", and the larger the value of this power, the greater the influence of the nearest neighbors (z_i) in the estimated value for the attribute of point z will be. The highest power used is 2, for which reason this interpolator is often called the "inverse square of distance" interpolator. It is also known as IDW (Inverse Distance Weighted).

9.3.4 Kriging

Kriging is the only interpolation method which sets a model for the behavior of the spatial variance of the raw data and uses this model to estimate the values of attributes in specific points in a regular grid. According to Vieira, (2000) it is necessary that the variables are spatially dependent for the making of maps by kriging. The study of spatial dependence is done by examining the semivariogram. For Vieira, (2000), this is the most appropriate tool to measure spatial dependence. The semivariogram is defined as the mathematical expectation of the squared difference between pairs of a variable in space, given by Equation 3.

$$\gamma(h) = \frac{1}{2} E\{[Z(x) - Z(x + h)]^2\} \tag{3}$$

Where:

g(h) = semivariogram,
Z(x) = value of the variable at point x,
Z(x + h) = value of the variable at point (x + h), and
h = distance between points x and x + h.

The semivariogram can be modeled in various computer programs. The most commonly used semivariograms are: Gaussian, exponential, spherical, circular, and linear. The following presents a methodology for mapping the spatial variability of attributes. The data are values of brix levels in sugarcane, obtained from Processing Plant Paineiras Inc., located in the city of Itapemirim, State of Espírito Santo. Thirty-six random points of brix values were sampled with average sample spatial resolution of 16.67 m in a parcel of sugarcane of about 1 ha of area. The sample values were interpolated by kriging on a regular grid of 1.0 m spatial resolution. The proposed methodology for the selection of the semivariogram model for kriging consists of two steps: modeling of the variogram and validation of the interpolation.

Modeling of semivariograms

In this example, the semivariograms were modeled using the software ArcGIS, version 9.2 (ESRI, 2008). Circular, spherical, exponential and Gaussian models were tested. The results of modeling of semivariograms are shown in Table 9.3.

Validation of interpolation

In this step, the previously adjusted semivariogram models are used to make predictions for values of attributes in points having the same coordinates of the observed values. Subsequently, a first-degree linear model of the predicted values is adjusted according to observed values. The parameters of this regression are evaluated according to the methodology described by Graybill (1976). The null hypothesis of the test is that the intercept is zero and the slope is one. When the test is not significant at 5%, the null hypothesis is accepted and predicted values are considered statistically identical to those observed. According to the results presented in Table 9.4, only the Gaussian model was non-significant in test F. Consequently, the semivariogram model that is used for the interpolation method of kriging is thus selected.

Table 9.3: Results of modeling of semivariograms using the software ArcGIS 9.2.

Parameter	Circular	Spherical	Exponential	Gaussian
Reach	81.40	81.40	81.40	81.4047
Threshold	0.232	0.211	0.194	0.29347
Nugget Effect	0.000	0.000	0.000	0.00029
Step	6.868	6.868	6.868	6.8677
Number of steps	12	12	12	12

Table 9.4: Results for the validation test for kriging for each of the semivariograms models tested in this example.

Item	Semivariogram	Intercept	Slope	F(H0)	F (5%, 2, 34)
1	Circular	1.756	0.861	3.8602*	3.2759
2	Spherical	1.772	0.859	3.8617*	
3	Exponential	2.215	0.824	6.3278*	
4	Gaussian	0.701	0.944	0.0705ns	

*Significant at 5% probability; ns, not significant at 5%.

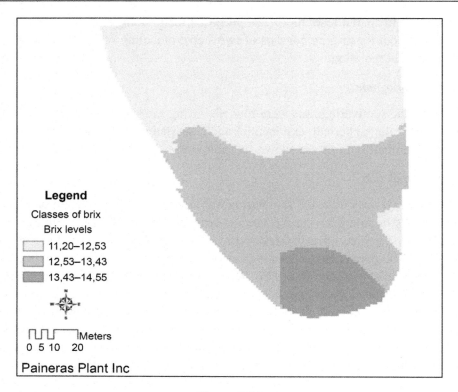

Figure 9.14
Map of spatial variability for brix in sugarcane collected in a parcel of about 1 ha.

9.4 Spatial Variability Maps

There are several computer programs that can generate maps of the spatial variability of attributes. In this example, we used the software ArcGIS 9.2. Figure 9.14 illustrates the map for the spatial variability of brix in sugarcane from a parcel of around 1 ha in area. It was possible to adjust a Gaussian model and validate the interpolation of brix values predicted by kriging in ArcGIS 9.2. One can conclude that the Gaussian semivariogram model was found to be adequate to estimate the spatial variability of the levels of brix through the kriging method, and that an irregular sampling grid of 36 points per ha is sufficient to map the spatial variability of brix in sugarcane. It is recommended to validate the interpolation before that particular semivariogram model is used for kriging. This way, the generated maps will present lower estimation errors for that attribute.

Bibliography

Bunnik, N.J.J., 1978. The multispectral reflectance of shortwave radiation by agricultural crops in relations with their morphological and optical properties. Thesis. Agricultural University of Wageningen. 175p.

Crósta, A.P., 1992. Processamento digital de imagens de sensoriamento remoto. Campinas. 170p.

Curran, P.J., 1985. Principles of remote sensing. London. 260p.

ESRI. Training and Education. Disponível em: <http://training.esri.com/gateway/index.cfm>. Acesso em: nov. 2008.

Graybill, F.A., 1976. Theory and Application of the Linear Model. Ouxburg Press, Massachusetts. 704p.

Huete, A.R., 1988. A soil-adjusted vegetation index (SAVI). Remote Sens. Environ. 25 (3), 295−309.

Hurcom, S.J., Harrison, A.R., Taberner, M., 1996. Assessment of biophysical vegetation properties through spectral decomposition techniques. Remote Sens. Environ. 56 (3), 203−214.

Kimes, D.S., Kirchner, A.J., 1983. Directional radiometric measurements of row-crop temperatures. Int. J. Remote Sens. 4 (2), 299−311.

Machado, H.M., 2003. Determinação da biomassa de cana-de-açúcar considerando a variação espacial de dados espectrais do satélite Landsat 7 − ETM + . Dissertação (Mestrado), Universidade Estadual de Capinas. Campinas. 61p.

Molin, J.P. 2008. Agricutura de precisão em cana-de-açúcar é mais do que uma realidade. Disponível em: <http://www.coplana.com/gxpfiles/ws001/design/RevistaCoplana/2007/Dezembro/pag22.pdf>. Acesso em: nov. 2008.

Moore, M. 1998. An Investigation into the accuracy of yield maps and their subsequent use in crop management. Tese (Doutorado), Cranfield University, Silsoe College, Disponível em: <http://www.cpf.kvl.dk/Papers/Mark_Moore_Thesis>. Acesso em: nov. 2008.

Moreira, A.M. 2001. Fundamentos do sensoriamento remoto e metodologias de aplicação. São José dos Campos. 250p.

Plant, R.E., 2001. Site-specific management: the application of information technology to crop production. Comput. Electron. Agric. 30, 9−29.

Pocknee, S., Boydell, B.C., Green, H.M., Waters D.J., Kvien, C.K., 1996. Directed soil sampling. In: Roberts, P.C., et al. (Eds.), Proceedings of the Third International Conference on Precision Agriculture Queensland Sugar Corporation, Queensland sugar in focus, Brisbane, Australia.

Ponzoni, F.J., Shimabukuro, Y.E., 2007. Sensoriamento remoto no estudo da vegetação. São José dos Campos. 127p.

Stafford, J.V., 2000. Implementing precision agriculture in the 21st century. J. Agric. Eng. Res. 76 (3), 267−275.

Stewart, C.M., McBratney, A.B., 2000. Development of a methodology for the variable-rate application of fertilizer in irrigated cotton fields. In: Robert, P.C., Rust, R.H., Larson, W.E. (Eds.), International Conference on Precision Agriculture, 5, Proceedings. Madison, ASA, CSSA, SSSA, CD-Rom.

USACE, 2003. Remote sensing − Manual of U.S. Army Corps of Engineers, Washington. 217p.

Vieira, S.R, 2000. Geoestatística aplicada á agricultura de precisão. In: Borém, A., Giudice, M.P., Queiroz, D.M., Mantovani, E.C., Ferreira, L.R., Valle, F.X.R., Gomide, R.L. (Eds.), Agricultura de Precisão. Editora UFV, Viçosa, pp. 93−108.

CHAPTER 10

Stalk Harvesting Systems

Tomaz Caetano Cannavam Ripoli[1,†] **and Marco Lorenzzo Cunali Ripoli**[2]

[1]*Universidade de São Paulo — Escola Superior de Agricultura "Luiz de Queiroz", São Paulo, SP, Brazil* [2]*John Deere — Latin America, Indaiatuba, SP, Brazil*

Introduction

From the point of view of selection and operationality of a harvesting system, whatever the crop is, the analysis must not be limited to aspects related to machinery or the workforce involved. A deeper study is necessary, which must take into account four main groups of conditioning factors: social, physiological, technological, and economic.

In the case of sugar crops, the harvesting of raw materials, which include manufacturable stalks, "thief" shoots, mineral (dust, metals), and vegetal (hay, green leaves, shoot tops, crop remnants, weeds) extraneous matter, must reflect the entire work developed in the planning and implementation of the crop, from the periodic preparation of the soil to the operations included in harvesting and extracting the product from the farm.

This universe of planning and execution actions must begin by the correct selection of varieties, according to local edaphic and climatic conditions, and persist all the way through choices regarding adequate conditions of the road network, of the transportation subsystem, of the reception subsystem for raw materials in the industry, and the potential and appropriate qualification of the available staff (manual laborers, machine operators, mechanics, technicians with higher education, area managers, etc.).

From the physiological viewpoint of sugarcane crops, the harvest represents the end of the growth and maturation cycle, the moment at which it reaches its maximum levels of agricultural productivity of stalks (as allowed by the edaphic and climatic conditions of the area, as well as by the agricultural technology and variety utilized), besides marking the ceasing of any production techniques. It is essential to have adequate planning and selection of varieties according to their Optimum Industrialization Period (PUI in Portuguese), so that one can obtain, throughout the harvest period, sugarcane crops with desirable levels of maturation and which are adequately distributed through the area belonging to an agroindustry or a supplier.

---

† In memoriam

Sugarcane. DOI: http://dx.doi.org/10.1016/B978-0-12-802239-9.00010-4
205

Even not bearing direct participation on the agricultural productivity of stalks, the cutting subsystems, as well as the raw materials transportation subsystems, can compromise its quality, as well as the productivity of stalks in subsequent cuttings if those are not carried out in compliance with adequately defined and implemented technical principles. Those will vary according to a countless number of agronomic, environmental, technical, and managing conditions, mainly. Consider, for the instance, the case of pre-harvest burning: performing it too soon before the programmed cutting, that is, in such a way that there is a period longer than 24−36 h after burning during which there is no manual cutting, loading, or mechanical harvest, leads to significant losses in sucrose, by inversion, due to more consistent attacks from *Leuconostocus* bacteria. Also, loading operations for raw materials with excessive dragging of the soil, stomping, and destruction of stubbles by transshipment and/or transportation vehicles are a few serious implications of operational and management inadequacies, which can lead to serious and costly implications in the industrial processing of the harvested material.

On the other hand, one must bear in mind that the harvest period for sugarcane crops causes a notable and sensitive modification in the socioeconomic environment of the regions in which this crop is predominant, with the appearance of a mobile workforce that is, to a very large extent, unqualified and seasonal, which can lead to implications in the social promotion and assistance sectors of the agroindustries and of sugarcane crop municipalities. In this early 21st century, it is still common to see thousands of manual laborers from the Northern regions of the state of Minas Gerais come to the São Paulo sugarcane crops looking for work.

Besides the intensive use of manual labor for the cutting of sugarcane, the harvest period brings about a "war operation", with the synchronized mobilization of tractors, harvesters, transshipment vehicles, loaders, transportation units, and reception subsystems, with a view to ensuring a continuous, 24-hour-a-day flow of the raw materials for their adequate and programmed supply. All this complex structure of equipment and technical−managerial actions must lead to a supply of raw materials which maintains its quality features preserved and, ideally, enhances them.

Finally, there is the economic aspect, which, due not only to the agricultural production and productivity of manufacturable stalks but also to the need for an adequate harvesting system as defined by the agroindustry, will be the defining factor for the harvest period, resulting in the success or failure of the agricultural activity. Once all desirable conditions for implementation and administration of the crop are seen to, the harvest period requires complex planning and management, through the work of a highly qualified workforce composed mainly of agricultural technicians, agricultural engineers and mechanical engineers, who must possess appropriate levels of knowledge and technical−academic expertise.

10.1 Types of Harvesting Systems

The cutting, loading, transportation and reception operations for raw materials present countless options (Figure 10.1).

The harvest subsystems currently being utilized in Brazil and in the world can be subdivided into three major groups:

- *Manual system* — in which the cutting and loading subsystems are manually processed. There may be an intermediate transportation subsystem, by animal traction or transshipment with specific devices. Despite being an apparently archaic system, it is still broadly utilized in declivitous areas of the Brazilian Northeast, mainly in Alagoas and Pernambuco, where sugarcane crops are planted in topographies that surpass 100% declivity levels.
- *Semi-mechanized system* — involves a manual cutting subsystem and a subsystem of loading, in the transportation units, which is handled by mechanic loaders. It is the most broadly utilized system in all sugarcane crop regions in Brazil, where topography does not exceed 20–25% in declivity.
- *Mechanized system* — a system that uses a mechanized subsystem with cutters of several types, according to the classification in Figure 10.1, or with whole sugarcane

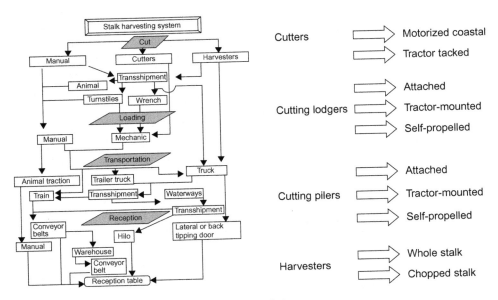

Figure 10.1

Options for harvest systems for sugarcane stalks in Brazil (updated from Ripoli and Paranhos, 1987), and classification for cutting and harvesting machinery for sugarcane (updated from Ripoli, 1974).

stalk harvesters with mechanical loading subsystems, or even a subsystem composed of combine harvesters (which cut, slice and partially clean the raw materials and load them into transportation units). The utilization of this system is feasible in topographies with 15−17% declivity levels (depending on the quality of the systematization of the parcel to be planted and the center of gravity of the machinery). Above those levels, because of dynamic stability issues with the equipment, work is compromised by risk of tipping.

The first machines to be used for sugarcane cutting were the so-called cutters. After those came the cutting lodgers, the cutting pilers, and finally the harvesters. As those machines developed, their operational capacity increased, and today, depending on their general work conditions, they cut, chop, clean and load around 30−70 tons of crude and/or burned sugarcane per operational working hour.

Ripoli (1974) presents a classification of machines for cutting and harvesting of sugarcane (Figure 10.1) and describes them as such:

- Cutters − machines that only perform the base cutting of the stalk, leaving the cut material on the terrain (some of those also perform apex cutting). This is the case with the Cameco model, of American manufacturing, which operates two rows at a time. It is a track tractor mounted on an adapted Caterpillar chassis model D-5. There is also the *push-rake* type, in use in Peru and Hawaii.
- Cutting spreader machines − besides performing the base cutting of the stalks, they also cut the shoot tops and subsequently place the stalks on the terrain in a threaded way so as to facilitate mechanical loading. This is the case with the Santal, CTE model, no longer manufactured.
- Cutting piler machines − they are similar to the aforementioned types, but instead of placing the stalks in a threaded way, they pile them in mounds at regular distances. They were manufactured by Santal, E. Artioli and Dedimac.
- Chopped sugarcane harvesters − also called combines, they perform base cutting and promote the partial elimination of vegetal and mineral extraneous matter through the action of gravity, fans, or exhaust pipes. They fraction the stalks in pieces of around 15−40 cm long, unloading them over a transportation or transshipment unit. They are manufactured, in Brazil, by Santal, Case-CHN, John Deere, Star, Civemasa, and Clima. The last one is a harvester attached to a tractor.
- Whole stalk harvesters − perform base and top cutting of the stalks, carry out the partial elimination of vegetal extraneous matter, store the stalks in a tilting deposit and, after moving to the outside of the parcel, deposit the collected material into a dirt road for posterior loading. Manufactured in Brazil by Motocana.

Ripoli (1974) outlines another classification of the main machines involved in the sugarcane cutting process. This classification is based on the machines' design in relation to the aspects that more directly influence cutting and handling of raw materials in their processing by the machinery:

- *Regarding the power source* — mechanic or hydrostatic transmission.
- *Regarding traction* — tires (tricycle, four supports or six supports, in tandem), semi-tracked, or tracked.
- *Regarding the number of rows cut at a time:* one, two, or three.
- *Regarding the lifting system for lodged stalks:* shaft, chains, fixed cones, or rotating cones with a spiral.
- *Regarding the number of base cutting discs* — one, two, or two with support for four blades.
- *Regarding the conduction mechanism for the sugarcane once inside the machine* — rotating serrated rollers, conveyor belts, fixed conveyors, and chains with conveyor plates, by rotor.
- *Regarding the chopping system for the sugarcane pieces* — two horizontal cylinders with blades, at the machine's entrance point or in a point intermediate to it; vertical disks with blade, at an intermediate point in the machine; or cylinder horizontally placed, with the blade in an intermediate point in the machine.
- *Regarding the type of raw materials supplied* — whole stalks or *rebolos* (Portuguese term for the pieces from chopped stalks).

Whichever harvesting system is adopted, there will always be so-called visible losses, that is, stalks, or fractions of such (including stubbles) that will remain in the area after the operations involved in the extraction of the raw materials from the land, besides residues and trash.

Ripoli and Ripoli (2001) presented the results of a five-year study on the effects of preharvest burning in stalk exudation. The average results obtained in numerous varieties and times of analyses showed loss of juice by exudation of the order of 5−130 liters of ethanol/ha. This high level of variation between samples was due to the diverse farm conditions found before, during and after the burning (variety, age, order of cutting, size of the crop, air humidity, wind speed, quality of burning, time elapsed between burning and harvest, and others).

Roughly estimating 4.5×10^6 ha of burned sugarcane harvested per season in Brazil, and taking the average rate of loss due to exudation as calculated by Ripoli and Ripoli (2001) in liters of ethanol (67.5 L/ha), we come to a final amount of the order of 303.75×10^6 L of ethanol theoretically not produced due to exudation. Not a value to be taken lightly in any scenario in which one wishes to minimize costs and waste in the sugar−ethanol industry.

After burning, the sugarcane must be cut, transported and processed as quickly as possible, deadlines between 24−36 h being considered satisfactory. Within this period, losses will not be very significant. The burned and cut sugarcane which is exposed to the weather will suffer dehydration, with loss of weight; there will be an intensification of respiratory processes for the stalk, with loss of sugars. After the aforementioned deadline, quite often, deterioration will reach elevated levels very quickly, completely compromising the quality of the raw materials. If rain occurs after the burning and prior to the cutting, or even after the cutting and before transportation to the industry, losses will be severely worsened.

Following the completion of burning, cutting operations commence. Either mechanical or manual, those are administratively organized into "cutting fronts" (one to four or more), with a staff of cutters or machinery for cutting pre-sized to supply the responsible industry with a pre-established amount of raw materials for crushing and stock supply. These fronts also count on specific fleets for loading (loaders) and transportation, besides supervision, control, maintenance, supply, and mechanical assistance personnel.

10.1.1 Cutting Subsystem

The choice of type of cutting of sugarcane stalks (manual or mechanic) will depend on a number of factors, such as: availability of staff, socioeconomic aspects, conditions of the farm where the sugarcane crop is implemented, the loading subsystem to be utilized, etc.

In Brazil, workers involved with manual cutting comprise a working class burdened with several shortcomings, including nutritional, health, and educational hurdles, besides qualification for the job itself. As a consequence of this, when the daily capacity of these workers is compared against that of other countries, like South Africa and Pueorto Rico, one observes that in those places they are able to cut, on average, 12−14 tons of sugarcane/day of work, being qualified, healthy and well-nourished workers, besides having at their disposal tools which are ergonomically adequate for their physical features. As a rule, only men take part in the job.

In Brazil, the mass of workers working the sugarcane field includes men, women, children and the elderly. But for a few self-taught exceptions, a good part of those are malnourished, illiterate, or semi-illiterate, and without tools that are adequate to their biotypes. Due to this scenario, the energy spent by each of the workers is above their own capacities, which reflects in low daily productivity, from 7 to 10 tons/day, and in a gradual loss of their bodily resistances. In national agro-industries that have already implemented feeding and training programs, among other measures, for these workers, their productivity has already reached 12 tons/day in scenarios where the stalks are heavier and more erect and the quality of the burning was good. An example of a successful training program is that implemented by the COSAN Group.

10.1.2 Manual Cutting

Manual cutting is characterized by a series of events that a manual laborer, in possession of a tool (a sort of scythe called leaf, podão, etc., depending on the region), utilizes to cut and eliminate from the stalk vegetal material of no interest for the production of ethanol or sugar. However, with the advent of new techniques that take advantage of the residues for co-generation, only the stalk tops are eliminated.

These events will depend on an initially existing condition and on a desired final condition with regard to the stalks *per se*. The initial condition allows for two possibilities:

- stalks, with leaves, straw and tops in natura (crude sugarcane); or
- stalks that have suffered the action of fire, with almost entire elimination of straw and partial elimination of green leaves, as well as the maintenance of tops.

In turn, the desired final condition also comprises two possibilities:

- stalks cut and spread randomly over the terrain; or.
- stalks cut and not randomly spread over the terrain, clustered in mounds, or threaded, forming the large beds of cut sugarcane.

However, the tendency in the Center-Southern region of Brazil is for a drastic reduction in the use of manual cutting, for socioeconomic and environmental reasons and because of demands from the international market (which will impose limits to products originating from practices that are harmful to the environment). Mechanical solutions are sought to replace manual laborers, who nowadays will work mainly with small and medium producers whose lands are placed in areas of higher declivity levels.

In terms of equipment to boost the operational performance of manual laborers, the company Agria ESM Cane Thumper offers a simple and easily handled cutter, developed from a lawnmower utilized in Germany. The Serra Grande plant (in Pernambuco) is already using it. Weighing a little over 200 kg, the machine works with a 7-hp engine with a four-gear shift. The speed of work can be changed: 1, 1.5, and 2 km/h. The fourth gear is for movement, when the machine is not cutting, and can reach 7 km/h. The engine is a Lombardin, produced in Italy and with its cutting system adapted in Germany by the company ESM. Its consumption of diesel is of the order of 0.6 L/h. On average, the 5-L fuel tank allows for 8 h of operation. The estimated cost of the machine is US$ 2700. At the intermediate 1.5 km/h speed and with a spacing of 1 m between rows, it reaches the rate of 1 ha cut every 8 h of work. Five people work at the front of the machine, embracing the bunch grass, and two remove the top of the lined sugarcane stalk, allowing for it to be cut near the stubble.

10.1.3 Mechanized Cut

The mechanization process of the harvesting of sugarcane is not a simple substitution of manual labor by machines. It reaches the dimensions of a system whose limits are ample enough to include the entire problem of transferring raw materials from the farm to the industrial unit. According to Mialhe and Ripoli (1976), three subsystems can be visualized within this system, namely:

- the cutting and loading subsystem;
- the transportation subsystem; and
- the reception subsystem.

It is considered that the subsystems, although they contain a specific part of the global problematic, present interfaces that include aspects of common interest. Thus, we have the formation of a binding chain between the farm and the factory (Figure 10.2) through which the flow of raw materials that feeds the plant is established. Therefore, the fundamental goal of the studies and research developed on the mechanized harvesting system of sugarcane is, ultimately, the optimization of this flow for the particular features of each sugar and/or ethanol production company, with a view to:

- improving raw materials quality, in terms of the maintenance of the level of sugar in the original levels of the farm and reduction of deterioration levels during the flow;
- performing the cleaning of the raw materials, in terms of reduction of extraneous matter; and
- reducing the cost of transfer of raw materials from the farm to the plant, in terms of reducing the cost per unit during the flow.

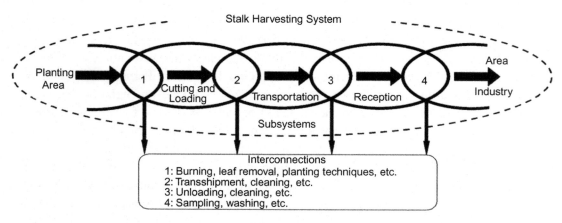

Figure 10.2
Flowchart of sugarcane harvesting systems, with their subsystems and interfaces. Without adequate synchronicity between the subsystems, costs are almost prohibitive.
Source: Mialhe and Ripoli (1976).

10.2 Cutters versus Harvesters

The adoption of combine harvesters, commonly known as "chopped stalk" machines, or the other type of machinery, commonly called "whole stalk" machines, will depend on a countless numbers of agro-industry-specific variables. Many are favorable of the former, claiming that they are more efficient at work, providing raw materials of better quality; others believe that the adoption of combine harvesters leads to a need for elevated investment, with drastic changes in the transportation and reception systems, besides leading to a subsequent reduction of agricultural output in subsequent cuttings. With a view to obviating such differences, Paranhos (1974) and Ripoli (1974) mention the main advantages and disadvantages of the two basic types of machine for sugarcane cutting.

10.2.1 Cutters (whole stalks)

Advantages:

- they can be easily introduced with any transportation system;
- cutting and loading are independent operations;
- whole stalks do not deteriorate as quickly as chopped stalks (which therefore cannot be stocked) and can be stocked for longer periods;
- no special recipients are necessary for the stocking of whole stalks but chains and steel cables, already present at the plant.

Disadvantages:

- there is a need for loaders, since the cutters deploy the cut sugarcane stalks into the terrain (in mounds or spread over it);
- any interruption in the subsystems of transportation, loading and reception at the plant can result in the cut sugarcane remaining on the farm for longer periods, with all the inconveniencies that such an event could cause;
- whole stalks produce cargos of higher densities to be loaded in the transportation vehicle, which, after loaded, will have a higher center of gravity and, therefore, be more unstable;
- the use of chains and cables is costly and time-consuming;
- the transportation system is not effectively used due to the high levels of variation found in the density of the cargos;
- the losses (sugarcane stalks that fall during the farm-processing plant trip) are significant;
- due to their design features, cutters have high centers of gravity, making them unsuitable to operate in topographies with declivity levels higher than 15–18%;

- the quality of the raw materials that come to the plant is harmed by the need to use loaders, which drag, together with the sugarcane, mineral and vegetal extraneous matter;
- cutters of simpler design, that is, which only cut, without piling them up in any way, leave the cut stalks spread over the planting rows and longitudinally to them, which makes it significantly more difficult to carry out the loading (and top removal, if it is the case) operations. This condition possibly was, and still is, the main cause of non-acceptance of these machines by sugarcane producers, despite their low purchase value;
- large difficulties in harvesting lodged cane.

10.2.2 Harvesters (chopped stalks)

Advantages:

- they are self-propelled machines, mounted on or attached to tractors that eliminate the need for loaders, depositing the chopped sugarcane directly into the transportation system;
- they cut all types of sugarcane (erect or extremely lodged), promoting a partial cleaning of the planting area;
- a higher level of density is obtained for the transported cargos (an average of 500 kg/m), allowing for a more realistic control of transportation;
- stalks will seldom fall on the roads during the farm—plant trip;
- countless gains are obtained with the crushing of fresh sugarcane, with no stocking required;
- their use results in a more efficient and well-programmed transportation system, since the chopped sugarcane must be delivered before deterioration occurs;
- interruptions in the plant or the transportation system do not in result in sugarcane being cut and left in the farm;
- they operate with crude sugarcane.

Disadvantages:

- the cutting and transportation operations are strictly connected;
- it implicates costly changes in the transportation system, since the chopped stalks need special transportation features (closed carts);
- special packaging would be necessary for an occasional need for stocking in the plant, which is not recommended;
- a more efficient team and a more perfected synchronicity of transportation are necessary to ensure a more rational use of the harvesters;
- if the chopping device is not efficient, or is inadequately used, an incorrect shearing of the stalks may result in imperfect pieces;

- in lodged cane, shoot tops are frequently included in the raw materials sent to the plant;
- depending on the distance from the planted sugarcane to the plant, there will be need for transshipment vehicles.

It is evident that, in this didactic comparison, depending on technical and economic conditions of each plant, a feature that could be an advantage for one plant may be a disadvantage for another. Therefore, it must be highlighted that a study must be carried out individually, involving all the aspects here presented, for it to be able to arrive at a more objective and rational conclusion.

10.3 Factors Involved in the Selection and Operational Capacity of Harvesters

Grounded on what Ripoli (1974) already postulated on the factors which should and must, still, be taken into account and that interfere in the operational capacity and utilization of cutters and harvesters, the following new considerations are presented, namely, on: "machine factors", which regard their project characteristics; "planting area factors", which regard the conditions of the farm in which the machine will operate; and "administrative factors", which regard managerial and planning aspects. It is regrettable that, even in this day and age, many plants and distilleries acquire harvesters and do not adapt their operational, managerial and planting area structures to their new systems. Thus, the costs per harvested ton become elevated in comparison with those of manual cutting.

10.3.1 Machine Factors

Center of gravity

Center of gravity interferes in the utilization and operational capacity of these machines, as with any other source of power. The more elevated the center of gravity is, the less the machine will be utilized according to the topography of the region. As to operational capacity, there will be a decrease, for there is the tendency to slow down movement speed if the center of gravity is higher, which creates more critical instability conditions and consequently makes operation more difficult. The only work published in Brazil on the determination of the center of gravity in harvesters in static equilibrium conditions was that of Ripoli et al. (1974). The machine was a Massey Fergusson 201 wheeled tractor, of Australian origin. That work presents a curve of influence of declivity on the safety of the operation. It is observed in the study that the safety of the operation rapidly decreases after a given level of declivity. This declivity does not refer to the planting area as a whole, but is limited to the equivalent distance between the extremes of the machines' front tires. The ordinate axis provides the level difference in cm, and in percentage of corresponding declivity, while the abscissa provides the margin of safety operation (in %).

The study reinforces the need for an adequate systematization of planting areas with regard to the elimination of depressions or elevations in the micro-topography of the terrain. Otherwise, one runs the risk (even in an area with declivity of the order of 12%) of the tires on one side (left, for instance) being over a depression while those on the right side are on an elevation, which could lead to a dynamic imbalance of the machine and to its tipping, even in a level of declivity that is considered safe.

Declivity in itself is not enough to ensure the possibility of mechanized harvesting of sugarcane, since there are a few scenarios which are critical for the active parts of these machines, such as outcropping rocks, stubbles in areas where there has been recent deforesting (reversion of areas that had previously been reforested or of natural formations), areas with low resistance to settling (swamp formations, turf, etc.) and areas with superficial waterbeds (lowlands without drainage, flood plains, etc.). Another aspect to be considered is the capacity for optimized management of the harvesting system, a factor that determines the return conditions for the capital invested in the mechanized harvesting system for crude sugarcane.

The maximum level of declivity that allows for the use of machinery for harvesting (or for any other mechanized operation), as defined by Mialhe (1966), is limited solely by the so-called "operational stability conditions of the specimen", which, in turn, are determined by the reactions of forces in equilibrium which act on the specimen (machinery, implements, etc.) that is being considered. Contrary to what has been reported in technical articles by authors who are not specialists in agricultural mechanization, the core of the issue of maximum declivity for utilization of machinery (Agricultural Mechanization) does not dwell solely on the declivity of the terrain, but also on the height and weight characteristics of the equipment, as has been clearly evidenced by Mialhe (1966) and several other authors, such as Chudacov (1977) and Gill and Vanden Berg (1967). If the use of machinery beyond certain limits is harmful or not from the perspective of soil conservation, fertility management, etc. is a different story. When defining 12% as the limit value for declivity for areas eligible for mechanized harvesting of sugarcane, the legislator did not take into consideration only the harvester machine factor, for those can easily operate in more declivitous terrain when its dynamic equilibrium condition is considered (mass distribution, position of center of gravity, etc.).

Capacity of the active cutting and internal conduction mechanisms

Depending on the characteristics of the internal conduction mechanisms for sugarcane (dimensions, rotation or speed), certain types of machinery may or may not be able to cut varieties of higher agricultural output. With regard to the cutting system of the stalks (which cuts them into smaller pieces), there must be some reasonable level of uniformity in their size. The cut must be shearing to avoid dilaceration of the stalk, which would lead to the wasting of raw materials and the aforementioned invisible losses. The conduction implements for the stalks inside the harvester must also refrain from causing sensitive damage to them. Inadequate maintenance of the base cutting and stalk fractioning blades, associated with other variety-specific features, can significantly increase the percentage of cracked, crushed, or incorrectly sheared stalks, which will increase levels of invisible losses.

Moving speed

Moving speed is directly influenced by the condition of the crop and the terrain. However, a machine at elevated speed will, of course, have a theoretically higher cutting capacity per period of time. Usually, those machines, according to specification of the manufacturers, can work at speeds up to 9 km/h. On the other hand, the harvesters and cutters in the São Paulo crops have not been able to go beyond average speeds of 4–6 km/h at work, possibly due to the lack of systematization of the parcels used for mechanized harvesting. Higher speeds in non-adequately systematized parcels, mainly as to what refers to the condition of stubbles, will lead to higher losses of raw materials. The adequate speed must be adjusted according to the characteristics of the parcel (systematization, size of the crop, and estimated agricultural productivity).

Characteristics of the mechanisms for lifting and chopping of lodged stalks and for ventilation (cleaning)

Certain machines do not possess mechanisms for the lifting of lodged stalks. This makes their usage limited (they will only work with assured efficacy in erect stalks). The harvesters produced in the country which possess this mechanism operate satisfactorily. With regard to the fractioning of the stalks into smaller pieces (rebolos), those must present some standardization of size, so that when the pieces go through the cleaning mechanisms (which also operate through fans and/or high power exhaust pipes), the separation between those and extraneous matter (by difference of density) does not come about in an inadequate way.

Without this standardization, it may happen that stalk pieces are eliminated and tops are placed together with raw materials for transportation. Accordingly, if the fans or exhaust pipes are not appropriately directed and regulated in their rotations this situation will be worsened, not to mention the interferences caused by the variety of sugarcane being cut. State-of-the-art harvesters allow for the variation of rotation in the exhaust pipes and fans, with a view to reducing losses (visible and invisible) and/or reducing the amount of extraneous matter carried together with the raw materials harvested.

The significant presence of cracked or badly sheared stalks may be due to the cutting mechanisms of the harvesters being deregulated or under inadequate maintenance. Also, it may be caused by excessive rotation of fans and/or exhaust pipes. The brittlement of stalks (variety-specific characteristic) can also influence this variable.

Mello (2002) presented results from the São Martinho plant (SP), which is the agro-industry with the highest percentage of chopped and crude sugarcane harvested in Brazil. This type of harvesting system, in the 2001–2002 harvest, reached 3,893,521 tons, being 94% of this amount harvested in sugarcane crops without pre-burning. Visible losses were of 2.52 tons/ha, soil percentage in raw materials samples was of 3.65 kg/ton of sugarcane harvested and extraneous vegetal matter reached 4.30%. The daily average harvest, per machine, during the 167 days, was 603 tons.

Pearce and Gonzales (2003) presented a few results obtained by the São Martino plant and by CTC-COOPERSUCAR regarding the influence of exhaust pipe rotation on crop losses, quality of raw materials, density of cargo and on other variables, using CASE-CND harvesters model A7700, 2003 version. In this study, an average cargo density of the order of 400 kg/m was obtained, with stalk pieces of average size of 170 mm.

Power

Today's harvesters have gross power levels of the order of 280–330 hp. This elevated power level is due to the demands of the agricultural productivity of the crop, of its bearing (erect, lodged, or sprawled), the cleaning and cutting mechanisms (which consist of several engines, pumps, and hydraulic cylinders) and of the devices for transportation of raw materials, from base cutting (beginning of operation) to the placement of the sugarcane in the transportation unit (end of operation).

In the expectedly not so distant future, full harvest (that is, the machine will perform a partial separation of extraneous matter and, therefore, everything it processes in its cutting and conduction mechanisms will go to transshipment) will be a routine operation (for the agro-industry which wishes to aggregate value through the usage of sugarcane crop residues), and harvesters will start using engines with lower power levels, since they will no longer require cleaning mechanisms, which nowadays are responsible for the consumption of more or less 30% of the available gross power.

Wheeling

There are harvesters which use tires (simple or in tandem), and of the track and semi-track types. In Brazil, only the first two are available, with track harvesters being predominant. Track machinery can operate in more severe topography scenarios and on soils with higher levels of humidity. Depending on the type of soil, the wheeling of the machine and the space left between rows, there will be different levels of compacting of the soil and destruction/crushing of stubbles, which will reflect negatively in the productivity of the area for the next harvest.

Beginning in 2000, several plants started developing integrated programs with a view to obtaining better-quality raw materials. As a rule, these are called "Clean House" programs.

The COSAN and João Lira groups, among others, are strongly engaged in this process. The basic goals to be fulfilled are the following: staff training (cutting fronts) for manual cutting and/or mechanized harvest (involving operators, repair and maintenance staff); decrease of the time between burning and cutting the crop; optimization of the transportation subsystem (improvement of synchronicity between harvest, transport and reception subsystems); new layout of parcels due to the need to adapt to the need for daily burning and cutting (which is a variable of the crushing capacity of the plant); better efficiency in the criteria for control of visible losses in the farm (reduction in the need for "bituqueiros" — workers who, after the work of loaders or harvesters, pick up stalks and/or their cut fractions and place them on some form of available transportation vehicle or group them in mounds for later loading); constant supervision of the conditions of the base cutting and chopping blades of the harvesters; avoidance of the leaving of burned erect sugarcane to be harvested the next day; and carrying out of pre-analysis of areas to be allotted for cutting, and subsequent releasing of such areas in the apex of the pol curve (ART/hectare), among other factors.

10.3.2 Farm-Specific Factors

Variety

The morphophysiological characteristics of the varieties chosen interfere heavily in the mechanical cutting of sugarcane. As a rule, both harvesters and cutters operate better on erect, robust sugarcane with deep root systems. Erect sugarcane stalks facilitate base and top cutting, with a consequent effective gain in the effective capacity of the machines (they will be able to work without larger interruption periods), decreased losses due to uncut stalks and better cleaning. Robust stalks with deep root systems resist mechanical base cutting performed by one or more rotating blades. There must be anchoring resistance in the stalks for the shearing to be made adequately. In case the sugarcane stalks have superficial root systems and are not robust, an imperfect or dilacerating cut may occur, which practically destroys that internode. Consequently, we have an increase in the infection area and higher deterioration levels, as well as higher likelihood of infection or destruction of the stubble, and the consequent reduction of germination in the next season.

Varieties that have already been abandoned by producers, such as Co419 and IAC52-326, were more shearable (lower fiber rate), which allowed for perfect base cutting, but, at the moment the stalks were crossing the machine internally (first stage of cleaning), they broke into small pieces and got lost under the harvester (machines that in the first stage had rollers and not tracks for transportation).

The varieties such as those developed in Australia, which have small tops, are more desirable when one decides to use mechanical cutting. They also occur in Puerto Rico and Hawaii, and have tropical characteristics. On the other hand, the Brazilian varieties, as a rule, have longer tops, which implies difficulties for the exhaust pipes and fans (cleaning devices) to perform separation by difference of density, since the biggest sugarcane hearts may be equivalent, in weight, to the cut-down pieces. Increasing the rpm rate of the exhaust pipe will eliminate usable sugarcane pieces, while decreasing it will include tops in the raw materials to be transported.

The sanitation of the crop can also interfere with the quality of mechanical cutting. For instance, stalks that are constantly attacked by the sugarcane borer (*Diatrea saccharalis*) present gaps within their internodes, which makes it more likely for those to get broken in the first stage of cleaning (in certain machines). With that, manufacturable raw materials can be lost from under those machines.

Sugarcane with good combustibility levels will present lower levels of vegetable extraneous matter after burning and will, therefore, offer greater ease of cleaning by the machine. A well-performed burning can eliminate up to 90% of the vegetable extraneous matter found. Besides, there is also a concern regarding the quality of the raw materials that get to the plant, and when one thinks of cutting or harvesting mechanically, this aspect should be considered.

Despite the existence of legislation regarding burning procedures in São Paulo crops, similarly to what happens in other countries, in most sugarcane-producing Brazilian states this practice is broadly adopted. The reasons for this are the significant increase in the performance levels of manual or mechanical cutting; the reduction of vegetable extraneous matter to be sent to the industrial processing unit; and the elimination of venomous animals or of insects that may attack workers. In this case, varieties which present leaves with good combustibility will facilitate the work of cutters and harvesters, and consequently offer raw materials of better quality.

It is advantageous for the harvesting machines that the parcel to be covered be homogeneous with regard to the height of the sugarcane and the lining up of stubbles. In Brazilian crops these conditions are not very commonly found, since the currently used varieties in Brazil, as a rule, have not been developed to be cut or harvested mechanically. Besides, large differences in the planting techniques adopted, with irregular distribution of fertilizer, makes this homogenization more unlikely. With the adoption of precision agricultural techniques in the sugarcane sector, this problem tends to become controllable.

The reflection of this non-homogeneity is that the machine is not uniformly fed, working without a constant flow of sugarcane to be harvested, which diminishes its operational capacity. As to the unevenness of height, there is also some form of harm to the quality of the work performed, for the operator will hardly be able to control, with perfection, the cutting of the tops. That causes the occasional elimination of internodes and non-elimination of tops, which makes it harder to regulate ventilation and/or exhaust pipe systems at the machines.

High stalk productivity, that is, beyond 130—150 tons/ha, may reduce the effective capacity of some harvesters due to the necessary speed reduction and because the sugarcane may be lodged, recumbent, etc., which, besides accounting for bad burning, makes it harder to clean and causes an increased likelihood of bushing. If the harvest is one of crude sugarcane, this negative effect will be stronger.

Campanhão (2000) presented the desirable characteristics in a variety for harvesting crude sugarcane: erect stalks, easy pealing, short sugarcane heart, medium fiber rate, good germination capacity under residues, uniform distribution of stalks, resistance to leafhoppers, average Pol of 16 and agricultural output above 88 tons/ha.

Status of the crop

As seen in the previous item, the more the crop finds itself more uniformly distributed, with more erect and more uniform stalks in terms of height, the better working conditions the machines will have, being then able to develop better speeds and offer raw materials of better quality. Those factors being confirmed, the machine can then develop higher speeds and offer raw materials of better quality, with a decreased likelihood of bushing.

Ripoli et al. (1977) proposed a set of criteria to define erect, lodged and sprawled stalks. Erect stalks are those which present a degree of alignment with the ground of 45° or higher. Sprawled are those which are found between 22.5° and 45°. Lodged are those which are found under 22.5° of inclination (Figure 10.3). For the characterization of a parcel, one can take 20 samples/ha of linear m of furrow, counting the stalks found in each sampling, and determining percentages. For that, one uses a rectangle triangle made of iron or wood, with 1-m-long catheti.

Soil preparation, planting system, and stalk spacing

Base cutting is performed by one or more rotating disks containing blades (knives), both on cutters and harvesters.

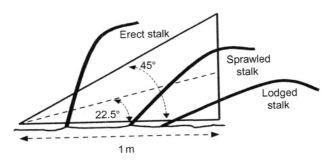

Figure 10.3
Criteria on defining the port of stalks in a sugarcane crop. *Source: Ripoli et al. (1977).*

Despite the fact that the set has hydraulic controls, allowing for it to vary the height of base cutting, the machinery is projected to cut the stalks at a point little above terrain level, that is, the bunch grass should be on top of a small ridge.

It is worth noting that certain cutter and harvester manufacturers have already developed devices which allow their machines to perform base cutting according to the imperfections in the micro-topography of the terrain, or even to operate on bunch grass that is below the level of the terrain.

John Deere has developed a system called (in Portuguese) CACB (which stands for automatic base cutting control), which is considered, in the sugarcane harvesting field worldwide, the most advanced system currently available for mechanized harvesting with machinery. It is composed by transducers, a main control (the system's "brain") and software specially designed to analyze the data collected. The transducers receive info based on pressure variation at the base cutting and the lifting roller, transforming those signals into electric signals. With that info, the base cutter will, automatically, compensate for changes in terrain declivity, with a view to: making the job of the machine operator easier; reducing losses of material on the farm; improving shearing quality in base cutting without causing damage to the stubble; reducing the amount of extraneous mineral and vegetable matter found in harvested material; and increasing the life span of the internal components of the machinery. CASE-CNH also presents a similarly built mechanism.

Due to the fact that the soil preparation techniques adopted in Brazil are not commonly oriented to mechanized cutting or harvest, one rarely finds ideal conditions in the plant-cane, that is, the first-cut stubble is seldom already leveled in relation to the terrain. With these conditions, if there are ratoons (older cane stalks), one runs the risk of damaging the stubble upon cutting, causing issues in the next germination period. There is also the inconveniency that the machinery will operate with tires at uneven heights, which will lead to unsatisfactory service levels. In case the blades will work under soil, their wearing will obviously be higher, due to abrasion, as well as that of all the conduction mechanisms for the cut material placed inside the machinery.

The blades of the chopping device have better resistance levels over time if compared with the base cutting disks, provided that their periodic replacement is made according to the data sent by the quality lab which tests the raw materials in the processing plant. Figure 10.4 shows adjustment needs for spacing and tire distance in machinery.

A lot of processing plants prefer to produce their own blades, using discarded springs from truck suspension systems. It is arguably a good option which is still lacking in evidence in terms of good cost to benefit ratios, if those are compared with the blades offered by professional manufacturers which are concerned with offering more than a sharpened piece of steel, but a product with reliable technical characteristics. Duraface, among a few other

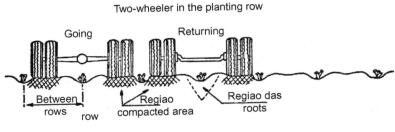

Two-wheeler in the planting row

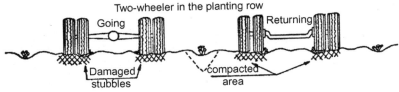

Two-wheeler in the planting row

Truck with one wheel in the center of the between-rows area

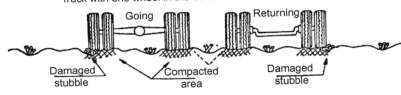

Figure 10.4

Relative positions between tractor tires and row spacing, which can cause higher or lower levels of soil compactation near the stubbles and/or their crushing, and reinforces the need for adequate adjustment of spacing between rows, of the distance between the tires of vehicles and of the machinery which will operate in the parcel. *Source: Mialhe (1980).*

manufacturers, produces blades both for the base cutter disks and for the chopping cylinders of combine harvesters.

Conversely, if the planting rows are in a furrow, there will be no way for the machine to cut the stalk at the desired height (stubble level), which leads to more dilaceration of the internode than shearing *per se*, which would be the expected result. This fact is due, mainly, to the lower anchoring resistance of the stalk upon cutting. Such a scenario leads to visible losses of raw materials.

Accordingly, preparation of the soil arises as one of the most important aspects with regard to mechanical harvesting. The higher the number of stones, wood stubbles, holes, uneven spots, etc., in the area, the higher difficulty levels one will face when operating the machines. This will lead to higher losses of raw materials, increase in the percentage of extraneous matter found, and a need for more maintenance of the machinery. These machines were built to operate in terrain that has been systematically prepared.

Furthermore, the spacing between planting rows also interferes in the quality of the mechanized harvest process. The machines currently available in the market demand an ideal spacing of 1.5 m between planting rows (due to the distance between their tires). There are a few models in other countries which cut two rows simultaneously (and for which spacing is not so relevant). In rows with smaller spacing, the machine, when operating on a row of stalks, will have their track over the next row to be cut, lodging stalks, crushing them, making their cutting more difficult, harming the stubble and favoring the compactation of the soil.

Even at ideal 1.5-m spacing levels (for mechanical harvest), one must control the operation of furrow formation strictly, to maintain good levels of parallelism between rows. The absence of such care leads to elevated losses of harvest and reduces the performance of mechanized operations.

Planting row length and state of dirt roads

Inadequate length of planting rows will directly influence effective harvest time. If planting rows are too short, from 80 to 100 m, the machine will necessarily have to perform a much larger number of u-turn maneuvers (besides demanding the same maneuvers from the transportation vehicles). In such cases, their operational working capacities will be harmed, significantly increasing the cost of the harvesting system.

The opposite situation (that is, rows of 1500−2000 m in length) may make it harder to obtain synchronicity between harvesting machinery and transportation vehicles, thus leading to higher management needs. At a given moment, the transportation vehicle will already be at its maximum value and the row will not have finished harvesting yet. The ideal condition for in-parcel transit would be exchange of transportation vehicle at the heads of the planting rows. Otherwise there will be a higher traffic of vehicles on the parcel itself, which leads to higher levels of soil compaction and destruction of stubbles. In this case, the recommended scenario is the creation of temporary dirt roads. Experience has shown that 500−1000 m is a satisfactory number for planting row length. However, with good management, use of automatic pilot and DGPS, 3-km-long parcels can be handled, which already happens in sugarcane crops in Australia.

The length of the dirt roads will influence the speed of maneuver of machinery and the transportation mode used. Machinery will, in general, be able to perform a complete u-turn maneuver within a minimum 7−10 m radius. If dirt roads are wider, maneuvering will be faster. Of course that conservation of dirt roads is a factor that should be observed, and differences of level between those and the parcel must be eliminated whenever possible. For the dirt roads, the demand is different and machines with a narrower turn radius will be better.

Layout of the parcels

In general, the sugarcane crops are formed based on a very wide range of parcel layouts, and wasted room in planting rows (the so-called "dead rows") will often occur due to excessively conservative practices being adopted.

It has already been proved that such scenarios cause losses in the mechanized operations, be they of preparation of soil, furrow digging, planting, culture treatments, etc., especially with regard to loss of time due to excessive numbers of maneuvers. This loss is aggravated in harvesting with combine harvesters, when there is a constant need for the transportation vehicle to be next to it. The desirable scenario would be rectangle-shaped parcels without dead rows or parcels that keep up with topography unevenness.

A possible scenario for systematization is the implementation of culture in ribbons, following topography levels, with lengths that can range from 2 to 3 km. Every 500—600 m a dirt road must be placed transversally to level curves, which will disappear in soil preparation operations. With this system, one can obtain increased operational capacity for all machines involved in the agricultural production process. A topography that allows for the rectangular shape is ideal for all mechanized operations, including harvest.

Terrain declivity

Terrain declivity is one of the limiting factors for the use of cutters and combine harvesters, because these machines have actually been built to operate in plain terrain (such as those in the sugarcane crops in Australia, Florida, etc.). Because of their relatively high center of gravity, those machines can only operate at certain levels of declivity, beyond which there is a risk of accident by tipping. Some tractor-mounted harvesters have higher tire distances, which offers them higher stability levels and allows them to safely work in terrain with declivity levels of up to 20% (manufacturer information), but most models should not work in declivity levels higher than 12—15% (wheeled vehicles) and 15—20% (tracked vehicles). Systematization conditions of the parcel are also worth highlighting. Above those declivity levels, besides the risk of tipping, there are difficulties for the transportation vehicle to follow the machinery, due to the tire—terrain ratio, as clarified by Beckker (1964).

In sum, an adequate systematization of the parcel must contemplate agricultural practices that refer to:

- better geometry of each parcel, due to the declivity levels of the terrain;
- desired parallelism between rows (to be obtained during planting);
- for simple furrows, 1.5 m spacing;
- better leveling of the terrain "brain".

This system demands high levels of synchronicity of operations between the subsystems involved so that the machine will not have to interrupt its work due to lack of transportation vehicles, besides reducing the number of transportation units per harvester. This will lead to favorable results in the cost of the system as a whole. The use of transshipment units (which follow the harvester while it operates) is more frequently found in Brazil. These units are also called intermediate transshipment units.

They present, in general, load-bearing capacity of 6—10 tons and lateral hydraulic tilting, besides high-flotation pneumatic tires (simple or double in tandem), which aim at reducing soil compactation and damage to stubbles They are mounted on tractors with 4×4 or 4×2 TDA wheeling.

After reaching their bearing capacity, the set will move to reach a dirt road, where a transportation vehicle of higher capacity (30—45 tons) is placed, which will then promote the transshipment of the cargo. However, the concatenation of aspects such as inadequate load for the type of transportation, excessive speeds and ill-maintained transit areas may lead to accidents.

Coordination and synchronicity of subsystems

The mechanized harvesting system involves, as has already been mentioned, the cutting, transportation, and reception subsystems. When harvesters are used in the cutting subsystem, the coordination and synchronicity between those three subsystems must be more efficient, otherwise there will be the risk of the harvester remaining idle for lack of transportation or transshipment vehicles in operation, which increases costs for the system's operation. Another aspect to be approached is the road system of the processing plant or distillery, which should obey criteria that lead to an agile circulation of vehicles, that is, the status of the traffic area, its width, acclivity and declivity levels, bridge widths, etc.

Management of the system

According to Mialhe and Ripoli (1999), the results obtained in the standardized tests in agricultural equipment in general have been seen under a new light: that of the reference aspects for the decrease of the size of the managerial challenge in mechanized agricultural systems. For instance, the operational capacity of sugarcane harvesters during a harvest, which directly affects the costs of the system, presents enormous discrepancies between test values and actual farm work values, which caters for very low efficiency-in-farm values. Increasing such efficiency is the great challenge currently faced by the professionals which are responsible for machinery management.

The diversity of socioeconomic and technical—managerial conditions that is found typically in Brazil has created enormous levels of confusion with regard to the criteria for the comparative evaluation of sugarcane harvesters (something that does not happen with grain harvesters). As a rule, the two main critical points regard the following aspects: (a) non-distinction between the usefulness of the data obtained in controlled tests and those obtained in the harvest (averages) through operational control; (b) overvaluation of the general average data obtained, to the detriment of the contribution from "clean" data obtained in standardized experimental conditions. Thus, the challenge that presents itself to the mechanization managers is always that of seeking to bridge the gap between harvest data and standardized testing data.

10.4 Loading Subsystems

10.4.1 Manual

Manual loading is, nowadays, a rare practice in Brazil, and occurs in some regions of accentuated topography in the south of Pernambuco, the north of Alagoas and the Zona da Mata area in Minas Gerais. It occurs when the dirt road is in very different topographic levels in relation to the parcel. In this case, a wooden board is placed to serve as a gateway between the top of the cliff and the cart of the transportation unit. Another situation in which manual loading is employed is the transportation of raw materials by the use of oxcarts. This scenario occurs in small sugarcane mills in the Northeast.

10.4.2 Mechanized or Semi-Mechanized

The biggest boom in the use of mechanical loading subsystems in Brazil happened around the second half of the 1950s, in the Center-Southern region of the country, still using imported machinery, as reported by Azzi (1972). Nowadays, as long as the topography of the terrain allows for it and the producer plants an area large enough to justify the purchase of loaders, one can find these machines in all producing regions of Brazil. Ripoli (1986) estimated that, from the 3867.2 million hectares planted in 1986, only around 400,000 ha, located mainly in the Pernambuco, Minas Gerais and Alagoas states, were not suitable for the use of conventional mechanical loaders because of terrain limitations. This number has probably been reduced today due to the introduction of tricycle and self-propelled loaders, initially manufactured by Cemasa and now by Implanor, in Pernambuco, whose project originated from a South-African model from BELL.

The basic types of loaders currently found in the Brazilian market are: tractor-mounted and self-propelled (the last type introduced in Brazil in 1981). Both are hydraulically activated. In the former, the hydraulic pump is moved by a gear shaft connected to the crankshaft at the tractor's engine, and the latter is activated hydrostatically.

From an ergonomic point of view, these tractor-mounted machines are normally sub-par, not offering any protection to the operator with regard to tipping risks, noise levels, vibration, etc., besides the inadequate positioning of commands and cockpit. The sum of these aspects leads more quickly to operator stress, to an increased number of occupational accidents, and to the decrease in his/her efficiency as an operator of agricultural machinery. The manufacturers offer, as an optional, acclimatized cabins, which dramatically improve working conditions for operators.

Conventional loaders are basically constituted of a metallic structure fixed on tractors, a hydraulic pump, hoses, command controls and hydraulic pistons, filters and hydraulic fluid containers. The hydraulic system provides motion for the active organs, that is, the rake and the spear, arrow and grab set.

The rake has the function of anchoring or piling together the stalks which are already cut (in piles or threaded) in the terrain, so as to facilitate the work of the grab, which elevates and deploys the raw materials on the transportation units. Experience has shown that the best type of rake is the floating one, for it leads to smaller levels of mineral and vegetal extraneous matter carried along with the raw materials loaded.

In manual loading, the raw materials can arrive at the processing plants in clusters tied up with the apex of the stalk itself (top with green leaves), a scenario still commonly found in high declivity areas whose intermediate transportation process is still carried out on oxcarts. Thus, with the exception of the apex piece, the raw materials are presented with better quality, with almost no presence of mineral extraneous matter. The mechanization of the loading process resulted in a significant increase in the amount of both vegetal and mineral extraneous matter loaded. It is regrettable, however, that, for several years, despite the development of new types of rakes, there has been no consistent scientific work that has been developed to effectively know their effect in the decrease in the amount of earth dragged together with the stalks. This makes it quite hard to ascertain the efficacy of the current projects as compared to the older ones. What one has is data obtained by sampling at the probe; however, in general, the training of the operator, the type of handling, and other variables are not controlled for. Those become average values, which are not always statistically significant.

Quality of burning

It is not simple to accurately determine the quality of a pre-harvest burning of the crop. For practicality purposes, we suggest the adoption of the criterion proposed by Balastreire and Ripoli (1975), which takes into account two possibilities only: good or bad. The burning will be considered to have been *good* when only the stalks and respective tops are left in the crop after it. *Bad* burning, conversely, will be considered that in which, after the action of fire, there remains in the area, besides the tops, green leaves, planting remnants and invasive plants. The quality of the burning is influenced by several factors, namely: size of the crop; stalk productivity rates; amount of straw and green leaves presented by that variety at the moment of burning; direction and speed of wind; air humidity in the crop;

and room temperature. In manual cutting, the worker will, in general, top off the stalks and the tops will fall on the ground. The cut stalks will, in turn, be spread or piled on the terrain and on top of the remnants from cutting. In sugarcane crops where burning was not adequate, the amount of unburned leaves is higher. As a consequence, during loading, those tops will be carried together with the stalks, increasing rates of vegetable extraneous matter found.

Granulometry and soil humidity

These two variables deeply influence the mineral extraneous matter rates found. The higher the concentration of clay in the soil is, and the more humid it is at the moment of loading, the higher the rates of extraneous matter will be, because those characteristics make it easier for that type of matter to adhere to the stalks during the action of the rake and the grab on them, at which moment they tend to roll on the ground. Besides this aspect, and with higher intensity, the rake, upon moving, sometimes drives its inferior extremities into the ground, which concurs for the increase of accumulated earth on the pile of stalks that is lifted by the grab. In clay-rich and humid soil, adherence to the rake is evident. Another aspect that leads to the increase in the rate of mineral extraneous matter is the post-burning condition of the stalks, as to what refers to the rate of exudation in the stalks. In both manual and mechanized (by cutting piler) cutting operations, when the stalks are deposited in direct contact with the terrain or dragged during loading, exudation acts as an adherence factor for dust particles, which will be carried together with the stalks upon transportation.

Disposition of the cut stalks

Piles or threads of adequately placed stalks lead to lower rates of extraneous matter, because they will demand less movement from the loader's grab. Crops with erect stalks also favor adequate placing on the ground.

Types of rake and grab

Floating rakes are more efficient than conventional rakes due to the fact that they have hydraulic devices that limit their soil penetration. Their inferior extremities tend to follow the micro-topography of the terrain. In turn, grabs which have hydraulic devices that limit their closing, therefore avoiding soil penetration, are also reducers of the amount of dust dragged along with stalks. Cane loader manufactures have, throughout the years, developed new solutions to offer the market in terms of grab mechanisms and constitution. The purchase of a given type of rake must be a decision based on the predominant type of soil and rain conditions during harvest at that specific region. In lighter soils, there is no need for more sophisticated equipment, as long as operators are well-trained.

Operator aptitude

Operating a sugarcane harvester is quite different from operating a conventional tractor. Lack of qualification in operators is one of the causes for increased extraneous matter in the loaded cargo, visible losses in raw materials, and increased repair and maintenance costs. Training of operators must not be oriented simply to the handling and regulation of the machine, but must also be a means of providing the operator with minimum orientation and knowledge regarding the sugarcane plantation.

Types of parcel

Three-, five-, or seven-row parcels, piled up or threaded, must lead to different amounts of extraneous matter dragged along. Few studies have been performed on the influence of this aspect in extraneous matter incidence rates, therefore one cannot affirm, so far, which scenario is more interesting. Theoretically, seven-row divisions would lead to smaller rates of extraneous matter because fewer machine operational cycles would be needed to load the same amount of raw materials.

Due to better equipment technology, better training of workforce (both for manual cutting and for the operation of loaders and harvesters), and programs for quality improvement process-wide (from harvest to reception of raw materials) currently available, the rates of extraneous matter currently found in processing plants are below those mentioned by Monteiro et al. (1982).

Another aspect to be considered in the mechanical loading subsystem refers to the manipulation efficacy (EM, in Portuguese) rates of this operation, that is, to the amount of raw materials piled up but not loaded by the machines, which are called losses. With a view to reducing such losses, processing plants currently used the so-called "bituqueiros". The cost of such an operation is compensated by return, since economic losses would be higher should these losses not be somehow recovered. However, the raw materials resulting from this operation are often of poor quality, due to the high rates of dust which adhere to the stalks and to the fact that they are normally partially or completely crushed.

Finally, one must consider, for the mechanical loading subsystem, the operational and effective capacity of these machines, measured in tons/hour, tons/day, or tons/month of raw materials loaded. These parameters depend, mainly, on the condition of cut stalks (erect or lodged), the capacity of the loader's grab (by operational cycle), the number of rows which are cut and subsequently form the beds for loading, the aptitude of the operator, the synchronicity between transportation vehicles, and the efficiency of the refueling, repair and maintenance fleet at the plant.

The ideal scenario for the representation of visible loss and extraneous matter rates analyses is for the results to be expressed in percentages, and not units per ton/ha. In the case of tons/ha, it is necessary to know the agricultural output rate of that crop to have the correct reference.

10.5 Transportation Subsystems

The implementation of transportation subsystems should rationally start (theoretically) concomitantly with the implementation of the physical infra-structure of the agro-industry itself, so that, as time goes by, it does not become a bottleneck in the raw materials transfer process from the farm to the industrial unit.

Due to the large areas that often characterize sugarcane planting units in Brazil, road transportation was consolidated as being the best option, despite the fact that it is not always the most economically feasible one. This scenario is a result of a transportation policy triggered in the early years of the implementation of the automobile industry in Brazil. So as to encourage and favor the trading of road vehicles, new roads were built, which, in itsself, is not undesirable. However, there was a lack of interest, equivalent in force, in the developing, maintaining and broadening of railroads and waterways, transportation systems which have been proved to be more economically feasible.

Thus, what was seen was the deactivation of railroads, which, at the time, were also available at sugar processing plants. Currently, no more than two or three sugar processing units, in Rio de Janeiro and in Pernambuco, sustain railroad use in the transportation of their sugarcane.

Therefore, the transportation subsystems used for sugarcane in Brazil are: roads, railroads, and waterways, with large predominance of the first.

10.5.1 Road Transportation Subsystem

An estimated 95% or more of the transportation of raw materials performed in Brazil is made through the road networks of the sugarcane producing regions. Those roads are of four types: roads within the production unit *per se* (dirt roads and side roads), local roads, state roads, and federal roads.

Dirt roads are 5–7 m wide on average. Besides being the primary segment of the road network in a property, they also play two roles: separating and delimiting the planting rows and serving as room for maneuver of all the agricultural machinery involved in the process of production and transportation of the cane. The basic architecture of the dirt roads and side roads is determined by the layout of the parcels, which, in turn, is a variable of the topography of the area and other aspects connected to the organizational planning of the agricultural infrastructure.

Side roads have a transit area of 7–10 m in width, so as to allow for better movement of the transportation units and to join the dirt roads to other access ways to the industrial processing unit.

Types of transportation unit

The currently available options in Brazil for types of raw materials transportation units include vehicles ranging from oxcarts to semi-trailers towed by tractors; trucks with one or two gear shafts (double-clutched); tractor trailers towing one or more semi-trailers, containers and others.

The choice of one unit or another will depend on factors related to the distance between production areas and industrial units, transit conditions of the road network (width, type, state of the transit area, acclivity and declivity, special structures (bridges, etc.)), and amount of raw materials to be processed daily, besides the operational costs of each type of transportation. In this regard, it is worth mentioning that, besides the concern with deploying raw materials on the reception spot at the processing plant within as little time as possible, it is imposed upon the producer to seek transportation options that minimize ton/km transportation costs. This goal is only achieved through relatively complex studies of economic and technical nature regarding available equipment in the market, associating those with specific conditions of the road network to be utilized.

Animal

The use of animal-towed transportation for sugarcane is very common in the south of Pernambuco, north of Alagoas and the Zona da Mata area in Minas Gerais. That happens due to the use of high-declivity areas for the planting of sugar cane. In some areas sugarcane is planted in areas with 100% declivity levels.

The main ways animals are used for transportation include: oxcarts, with the use of float-boards made of wood or metal called zorras; mechanical turnstiles; or transportation on the back of mules, with special support devices called *cambitos*. With the exception of oxcarts, used only in small sugarcane mills and in small Panela craft factories in the Brazilian Northeast, the other forms of transport, called intermediate transportation, happen within parcels, and only up to a point where it is possible to reach engine-driven vehicles, where transshipment (manual or mechanic) happens.

Cambitos are devices placed upon the back of mules which are usually V-shaped and are tied together by leather or denim straps. On average, an animal is capable of carrying 300 kg of stalks downhill.

In turn, zorra is a type of trolley, the oldest models being made of wood and the more recent ones made of iron angles and plates. On average, they have a load-bearing capacity of 1 ton.

The mechanical turnstile comprises a rack (metal tube) which, through the saddling, is fixed to the joints of oxen. It has a manually activated turnstile with a steel cable that revolves around the pile of cane to be towed downhill. On average, this mode of intermediate transportation has a capacity of 800—1000 kg, depending on how the piles of cane are tied up.

Another practice for the removal of the harvested stalks from the farm which is commonly found in areas with more than 60% declivity levels is the so-called "tombo" ("falling" in English). Operators with an iron or wooden shaft push cut stalks downhill until they fall on a dirt road where the actions of loaders and engine-driven transportation vehicles is feasible. Both turnstile and "falling" transportation methods require the action of bituqueiros, since the material which is not manually taken to a dirt road can represent up to 20% of the agricultural output.

Tractors towing semi-trailers

For our conditions, practice has demonstrated that the use of agricultural tractors to tow one, two, or three truck trailers is feasible for distances no longer than 5 km between production and industrial areas, and in areas where the road network does not present high levels of acclivity or declivity, which would bring difficulties in handling of the produce and increased risk of accidents.

These semi-trailers, depending on the type of build, can transport whole stalks when they have wood or iron stanchions, and chopped stalks when the trailers are protected by nets. When only one trailer is towed at a time, its load-bearing capacity is approximately 10 tons. When more than one trailer is towed simultaneously, capacity is reduced to 4—6 tons per cart.

Truck

The dominance of trucks in the transportation of sugarcane is irrefutable, and the range of options available in the market is quite wide, going from medium-sized trucks with 8—10 tons of load-bearing capacity to the so-called "super heavy" models, with 45—50 tons of load-bearing capacity. Economically speaking, higher load-bearing capacity rates should be selected for longer distances.

The most traditional type of truck is the one whose trailers have iron or wood stanchions attached to it. Comparing a given model with simple or double clutching, the advantages of the second are as follows: increased load-bearing capacity, from 8 to 17 tons per trip; smaller compactation of the soil, because of increased areas of contact between tires and the ground, and, in rainy seasons, reduced risk of vehicles being stuck in the mud.

Depending on the way the stalks are placed in the loading unit and the transit conditions of the transit area, stanchion platform trucks will have the inconvenience of dropping stalks during the journey, even if they are tied up with steel cables. These losses are difficult to assess; however, it is known that they are significant during harvest and that they cause accidents.

Currently, the predominance is of trailers which are closed in their front and back parts and with large metal stanchions on the sides, of several models (semi-trailers, tow trucks, trailers, etc.) and of more careful loading actions, such as limiting the height of the cargo placed in compatible levels so as to avoid higher losses of stalks during the trip to the processing plant.

The use of more powerful trucks that can tow an extra trailer besides its original cart is already consolidated. Those are the Rodotrem ("Rodotrain") type, commonly known as "Romeo and Juliet", with a load-bearing capacity of 25−30 tons. Their bodies can be either closed or with stanchions. This option is recommended for distances between processing plant and farm of 25−50 km. For these vehicles to transit in a cost-effective way, it is necessary that the transit area be in a good state of conservation, so as to allow for higher speeds during transport (50−70 km/h), otherwise the cost of the ton per kilometer transported may become excessively high, compromising the use of such vehicles.

Finally, we have the "super heavy" vehicles, which can tow three or four trailers at a time, and are able to carry 60 tons of stalks per trip (there are restrictions on the National Transit Code regarding the transit of these on state roads). They are recommended for long-distance trips, above 30 km of distance, and can be used with stanchion-filled trailers or closed ones. In this case, the road network should, preferably, contain paved roads so as to allow for more agile transportation. Using these vehicles on narrow, ill-conserved roads or for short-distance trips is the shortest path to compromising the operational cost of these units and of the ton of cane transported per kilometer.

As the cost of transporting the raw materials is largely significant for the whole process, several sugarcane companies have been refining their planning in that regard. With that, the level of control has reached the point where the wearing of tires is verified and the engine and shift oils are physically and chemically analyzed for all working vehicles.

Another system, which has been only recently introduced and shows promise (despite not being very widespread yet) deserves a separate mention. It is the *containers* system, which uses standard crane-attached trailers or metal boxes which are movable and transportable separately, with 5−7 ton load-bearing capacity, in substitution of the conventional trailers which are fixed to the chassis. This system, in general, makes use of a tractor which tows an appropriate trailer for the conduction of the container, together with the loading unit, be it a conventional loader or a combine loader (this second one is the most common) in the parcel.

Once it is loaded, the container is transferred, usually by lift-trucks, into the transportation system. This transportation system will usually be able to carry two containers of 7 tons or three of 5 tons, depending on the type of transportation unit. The reception and unloading of the containers in the processing plant are specific for that type of unit, and performed by the use of lift-truck type equipment. Unloading can be made directly into the reception table or by simply piling up the containers, which will then work as temporary storage units.

Despite implications in the mandatory use of transshipment in the farm and of special units for the handling of the containers, this system has the great merit of speeding up the loading operation in the parcel, reducing compactation problems caused by the transit of very heavy transportation units in the field. Once the transportation vehicles no longer need to enter the parcels, they can be selected according to their best efficiency levels for transportation on roads. The reception also becomes independent from the crushing process, since the sugarcane can be stored in the containers themselves, even though this procedure usually leads to an oversizing of the transportation units.

Initially designed to be used with chopped stalks, the container system can also be re-sized to adapt for whole stalks, although in these conditions some of the advantages the system brings with regard to handling may no longer be available.

10.5.2 Railway Transportation Subsystem

As has been reported here, this subsystem has almost completely fallen into disuse in Brazil, even though it is widely utilized in countries such as Australia and Cuba, where the largest part of the agricultural output uses this form of transportation. Currently, railways, where available in Brazil, are part of the processing plant's transportation system, and are associated with road transportation. Either through tractors towing trailers or trucks, the raw materials arrive at the so-called transshipment stations, in which they are transferred to wagons and then taken to the processing plant units.

10.5.3 Waterway Transportation Subsystem

Beginning in 1980, in a pioneer experience (at least so far) in Brazil, the Diamante processing plant, nowadays part of the COSAN Group, located in the Medium Tietê River Basin (SP), implemented a waterway transportation system for sugarcane, taking advantage of the lock system implemented in the river's area. As happens with the railway transportation subsystem, the waterway system must also be conjugated with road transportation. Ripoli et al. (1984) reports that, in this plant, four transshipment points were built through skip hoists, in which the raw materials were transferred to the punts. The maximum distance for which this subsystem is used is 35 km. If the same raw materials were to be transferred by road, the corresponding distance to reach would be around 100 km. The towing sources for these punts are boats called "pushers", running on 270- to 340-hp engines, which tow two to three punts, carrying, per trip, 200−500 t of raw materials. Researchers also report that the use of waterway transportation systems resulted in a reduction of costs of around 53.1% per transported ton, and that, due to this association between waterway and road transportation, the current length of transportation made through roads does not exceed 10 km from farm to processing plant.

10.5.4 Transshipment Options

The transshipment operation is understood to be that of transferring the existing raw materials from one transportation vehicle to another. This operation can occur in any transportation subsystem, and the locations where it is performed are called transshipment stations.

In the railway or waterway transportation subsystems, transshipment happens from roads to wagons or boats, through mechanical or hydraulic hoists or by the use of skip-hoists.

In the road transportation subsystems, there are two types of transshipment systems to be considered: direct and indirect. Direct transshipment is used in harvesting systems that use harvesters (combine), so the raw material is usually found in chopped stalks (incorrectly called "toletes"; the correct name is "rebolo", referred to in this work as simply "pieces", since the term has no English translation). To follow the harvesters, trailers (also called transshipment vehicles) towed by wheeled tractors or special vehicles are used.

When the first type is used, there is a demand for the assembling of separate devices that allow for the transshipment operation to happen. In the second case, the transportation unit itself will have hydraulic devices which will cater for the unloading process. On average, these vehicles have a load-bearing capacity of 4−12 tons.

The use of these vehicles is justifiable in places where compactation of the soil is an issue, since they exert less pressure on the terrain than conventional double-clutched trucks.

Intermediate transshipment operations, on the other hand, are justified when the cutting takes place more than 25 km away from the processing plant. With that type of handling for transportation, the flow of vehicles can be sped up and the costs per kilometer of transported ton can decrease substantially, since small, slow units will operate close to the planting area, and faster units with more load-bearing capacity will run the larger distances.

With regard to the tires used in sugarcane transportation vehicles, it is worth mentioning that by 1988 Trelleborg (Swedish manufacturer) introduced in Brazil low-pressure radial tires (high flotation), especially for use in transshipment units, that is, those which transit inside the parcel. Studies have shown that these tires cause less soil compaction, reduce transportation costs, present higher durability (around 15,000 h) and lead to lower fuel consumption in the engines with which they are used.

10.6 Reception Subsystem

This subsystem involves the following stages: weighing of the transportation unit in a platform scale; taking of sample, by probe; unloading (directly at the reception table or in patios/warehouse) and weighing of the vehicle in order to determine its unladen weight.

10.6.1 Unloading

After the transportation unit has been weighed and a sample by probe taken of the raw materials for analysis of its quality, it may then proceed to two possible areas in the processing plant: warehouse, or direct unloading at the reception table. The definition of where the unloading operation will take place depends on the operationalization of the processing plant, the harvesting system, the amount of raw materials that arrive at the processing plant, and the crushing capacity of the molasses mill.

In the case of manual cutting and cutting by machine, the raw materials are found in the form of whole stalks. The cargo is removed from the transportation unit and deployed in the warehouse for subsequent cutting through traveling cranes. This type of unloading has been avoided by the processing plants as much as possible because the goal is to crush the raw materials as quickly as possible after cutting. Stocking sugarcane is only justifiable in nightly unloading operations. In case there is adequate planning of harvest and transportation, that is, if such planning has taken into account the daily crushing capacity of the plant, unloading is performed by a fixed hydraulic hoist system, called Hilo, which, through steel cables, laterally tips the cargo at the transportation unit, throwing it to the floor of the warehouse (less usual) or onto the reception table. In the first case, tractors with front rakes will perform the moving of these raw materials to the reception tables.

The direct unloading of whole stalks is that in which the transportation unit arrives at the Hilo and works by throwing the raw materials directly on the reception table at the plant. The transportation unit parks between the Hilo and the reception table. Steel cables are attached between the cable and the base of the trailer, being tied, at one of the extremities, to the crane of the Hilo, and at the other end to the top of the reception table. With the activation of the hoist's crankshaft, the cargo is tipped laterally, falling on the table.

When the raw materials are collected by combine harvesters and, therefore, are fractioned in pieces of around 15−30 cm in length, its unloading should not be indirect nor should it be washed, i.e., it should be thrown directly on the reception table. The reason for this is related to deterioration and sucrose loss aspects that have been discussed before. For its unloading, it is demanded that the reception tables and the transportation traveling cranes be assembled below the level of transit of transportation units, to allow for lateral or back unloading by tilting of the transportation unit's cargo. That leads the raw materials to fall on the reception table by action of gravity.

The transportation units have a pantograph linkage system for the opening of the lateral or back door proportionally to its lifting. This tilting can occur through traveling cranes or through hydraulic crankshafts fixed to the ground which push the tilting trunk laterally.

The reception subsystem also deserves attention in the management aspect, since its agility in receiving, unloading and releasing transportation units will define the adequate synchronicity level with the planting and loading operations, thus reducing lines of vehicles. Table 10.1 shows the outline percentages of extraneous matter within raw materials as an important point reference points.

Table 10.1: Reference criteria to outline percentages of extraneous matter within raw materials.

Terrain	Manual Cutting (%)	Mechanical Harvesting(%)
Excellent	Up to 0.66	Up to 0.60
Good	0.67−1.32	0.61−1.67
Regular	1.33−1.94	1.68−2.74
Bad	>1.94	>2.74
Average	1.32	1.67
E.V.M		
Excellent	Up to 2.34	Up to 4.29
Good	2.35−3.15	4.30−6.47
Regular	3.16−3.97	6.48−8.66
Bad	>3.97	>8.66
Averages	3.15	6.47

E.V.M = Extraneous Vegetable Matter.
Source: Ideanews (2003).

Bibliography

Azzi, G.M., 1972. Incidência de matéria estranha nos processos de carregamento de cana-de-açúcar. Tese (Doutorado), Escola Superior de Agricultura "Luiz de Queiroz". Universidade de São Paulo, Piracicaba, 112p.

Balastreire, L.A., Ripoli, T.C., 1975. Estudos básicos para quantificação de colhedoras e veículos de transporte. In: Seminário Copersucar da Agroindústria Açucareira, 2., 1975, Águas de Lindóia. Anais. São Paulo: Copersucar, pp. 345−353.

Barbosa, V., 2002. Efeito do pisoteio na produtividade da cana. In: Seminário de Mecanização Agrícola: Perda de Produtividade, 2002, Ribeirão Preto. Ribeirão Preto: Sociedade dos Técnicos Açucareiros e Alcooleiros do Brasil, 1 CD-Rom.

Bekker, M.G., 1964. Theory of Land Locomotion. The University Press, Ann Harbor, 520p.

Campanhão, J.M., 2000. Comportamento de variedades em áreas de colheita mecanizada de cana verde. Cana Crua 2000 Experiência Acumulada. UNESP/STAB, Jaboticabal, Piracicaba, CD Rom.

Chudacov, D.A., 1977. Fundamentos de la teoria y el calculo de tractores y automóviles. Mir, Moscou, 435p.

Gill, W.R., Vanden Berg, G.E., 1967. Soil Dynamics in Tiliage and Traction. USDA, Washington (Agricultural Handbook, 316). 511p.

Impurezas, sujeira, trash: é só prejuízo, 2003. IDEANews ano 4 (36), 66−72.

Mello, M.O., 2002. Colheita mecanizada de alto desempenho. In: Seminário sobre Mecanização e Produção de Cana-de-Açúcar, 4., 2002, Ribeirão Preto. Anais. Ribeiráo Preto: IDEA, 1 CD-Rom.

Mialhe, L.G., 1980. Curso intensivo de mecanização agrícolas na produção de álcool. ESALQ, Piracicaba, 174p.

Mialhe, L.G., 1996. Máquinas agrícolas: ensaios & certificação. FEALQ, Piracicaba, 722p.

Mialhe, L.G., Ripoli, T.C.C., 1976. Evaluación de cosechadoras automotrices de caña de azúcar. In: Seminário Internacional sobre Mecanización de la Cosecha de Caña de Azúcar, 1976, Caracas. Caracas: Dist. Venezolana de Azucares, SLR, pp. 189−204.

Mialhe, L.G., Ripoli, T.C.C., 1999. Critérios de avaliação em análises de confronto de colhedoras de cana-de-açúcar. In: Semana da Cana-de-açúcar de Piracicaba, 4., Piracicaba, 1999. Resumos, Piracicaba: STAB, p. 8.

Monteiro, H., Pexe, C.A., Bassinelo, J.L., Ripoli, T.C.C., 1982. Matéria estranha: custos e técnicas de sua diminuição na colheita. Álcool e Açúcar. 2 (6), 20−26.

Paranhos, S.B., 1974. Colheita mecânica de cana-de-açúcar. In: Seminário Agronômico de Pinhal, 4., 1974. Espírito Santo do Pinhal: Fundação Pinhalense de Ensino, Faculdade de Agronomia, 10p.

Pearce, J., Gonzales, C., 2003. Resultados de ensaios de qualidade de colhedoras série 7000 versão 2003. In: Seminário de Mecanização e Produção de Cana-de-Açúcar, 5., 2003, Ribeirão Preto. Ribeirão Preto: IDEA, 1 CD-Rom.

Ripoli, T.C.C., 1974. Corte, carregamento, transporte e recepção de cana-de-açúcar. ESALQ, Departamento de Engenharia Rural, Piracicaba, 52p.

Ripoli, T.C.C., 1986. Modelagem de desempenho operacional de uma carregadora auto-propelida para cana-de-açúcar (*Saccharum* spp.). Tese (Doutorado), Escola Superior de Agricultura "Luiz De Queiroz". Universidade de São Paulo, Piracicaba, 105p.

Ripoli, T.C.C., Paranhos, S.B., 1987. Colheita. In: Paranhos, S.B. (Ed.), Cana-de-açúcar: cultivo e utilização, vol. 2. Fundação Cargill, Campinas, pp. 517−597.

Ripoli, T.C.C., Ripoli, M.L.C., 2001. Effects of pre-harvester burning in sugar cane (Saccharum spp.) in Brazil. Rivista de Ingegneria Agrária. 32 (4), 202−210.

Ripoli, T.C.C., Franceschi, P., Mialhe, L.G., 1984. Transporte fluvial de cana-de-açúcar no Estado de São Paulo. Álcool e Açúcar. 16 (4), 18−22.

Ripoli, T.C.C., Paranhos, S.B., Santos, N.C. dos, Mialhe, L.G., 1974. Determinação de declividade limite para trabalho de colhedora de cana Massey Fergusson. In: Congresso Nacional de Engenharia Agrícola, 4., 1974, Viçosa. Anais. Viçosa: SBEA, pp. 27−31.

Ripoli, T.C., Mialhe, L.G., Novaes, H.P., 1977. Um critério para avaliação de canaviais visando a colheita. In: Separata de Congresso Brasileiro de Engenharia Agrícola, 4., Pelotas, 1977. Pelotas: SBEA, 10p.

CHAPTER 11

Breeding Program and Cultivar Recommendations

Márcio Henrique Pereira Barbosa and Luís Cláudio Inácio da Silveira
Federal University of Viçosa, Viçosa, MG, Brazil

Introduction

The cultivar is the most important technology and of lowest cost to the producer. It is the foundation which underpins all other production technologies. The successful production of energy, ethanol, sugar and other sub-products necessarily entails the production of quality raw material. In this respect, the cultivar takes a leading role in the success of the enterprise.

Since the early 1970s, the productivity of sugarcane stalks and sugar in Brazil has increased at a rate of 0.8 tons sugarcane stalks/ha and 1.8 kg sugar/ton sugarcane stalks. Obviously, these gains are due to the employment of technology both in the agricultural and industrial applications. In this scenario, cultivars are known to occupy a prominent position, although it is difficult to quantify the contribution of each production factor to global advancement.

Breeding programs are also the main methods used to control the major diseases affecting the culture, such as sugarcane rust *(Puccinia melanocephala* H. & P. Syd.), Coal *(Ustilago scitaminea* Syd.), leaf scald *(Xanthomonas albilineans* (Ashby) Dowson), among others.

This chapter aims to present an overview of the breeding program and management of cultivars of sugarcane.

11.1 Breeding Programs in Brazil

The first Brazilian sugar mills were concentrated in Pernambuco and Alagoas until the 20th century. However, with the expansion of coffee production and the abolition of slavery, the sugarcane industry declined, only recovering in the 1930s with the creation of the Institute of Sugar and Alcohol (IAA) in 1933. The Institute had as main objectives the regulation of the sugar market in the country and promoting the production of ethanol.

For the development of technologies for the ethanol sector, the Planalsucar was created, linked to the IAA. At this point, the Breeding Program of Sugarcane appeared and was later extinguished by the federal government in 1990, together with the IAA. However, after a year, staff and infrastructure of company departments and research stations of this program were absorbed initially by seven federal universities (UFPR, UFSCar, UFV, UFRRJ, UFS, UFAL, and UFRPE), which established the Inter-University Network for Development of the Sugarcane Production — RIDESA (www.ridesa.com.br). In 2004, the Federal University of Goiás (UFG) was also added to the RIDESA and in 2008, the Federal Universities of Mato Grosso (UFMT) and Piauí (UFPI).

The first cultivars developed by Planalsucar, called RB (Republic of Brazil), were released to the State of Alagoas in 1977 (Oiticica, 1977). Among all the cultivars developed, "RB72454" had a greater potential in the national scenery.

Currently, Brazil is among the pioneers in obtaining cultivars of sugarcane with market value, which are developed by three main programs: the traditional Breeding Program of the Agronomy Institute of Campinas — IAC (http://www.iac.sp.gov.br), the Cane Technology Center — CTC (http://www.ctc.com.br), extinct Copersucar, and the federal universities that make up the University Network for Development of the Sugarcane Production — RIDESA (http://www.ridesa.com.br). The youngest of them, the Canavialis (http://www.canavialis.com.br), appeared in 2004. In 2008, this program was acquired by the Monsanto Company.

Sugarcane cultivars are coded by letters and numbers, following an international standard, where the first two or three letters are the initials of the acquiring research institution. In the sequence, the first two numbers represent the year in which the hybridization was performed and the last numbers refer to the code that the clone initially received in the experiments. For example, in the case of Brazil, the cultivars IAC86-2210, SP80-1816 and RB867515 were developed by the IAC, Copersucar and UFV/RIDESA programs, respectively. The CTC does not follow this reported nomenclature.

The university members of RIDESA continue to produce cultivars with the initials RB (Republic of Brazil), similarly to the extinct Planalsucar. Currently, in almost 60% of the planted area in Brazil, RB cultivars are planted. Among the most planted cultivars in Brazil, RB867515 is one of the most relevant to the production of sugar and ethanol, as well as the production of forage. This cultivar was originally developed at the Federal University of Viçosa and is protected by the National Service of Cultivar Protection (NSCP) from the Ministry of Agriculture.

11.2 Strategies for the Breeding Program

11.2.1 Flowering and Hybridization

Sugarcane is an allogamous plant, belonging to the family *Gramineae (Poaceae)*, *Andropogoneae* family and genus *Saccharum*. The most accepted botanical classification as to the species is the one reported by Jeswiet (1925), modified by Brandes (1956), as cited by Daniels and Roach (1987). According to these authors, six species occur in the genus *Saccharum*: *S. officinarum, S. spontaneum, S. robustum, S. sinense, S. barberi*, and *S. edule*.

The allogamy is mainly due to male sterility. Therefore, it is important to determine whether the genotype will be used in the breeding program as male, providing pollen, or as female, receiving pollen. Depending on the potential of the female clone to release viable pollen, it must go through a process of emasculation. The emasculation is an important tool to avoid self-pollination in biparental crossbreeding. Among the various proposed treatments, the most frequently used is treatment with hot water at 50°C, for 4.5 min (Machado Jr. et al., 1995). McIntyre and Jackson (1996) estimated, using molecular markers, 2.8−17.6% of self-pollination.

To perform hybridization, the first requirement is flowering and its timing. These two factors were undoubtedly the major barriers faced by breeders of sugarcane. To solve this problem, some researchers studied the areas where sugarcane flowers regularly, as well as its center of origin. They concluded that the favorable agro-ecological conditions to their flowering is a minimum temperature of 18°C and a maximum of 32−35°C, high relative humidity and rainfall, lower natural fertility soils and latitude between 10°N and 10°S (Levi, 1983, 1992).

In Brazil, there are crossbreeding stations where the flowering of most of the genotypes is possible in natural conditions: (a) crossbreeding station of Serra do Ouro in Murici, AL, of the Federal University of Alagoas and crossbreeding station of Devaneio in Amaraji, PE, of Federal University of Pernambuco, which provides seeds to the University Network for the Development of the Sugarcane Sector (RIDESA); (b) CTC crossbreeding station (formerly Copersucar), located in Camamu, BA; (c) Experimental Station Mocambo, of the Bahia Company for Agricultural Development, in partnership with the breeding program of the IAC, located on the island of Itaparica, BA; and (d) Canavialis crossbreeding station acquired by the Monsanto Company in 2008.

In countries where the agro-ecological requirements are not satisfactory, flowering is only possible with the use of climatic chambers where the temperature, humidity, light, and nutrients are properly controlled.

Among the methods of hybridization of sugarcane, the most commonly used are multiple crossbreeding, which consists of bringing together large numbers of parents for the crossbreeding, and the simple breeding (biparental), in which male and female parents are known. This latter method is preferred, being the most frequently used by the breeders. In order to make breeding possible, an acid solution was developed, which keeps alive the sugarcane stalks removed from the field for the duration of the time required for maturation of the seed (Heinz and Tew, 1987).

Under the commercial aspect, flowering is undesirable, because it interrupts growth and consumes energy, being possible and can dry up the parenchymatous cells, a phenomenon called "pith process". Therefore, the reluctance to flower is a desired character in the selection of cultivars, which obligates breeding programs to locate the crossbreeding stations in locations where these genotypes, even the most reluctant, may flower regularly and have fertility.

11.2.2 Stages of the RIDESA Breeding Program

The starting point for the Sugarcane Breeding Program (SBP) of RIDESA is the germplasm bank located in the Flowering and Crossbreeding Station at Serra do Ouro (UFAL; lat 09° 13' S, long 35° 50' W and alt 515 m asl), in the municipality of Murici, State of Alagoas. In that bank, more than 2700 genotypes are gathered, among cultivars used in Brazil, clones, other species related to the genus *Saccharum*, and cultivars imported from different sugar regions of the world. RIDESA has also counted on the Crossbreeding Station of Devaneio, in Amaraji (lat 08° 19.8' S, long 35° 24.8' W and alt 514 m asl), Pernambuco, since 2007. It was established by UFRPE with the objective of complementing Serra do Ouro Station in seed production for the breeding network (Barbosa et al., 2012).

The seeds obtained in pre-established crossbreeding by the teams of the universities, which make up RIDESA, are sent to the respective States, where the seedlings are produced, which once transplanted to the field, define the first phase of selection (T1). RIDESA has produced annually about 2 million seedlings for phase T1. In some universities, the selection is made in two periods, April and July, in order to get at that first time, clones that have the important characteristic of precocity. More than 60,000 new clones have been generated annually by the universities. These clones are subsequently evaluated in experimental stations in the phase called T2. The clones are experimentally evaluated in plots of a furrow of 5—8 m in length, using the augmented block design.

In the second phase of selection (T2) the superior clones are selected from among the cane-plant and ratoon, which in turn are evaluated in phase T3. RIDESA has selected for phase T3 more than 7000 new clones per year. From this phase, the clones selected in each university are exchanged among them. At this phase, the new clones are multiplied and introduced in the mills and distilleries in agreements with the respective universities and working in different sugarcane regions of Brazil. In the lands of the mills and distilleries the promising clones have been evaluated through experiments for 3 consecutive years. This phase is called EP, i.e., experimental phase.

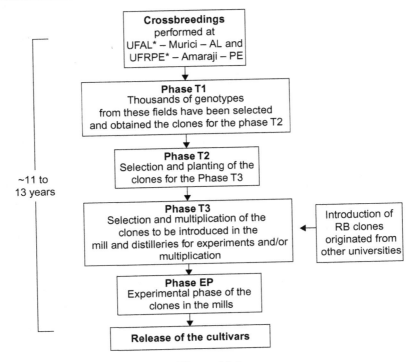

Figure 11.1

Flowchart of the phases of the breeding program of the universities which are members of RIDESA. *UFAL: Federal University of Alagoas and UFRPE: Federal University of Pernambuco.

The development program of new cultivars of sugarcane is, by nature, of long duration. Therefore, persistence is a virtue of the people involved in this process. The release of new cultivars has occurred after about 13 years of numerous evaluations of the clones, by experiments, observing the reaction of the clones to diseases and pests, as well as their productivity in different production environments. A flowchart on the previously mentioned case is presented in Figure 11.1.

11.2.3 Recurrent Selection

Recurrent selection is all cyclical process of breeding that involves obtaining the progenies, their evaluation and crossbreeding (recombination) of the best cultivars or clones. For this concept, virtually the breeding of sugarcane is, implicitly, a form of recurrent selection, as the superior clones and new cultivars will be used again in crossbreeding to generate new progeny.

In general, though not explicitly, intrapopulational recurrent selection (IRS) has been applied to the breeding of sugarcane. The superior clones generated at the end of the basic procedure (crossbreeding followed by clonal selection) are intercrossed (recombination) for

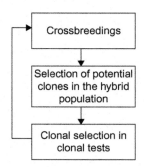

Figure 11.2
Implicit IRS outline for the breeding of sugarcane.

the generation of hybrid families of a new selective cycle (Figure 11.2). It is important to note that the sugarcane is semi-perennial and, therefore, there is an overlap of generations and clones of different generations are intercrossed, not just those of a given selective cycle. IRS is more effective in species that do not show high heterosis and/or genetic divergence in the material under breeding. Otherwise, the reciprocal recurrent selection (RRS) should be preferred, as will be reported later for the breeding of sugarcane.

Recurrent selection allows gradual increase in the frequency of favorable alleles through successive cycles of selection and recombination of the best individuals of the best progeny. This is relevant, since most characteristics of agronomic importance of the sugarcane are controlled by several genes. Therefore, the probability of a clone to come to possess all the favorable alleles is very low, hence the importance of recurrence.

Without doubt, the selection of superior individuals (obtainment of clones) will be more efficient if conducted in populations with a higher average or higher frequency of favorable alleles. Based on this premise some sugarcane breeding programs in the world have practiced routinely the selection of families before obtaining clones (Cox et al., 2000; Bressiani, 2001; Kimbeng and Cox, 2003; Barbosa et al., 2005), especially for traits whose heritability based on family averages has been higher than the heritability of individual plants, such as the production of sugarcane stalks.

The scheme (a), shown in Figure 11.3, has been the procedure adopted by RIDESA in recent years. From the crossbreedings are selected, considering the mass, the individuals being submitted for clonal tests. In other words, the information from the family is not used for the selection of potential clones. The criteria used for choosing the crossbreedings are:

- avoid crossbreeding between relatives;
- preferably, use in the crossbreedings elite clones and cultivars developed in the country or region; and
- associate important agro-industrial characteristics.

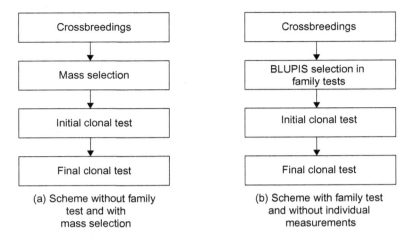

Figure 11.3
Basic schemes for the generation of superior clones in sugarcane.

The scheme (b), as shown in Figure 11.3, uses the tests of families in experiments with repetitions, but no individual measurements are made. Thus, it is not possible to predict values of individual genotypic of potential clones by individual BLUP procedure. However, you can use the BLUPIS (simulated individual BLUP) developed by Resende and Barbosa (2005b), which is an approach to individual BLUP and indicates how many individuals should be selected in each family and subjected to a clonal test. This scheme is ideal for use in sugarcane and forage (elephant grass, Brachiaria, *Panicum*), where the experimental plots are harvested in their totality.

Studies in quantitative genetics have shown that the additive genetic variance is more important than the non-additive genetic variance for most traits of agro-industrial importance of sugarcane. The main exception is for the production of sugarcane stalks, where both variances seem to be equally important (Bastos et al., 2003).

The results obtained by Hogarth et al. (1981) showed that the additive genetic variance was higher for brix, number, diameter, and height of stalks. They also reported the significance of non-additive genetic variance for all traits, except for brix and number of stalks. These results agree with those obtained by Bastos et al. (2003), who verified that additive genetic effects are as important as non-additive genetic effects in the expression of traits of economic importance of sugarcane. Therefore, for these characteristics, the specific combining ability (SCA) is as important as the general combining ability.

Considering the importance of exploring both the general and the specific ability of combination, the implementation of reciprocal recurrent selection (RRS) in the improvement of sugarcane is proposed. The RRS can be implemented in two levels: (i) populational, involving the crossbreeding of many individuals in a population with individuals from the reciprocal population; and (ii) individual (IRRS), involving only one individual from each population, which should yield an excellent crossbreeding with high total genotypic value and also high SCA. Those individuals that produce superior crossbreeding are auto-fertilized to produce two populations S_1, in which superior individuals are selected to integrate a program of reciprocal recurrent selection. The IRRS using S_1 is indicated for the improvement of sugarcane, which aims to explore the best of SCA from the prior identification of the best crossbreeding through the evaluation experiments of families. The use of inbred individuals selected from S_1 families is intended to eliminate the genetic load of population and explore for one more cycle the superior hybrid combination identified in advance. It is important to highlight that the process is cyclical, so new hybrid families will be identified and exploited by the process of IRRS-S_1. The general schema improvement via population RRS and IRRS is presented by Resende and Barbosa (2005a).

11.3 Desirable Characteristics in Cultivars of Sugarcane

11.3.1 High Productivity

Productivity, adaptability, and production stability of the clones have been quantified by evaluation in field experiments in different soil and climate conditions (Bastos et al., 2007). Based on these experiments and observations of local semi-commercial ranges (10–50 ha), the technical recommendations for the proper management of cultivars are prepared.

11.3.2 High Sucrose Content

The cultivar should have high sucrose content (PC) in the juice. Therefore, it is recommended to quantify the maturation curve of cultivars in different growing seasons in the property or location of cultivation. Associating this variable with the productivity of sugarcane stalks per hectare (TCH) in different harvesting times has resulted in TPH values ($\times$PC/100 TCH) in kg of sucrose per hectare at different harvesting times, in the harvest period. For a given harvest date, cultivars that provide greater TPH must be used, while taking into account other features mentioned in this chapter.

To raise the level of sucrose, especially at the beginning of harvest, many companies apply maturator, which in some cases also inhibits flowering and pith process, resulting in higher productivity of sugar per ton. The cultivars respond differently to the application of maturator, and soon tests will be undertaken to define the correct handling of this product.

11.3.3 Fiber Content

Typically, there is a negative association between fiber and sugar. The early maturation cultivars generally have lower fiber content. This can compromise the amount of available bagasse for firing at the beginning of harvest, and the stalks' productivity. The ideal amount of fiber cultivars varies between 12 and 13%. The fiber has been important in generating electricity to meet the needs of the mills and distilleries, as well as sale of surplus energy.

Certainly, the demand for electricity will influence the selection of clones for quality and quantity of fiber. Few studies have been carried out on this new scenario. Barbosa et al. (2004) have already begun the selection of specialized families in stalk productivity. However, further studies should be conducted to evaluate the use of fiber and its calorific value for cogeneration of electricity, as well as for its convertibility to produce cellulosic ethanol.

11.3.4 Sprouting and Longevity of the Ratoons

The productivity of stalks reduces every cut year. On average, the cane fields are renewed after four cuts. Therefore, it is highly desirable that the cultivars show good sprouting in the ratoon cane, especially those cut in mid-season of the dry season. Considering that about 80% of cane fields are ratoon cane, the importance of this trait in breeding programs is observed. Ferreira et al. (2005) demonstrated the importance of conducting three cuts in clone competition experiments to practice selection with predictability of the real value of the genotype above 80%.

In the case of mechanical harvesting, without previous burning, it is also important to evaluate the clones with regard to the sprouting of the ratoon cane under the residual straw. There are clones that do not sprout well under the straw.

11.3.5 Tillering and Characteristics of the Stalks

It is desirable that the cultivars show fast initial development, good tillering and appropriate closed interlines to minimize competition with weeds. The characteristic fast initial development and closed interlines is very important for areas harvested at the end of harvest and planting as cane-plant of year. The fast initial development is an important characteristic for sugarcane growing in areas with limited rainfall or very rigorous winters, factors which reduce the productivity of sugarcane. It is also interesting that the cultivars do not show late sprouting, which will undermine the uniformity of maturation of the sugarcane field.

Furthermore, it is desirable that they present uniform and average diameter of stalks that do not break easily. There are clones that have excessive broken tops due to winds.

The upright growth habit of stalks is important for both mechanical cutting and for manual cutting. However, the high productivity and the winds promote the lodging of the sugarcane fields. Therefore, higher productivity is not always associated with higher efficiency in crop yield.

Excessive lodging can cause stalk rooting when in contact with the soil surface. There is sprouting and pullout or ratoon lifting, or even movement of the root system by the action of strong winds. In addition, the lodging increases the vegetable and mineral impurities, increases the losses in manual and mechanized cutting, increases scavenging of butt after manual cutting, reducing cutting yield, worsening the quality of raw materials, reduces the charge density and increases the fiber content and losses in extraction.

It is interesting that the cultivars present easy straw removal. This has been especially important, given the need to help the harvester when cleaning the cane, causing minor vegetable impurities and smaller size of straw. If considered the manual harvesting of green cane prior, it is desirable that the cultivars present detrash easy or even natural.

11.3.6 Excessive Non-Flowering

Flowering can cause reduction in quality of raw materials due to the effects of the pith process in the stalks, increasing the proportion of fiber, causing the sprouting of buds of the standing stalks, reducing the juice extracted by the mills, and halting the vegetative development of flowering stalks, which may cause loss in productivity.

In the case of early maturation cultivars, flowering is not a problem because the negative effects mentioned above are not significant when the sugarcane harvest takes place in April and May in south-central Brazil. Moreover, the actual induction of flowering promotes improvement in the maturation of the cultivar.

11.3.7 Tolerance to Major Diseases and Pests

There was never and will never be the perfect cultivar. For the sake of probability, improvement cannot associate in a clone all the desirable characteristics of yield and tolerance to diseases and pests. Another issue is that the technological demands and even the pathogens change over time. Therefore, we have a cycle in which the cultivar has ephemeral life.

It is important to use cultivars that have lower risk as to expected productivity. The ideal is that they present good field tolerance to major diseases and pests.

Some diseases are more important in certain locations or situations of crop management. There are diseases where the primary method of control is the planting of tolerant cultivars, such as rust, coal and scald, while others can be managed with thermal treatment, such as stunting disease.

The breeding program aiming at resistance to pests, has been little studied. The main pests of sugarcane have been fought through biological control with great success, but it would be desirable that cultivars presented tolerance to these pests, for use in the practice of integrated pest management.

11.4 Management of Cultivars

11.4.1 Definition of Management, Number and Allocation of Cultivars

The management of cultivars can be defined as the process that aims at the use of the cultivar to maximize agro-industrial productivity.

In an extensive monoculture, as is the case of sugarcane, there are risks of decreased productivity due to biotic and abiotic factors. These conditions of environmental stress and biological interactions with cultivars cause decline in productivity of sugarcane. Man is responsible and should interfere in the process, in order to mitigate the risks of declining productivity of cultivars. Cultivars, in turn, will always be replaced by others with better productivity that meet the new demands imposed by changes in production technologies, or even epidemics of new diseases introduced from other countries.

In this sense, producers use some empirical recommendations, such as not exceeding the limit of 15% of the total area cultivated with only a single cultivar. In general, currently, it is recommended to diversify the number of cultivars used in the company. It is suggested to employ 10 cultivars with low degrees of kinship between them and which present lower risk due to their reaction to major diseases of the crop.

Table 11.1 presents a scheme dealing with the major cultivars of RIDESA that have been used in Minas Gerais to produce energy, ethanol and sugar. The allocation of the cultivars at each harvest period takes into account not only maturation, but also all the agricultural traits.

Table 11.1: Suggested management of RB cultivars in Minas Gerais.

Beginning of the Season			Middle of the Season			End of the season	
April	May	June	July	August	September	October	November
RB855156	RB835054 RB966928 RB855453	RB835054 RB966928 RB855453 RB937570	RB835054 RB966928 RB965902 RB855453 RB937570 RB855536 RB867515 RB92579	RB966928 RB965902 RB937570 RB855536 RB867515 RB92579	RB855536 RB867515 RB928064	RB867515 RB928064	RB928064

11.4.2 Grading of Cultivars According to Maturation

In sugarcane field planning, besides the production of stalks, one should take into account the maturation of the cultivars. As well as the productivity of stalks, maturation is also influenced by soil and climate conditions. In general, sugarcane requires 6—8 months with high temperatures, intense sunlight and regular rainfall, in order to have a perfect growing season, followed by 4—6 months in dry season and/or low temperature (these latter conditions are unfavorable to growth but beneficial to the accumulation of sucrose).

Maturation is the physiological process of transport and storage of sucrose in parenchyma cells of the stalks. The sugar concentration is higher from the base of the stalks to the top and from outside the stalks to the inside. The cultivars can be classified according to maturation as follows (Figure 11.4):

- Early — present a superior content of sucrose than other cultivars at the beginning of the harvest (April and May). Considering only the maturation, usually have long Working Time of Industrialization (WTI).
- Average — present a superior content of sucrose than other cultivars in the middle of the harvest (June, July, August), having average WTI.
- Late — present high content of sucrose in the mid to end of harvest, having short WTI (70—120 days).

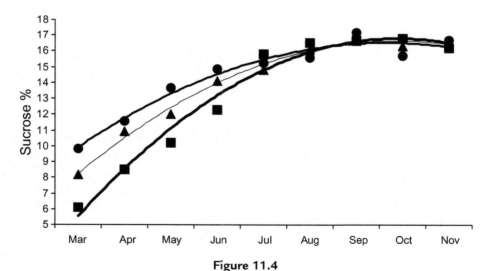

Figure 11.4

Maturation curves of early maturation cultivars (RB855156), average (SP81-3250), and late (RB72454) in south-central Brazil.

11.4.3 Planning of Planting and Allocation of the Cultivars

As mentioned earlier, one sees the need to know the past behavior of the cultivars for resistance to diseases, maturation and adaptation to handling and the harvest time.

The producer should plan the sugarcane field to offer quality raw material throughout the whole harvest period, in which case the center-south of Brazil covers the period from April to November.

As part of planning and considering the logistics of harvesting, the crop areas are divided into large blocks of harvest, aiming to: (a) improve the efficiency of harvesting and all the ratoon maintenance operations, such as cultivation and fertilization, application of vinasse, herbicides, and maturators; (b) reduce the need of use of board-trucks to make changes to the machinery, and (c) reduce costs due to adjustments of the operations mentioned in (a) and (b).

Thus, the fronts of the harvest will be directed to the large blocks. Usually the crop is operated with at least three cut fronts simultaneously and daily harvesting of sugarcane from the beginning to the end of season.

In general, we try to adjust the balance of early, middle and late maturation cultivars in all blocks of the harvest. We suggest the following proportions: 30% early, 50% average and 20% late. In other words, we seek to respect this proportion in each block of harvest.

An important block of harvest is represented by vinasse areas. So at the beginning of harvest, it is necessary to harvest and immediately release the areas for receiving the application of vinasse; this requirement will last until the end of the season. Therefore, in this block the early, middle, and late maturation cultivars should be allocated. On this basis, it follows a script to set the cultivars that will be allocated in large blocks of harvest, respecting their proportion within the blocks and in all areas. The first step would be the allocation of the secondary nursery. near the reformed areas. That is, in year 1 — planting nursery; in year 2 — reform the area after harvest, while developing the plants in an area close to the area of reform; and in year 3 — planting in the reformed area. This planning should be performed 2 years prior to the reform of the area. So, gradually, the blocks will have cultivars targeted for favorable harvest periods, in order to maximize agro-industrial productivity.

Besides the appropriate choice of cultivars, it is important to observe certain agricultural practices that could improve productivity, such as: soil fertilization based on its chemical analysis; use of limestone; organic fertilization; agricultural use of vinasse; planting period, cultural treaties; and creation of greenhouses with thermally treated seedlings, among others.

11.4.4 Validation of the Performance of Clones and New Cultivars

This point is crucial and explains why some companies exploit the potential of new cultivars and clones first. Therefore, it is important not to neglect the creation of nursery with superior genotypes, since the planting of cultivars and new clones is planned based on the availability of cuttings.

Superior clones are indicated by the results of experiments conducted in the lands of the mills and distilleries. The validation of new clones and cultivars occurs in ranges with 10–50 ha for at least three harvests in different soil types at different harvest times and with or without irrigation (with or without vinasse). Later, the growers could increase the new clone one for comercial production.

11.5 Final Considerations

The cultivars of sugarcane are to be the main factor for increasing productivity. It was reported that there was a more than 30% increase in sugar yield per hectare since the beginning of the program to encourage the use of ethanol in Brazil. Through the use of early maturation cultivars, the mills and distilleries could extend the harvest season and improve the quality of raw material. Breeding also contributes to the sustainability of the production system, since it completely excludes the use of fungicides in the crops, because the cultivars are tolerant to the major fungal diseases.

The main breeding programs of sugarcane in Brazil have as a strong point a partnership with the companies. The mills and distilleries have participated in the development of cultivars since the early stages of the program. Thus, the adoption and management of new cultivars happens seamlessly, in view of the fact that the producer has participated actively in the development of the technology.

Bibliography

Barbosa, M.H.P., Resende, M.D.V., Dias, L.A.D., Barbosa, G.V.S., Oliveira, R.A., Peternelli, L.A., et al., 2012. Genetic improvement of sugar cane for bioenergy: the Brazilian experience in network research with RIDESA. Crop Breed. Appl. Biotechnol. S2, 87–98.

Barbosa, M.H.P., Resende, M.D.V., Peternelli, L.A., Bressiani, J.A., Silveira, L.C.I., Silva, F.L., et al., 2004. Use of REML/BLUP for the selection of sugarcane families specialized in biomass production. Crop Breed. Appl. Biotechnol. 4, 218–226.

Barbosa, M.H.P., Resende, M.D.V., Bressiani, J.A., Silveira, L.C.I., Peternelli, L.A., 2005. Selection of sugarcane families and parents by REML/BLUP. Crop Breed. Appl. Biotechnol. 5, 443–450.

Bastos, I.T., Barbosa, M.H.P., Cruz, C.D., Burnquist, W.L., Bressiani, J.A., Silva, F.L., 2003. Análise dialélica em clones de cana-de-açúcar. Bragantia. 62, 199–206.

Bastos, I.T., Barbosa, M.H.P., Resende, M.D.V., Peternelli, L.A., Silveira, L.C.I., Donda, L.R., et al., 2007. Avaliação da interação genótipo x ambiente em cana-de-açúcar via modelos mistos. Pesqui. Agropecuária Trop. 37, 195–203.

Bressiani, J.A., 2001. Seleção seqüencial em cana-de-açúcar. Tese (Doutorado), Escola Superior de Agricultura "Luiz de Queiroz". USP, Piracicaba, 134p.

Cox, M.C., Hogarth, D.M., Smith, G., 2000. Cane breeding and improvement. In: Hogarth, D.M., Allsopp, P.G. (Eds.). Manual of cane growing. Bureau of sugar experiment stations, 436p.

Daniels, J., Roach, B.T., 1987. Taxonomy and evolution. In: Heinz, D.J. (Ed.), Sugarcane Improvement Through Breeding. Elsevier, Amsterdam, pp. 7–84.

Ferreira, A., Barbosa, M.H.P., Cruz, C.D., Hoffmann, H.P., Vieira, M.A.F., Bassinello, A.I., et al., 2005. Repetibilidade e número de colheitas para seleção de clones de cana-de-açúcar. Pesqui. Agropecuária Bras. 41, 761–767.

Heinz, D.J., Tew, T.L., 1987. Hybridization procedures. In: Heinz, D.J. (Ed.), Sugarcane Improvement Through Breeding. Elsevier, Amsterdam, pp. 313–342.

Hogarth, D.M., Wu, K.K., Heinz, D.J., 1981. Estimating genetic variance in sugarcane using a factorial cross design. Crop Sci. 21, 21–25.

Kimbeng, C.A., Cox, M., 2003. Early generation selection of sugarcane families and clones in Australia: a review. J. Am. Soc. Sugarcane Technol. 23, 20–39.

Levi, C.A., 1983. Floracion de caña de azúcar. Determinación de Requerimientos Inductivos. Revista Industrial y Agrícola de Tucumán. 60, 1–15.

Levi, C.A., 1992. La floracion de la caña de azúcar en tucumán. Revista Industrial y Agrícola de Tucumán. 69, 175–178.

Machado Jr., G.R., Walker, D.I., Bressiani, J.A., Silva, J.A.G., 1995. Utilização de água quente para a emasculação de flechas de cana-de-açúcar (*Saccharum* spp.). STAB. 13, 28–32.

Mcintyre, L., Jackson, P., 1996. Does selfing occur in sugarcane? In: Plant Genome IV Conference. San Diego, California. p. 165. (abstracts).

Oiticica, J., 1977. Sugarcane Experiment Station. Northeast Brazil. Sugar J. 39, 16–17.

Resende, M.D.V., Barbosa, M.H.P., 2005a. Melhoramento genético de plantas de propagação assexuada. Embrapa Floresta, Colombo, 130p.

Resende, M.D.V., Barbosa, M.H.P., 2005b. Selection via simulated individual blup based on family genotypic effects in sugarcane. Pesqui. Agropecuária Bras. 41, 421–429.

CHAPTER 12

Molecular Biology and Biotechnology

Aluízio Borém[1], Jorge A. Doe[2] and Valdir Diola[3,†]

[1]Universidade Federal de Viçosa, Viçosa, MG, Brazil [2]Texas A&M University, College Station, TX, USA [3]Department of Plant Physiology, Universidade Federal de Viçosa, Brazil

Introduction

Sugarcane is a perennial crop whose C4 metabolism makes it one of the most efficient species in carbon conversion. It is one of the most productive amongst all cultivated crops as well as the main feedstock for sugar production, accounting for nearly two-thirds of the world's production. The crop, which has gained attention recently due to its potential for biofuel production, has also engaged the interest of plant breeders and biotechnology companies, whose purpose is to obtain varieties that are more productive as well as more efficient and resistant to different types of stress. However, sugarcane has an extremely complex genome, which for a long time and for different reasons has limited the progress of breeding.

It belongs to the Sugarcane *Saccharum* L. genus, part of the Poaceae (grasses) family and the Andropogoneae tribe, one of the single polyploid species. Sugarcane, *Saccharum* spp. has been improved by many centuries of breeding. It derives from a series of inter-specific crosses between *S. officinarum* and *S. spontaneum*, which are polyploid species. The first has 2n = 80 and the latter two cytotypes, 2n = 64 or 2n = 112 (Bremer, 1961, Ewing et al., 1998). As a result, sugarcane varieties can be highly polyploid and aneuploid, ranging from 80 to 130 chromosomes. Most chromosomes derive from *S. officinarum* and 10−25% are inherited from *S. spontaneum* (D'Hont et al., 1996). The size of the non-replicated genome of the sugarcane somatic cell (2C) is estimated at 7440 Mb. The size of the complete non-redundant *S. officinarum* chromosome is approximately 930 Mb (D'Hont and Glaszmann, 2001). This value is comparable to that of sorghum (~760 Mb) and twice that of rice (~430 Mb). The level of heterozygosity of all sugarcane varieties is high (Lu et al., 1994) and the basic number of chromosomes for *S. officinaraum* is 10 ×. These variables make studying the molecular genetic of the species more difficult and complex.

---

† In memoriam

Sugarcane. DOI: http://dx.doi.org/10.1016/B978-0-12-802239-9.00012-8

With the advent of molecular genomics, the sugarcane genome has become less mysterious, although its complexity has been confirmed in many respects. Shortcuts have been identified for genomic analyzes thanks to the syntenic conservation between very close parental grass crops, particularly sorghum and rice. Over time, new tools have become more accessible, allowing researchers to understand the molecular basis related to productivity. Thus, interest in the study of these bases as well as in their application in the genetics and physiology of the crop has been renewed, as the goal of such research is to improve the cost—benefit ratio during sugarcane cultivation and processing.

The objective of this chapter is to present a general discussion of molecular bases and biotechnological processes as well as their application and prospective advances in the sugar-alcohol sector. The topics are not addressed in detail, as the theme is very broad, the generation of new information highly dynamic, and the amount of information exponential.

12.1 Molecular Bases of Biotechnology

The basis for the improvement of plants and the transmission of hereditary traits fall on the genetic material, DNA. This stable molecule is formed by a double helix in opposite directions of the sequence of nucleotides adenine, cytosine, thymine, and guanine. Practically more than 98% of the genome sequences do not encode expressed genes. A gene is a DNA fragment that contains specific sites: the promoter, the sequence that encodes a protein (the gene itself), and the terminator sequence (Figure 12.1). By specific environmental stimuli, the specific gene promoter is activated and genetic material transcription enzymes synthesize a RNA strand complementary to DNA. This transcript undergoes the process of intronic removal, forming the mRNA which is then transported outside the nucleus cell and translated into protein by ribosome in the endoplasmic reticulum. Each triplet of nucleotides encodes for an amino acid. An extended sequence of amino acids forms a protein, which undergoes various degrees of development. A more compact form of the protein is usually the inactive state. Environmental stimuli alter the protein conformations, making them active or, in exceptional cases, inactive. This process promotes a noticeable change in phenotype, evidenced, for example, by the morphological variation within species or between different species.

The morphological aspects are controlled by different points of gene regulation (pre- and post-transcriptional), called epigenetic changes, or by the combination of phenomena and the interaction of several proteins involved in the process (post-translational regulation). The expression of a trait is dependent on the intensity of expression of many genes, usually specific to environmental stimuli.

In possession of knowledge accumulated over time, from Mendel (discovery of Mendelian segregation of characters) to Watson and Crick (DNA structure) up to current times when

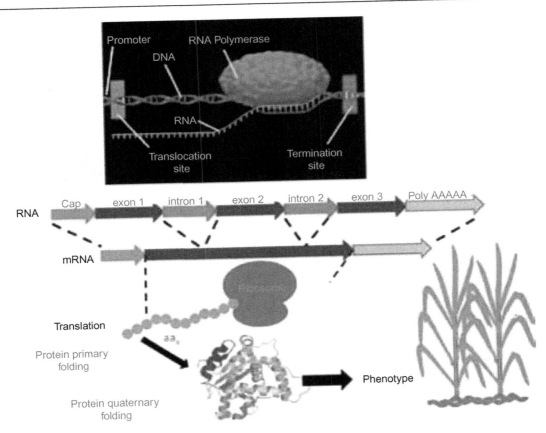

Figure 12.1

Aspects of DNA transcription, forming a single strand of RNA and in the mRNA synthesis, following towards the translation of nucleotides into amino acids and the formation of the protein with the primary configuration (linear) and quaternary conformation (maximum development).

the role of hereditary material is being studied, little is known about the interaction of genetic material and evolution of the species.

As the genetic material of different species is encoded by four nucleotides, in general, the shape of the arrangement of these nucleotides differs little or a lot according to the degree of morphological similarity of the species. Since sugarcane has its closest relatives in grasses, the gene synteny, i.e., the homology among the correlated genes of these species is high. Conventional breeding enables the transfer of genes by sexual processes within species, which limits most inter-specific crosses, due to genetic barriers. When this mismatch is implemented, there is the creation of an inter-specific hybrid, or an individual usually sterile, which in most cases in undesirable. The transmission of traits of interest among different incompatible species can be overcome by the use of biotechnology tools, assisting in the transfer of genes by different gene pools, or through direct transfer by recombinant DNA technique.

Genetic engineering is a branch of biotechnology, which is detailed below. Modern biotechnology includes — in its definition — from the manipulation of organisms to overcome natural barriers, be it cellular or molecular in nature, to more radical procedures such as genetic recombination between similar species or distinct and different techniques that use DNA to generate recombinant products or services. The recombinant DNA was first discovered by the natural presence of circular segments of genomic DNA, called plasmids, which colonized its transient host in a single copy, usually a specific bacterium called *Agrobacterium tumefaciens*. These segments are circular, independent of bacterial chromosomal material and contain specific genes, which in most cases carry transposable elements capable of inserting them into the secondary genomes of host organisms, usually a plant.

Understanding this phenomenon has raised enormous interest because if the exogenous genetic material could be inserted into the genome of a plant and express foreign genes, the same process could be used to introduce genes of interest.

Thus, biotechnology is an important tool for obtaining new varieties, because it can contribute to conventional breeding, making it more efficient, fast and accurate. Among the various procedures using biotechnological potential in the biofuels industry, we can highlight tissue culture, genetic engineering, genomics, transcriptome, proteomics and metabolomics, among other omics.

12.2 Tissue Culture

The totipotency of plants was recognized from the observations of regenerating tissues of damaged tissues of plants. The recognition of this biological principle is credited to the German plant physiologist Haberlandt, who in 1902 announced that each plant cell had the genetic potential to reproduce a complete organism. This physiologist predicted that tissues, cells, and organs could be maintained indefinitely in culture. Somehow, the concept of totipotentiality was already inherent in the cell theory of Schleiden and Schwan (1838 *apud* Vasil et al., 1979) when they postulated that some cells were capable of being separated from the body and continue to grow independently. Tissue culture can be defined as the development of tissues and or cells in *in vitro* systems, i.e., separated from the body's source material, in culture medium containing carbohydrates, vitamins, hormones, minerals, and other nutrients essential to the growth of cultured tissues. It can be classified as follows.

12.2.1 Disorganized Tissue Culture

- Corns — cultivation and maintenance of disorganized cell masses that originate from the uncontrolled proliferation of tissue or organs cultured *in vitro*.
- Suspensions — proliferation of isolated cells or small clusters dispersed in a liquid medium, under agitation.

12.2.2 Culture of Organized Structures

Organ culture — organized forms of growth can be maintained continuously *in vitro*. Includes aseptic isolation of structures defined as primordia and leaf segments. For the purposes of micropropagation, the most important types are:

- Meristem culture — meristems consisting of free apical dome or containing one or two leaf primordia give origin to unipolar stem axes.
- Shoot tips cultures — initiated from shoot tips or lateral buds. Used to establish cultures that originate multigems, gems, axes, or multiple budding.
- Node culture — contains stem segments that carry single or multiple buds.
- Embryo culture (or redemption) — zygotic embryos are excised and cultured to give rise to seedlings.
- Culture of isolated roots — roots grown without connection to the stem axis, producing new roots.

Another technique that can assist in improving sugarcane is the production of cell suspensions from haploid cells of pollen or other tissues. This area needs to be further developed in order to have applicability. Haploid cells or tissues with half the genetic makeup may undergo the process of chromosome duplication, generating the double haploid equivalent to isogenic lines obtained in conventional breeding after five or more generations of selfing. This technique enables substantially reducing the time required for developing a new variety. A successful example was published in an editorial in Current Contents in 2002 on the recovery of the juvenility of sugarcane. Another technique that has enabled significant advances in obtaining genetic variants in sugarcane is the fusion of protoplasts. It is the fusion of two cells in order to obtain a unilateral exchange (hybrid) or reciprocal (cybrid) extranuclear genetic material within the species *S. officinarum* or even within the genus *Saccharum*, overcoming incompatibility barriers. Protoplasts were obtained successfully by Falco et al. (1996), paving the way for this possibility.

The major contribution of tissue culture is undoubtedly its use in the regeneration of whole individuals from genetically transformed cells (Figure 12.2).

12.3 Genomics, Transcriptomics, Proteomics, and Metabolomics

In the biological expression of morphological features, it becomes extremely difficult to separate biological processes, as all are interconnected, starting with the DNA sequence to the protein or specific metabolic. It is worth mentioning that the study tools contribute immeasurably to the genetic improvement of plants, besides facilitating the study and application of biotechnology processes.

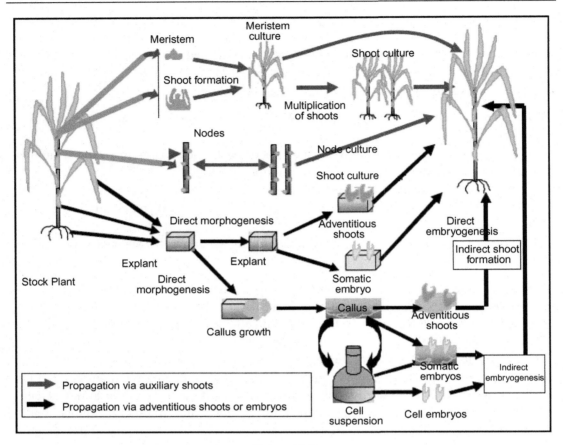

Figure 12.2
Main micropropagation methods and plant growth routes of explants.
Source: Adapted from George (1996).

Genomics is based on modern technologies for obtaining, analyzing, and interpreting large amounts of data from DNA sequencing. It provides phylogenetic studies, gene synteny, molecular markers, physical mapping, annotation of sequences, etc. The transcriptome, as suggested, studies the transcription process by means of techniques that analyze the expression of genes at any given time by a specific stimulus. The main techniques for gene expression analysis are based on the extraction of total RNA and mRNA isolation, followed by techniques that use hybridization or amplification of the molecule for quantitative analysis of the transcript. Proteomics, in turn, uses the protein extract as a sample, determining which proteins are present and the intensity of their synthesis with a specific environmental stimulus. The crude protein extract is usually separated by gel electrophoresis of first or second dimension, retrieving specific spots and amino acid sequencing to identify the primary structure of proteins. Metabolomics, in most cases, studies the final stage of this chain, i.e., the production of metabolics. Some specific examples of sugarcane which have aroused most interest are composed of cell wall synthesis, antimicrobial compounds, precursors of sucrose, among others. Metabolomics has greatly benefited the study of plant—microorganism interaction and assisted in the elucidation of several compounds that are synthesized from this interaction. Some classic examples include the sugarcane—diazotrophs interaction, which helps in increasing plant uptake of soil nitrogen by raising the level of metabolic disorders, such as hormones, an example of auxin.

All this genetic knowledge generated must be stored in databases, some of public domain and free access, others restricted and of private domain. These areas of bioinformatics become the most important tool for modern researchers. Currently, the largest public database freely accessible on sugarcane is the NCBI (*National Center for Biotechnology Information*), which includes 275,000 nucleotides sequences. Of these, 2829 are from genomic sequences, 26,2357 of EST (*Expressed Sequence tag*) and 9515 GSS (*Genome Survey Sequence*), in addition to 6545 proteins. The volume of information stored in that database increases nearly two-fold every 18 months.

Among the express sequences (EST — *Express Sequence Tag*) stored in the databases, special mention should be made of those related to the expression of genes associated with the presence of the *Leifsonia xyli* subsp. *Xyli* str. CTCB07 (4060) bacterium, the yellow leaf virus (447), the mosaic virus (314), and the *Leifsonia xyli* subsp. *Xyli* (4332) bacterium.

The main programs in genomic sequencing and annotation of sugarcane genes are:

- Sugarcane BACs: Plant Genome Laboratory, State University of Campinas. Construction of a library of bacterial artificial chromosome for genomic studies.
- Research programs in Mauritius: Sugar Industry Research Institute (Msiri).
- Project for the genetic mapping of sugarcane — United States: The Plant Genome Data and Information Center (PGDIC), in conjunction with the USDA, is developing the genomic map of this species.
- SUCEST — Sugarcane EST Genome Project: it is part of ONSA, a network of research laboratories in the State of São Paulo. The project identified 50,000 sugarcane genes.

In Brazil, the EST sequencing project, the SUCEST (Sugarcane EST), was coordinated by the University of Campinas (UNICAMP), with the support of the CTC, then Copersucar Technology Center. More than 260,000 complementary DNA (cDNA) clones from 26 cDNA libraries of plant material originated from different tissues were partially sequenced. A comprehensive analysis of the set of SUCEST data indicated that 14,409 sequences (33% of total) contain at least one full-length cDNA. The annotation of 43,141 sequences associated almost 50% of putatively identified sugarcane genes with protein metabolism, signal transduction and cell communication, bioenergetics, and responses to stresses.

The comparative mapping within the *Andropogoneae* tribe (Figure 12.3), which sugarcane belongs to, has advanced recently with the development of various molecular markers for maize, which also apply to sorghum and sugarcane (Grivet et al., 1994). These authors used RFLP markers from maize and the technique of gene synteny to locate the loci of association between species, based on the genetic map of maize. Syntenic regions were identified in three species, establishing a pattern of pairing between the genomes. In many cases, the two arms of a single maize chromosome correspond to at least two syntenic clusters of sugarcane. A more detailed analysis of two syntenic groups showed that sugarcane has the lowest rate of recombination among the three species. All three genomes showed a linear relationship, and distances between genes were similar to those observed in maize and sorghum.

Among the omics, genomics is the one that has generated more results applicable to sugarcane. Some of the most important include RFLPs, RAPDs, AFLPs markers,

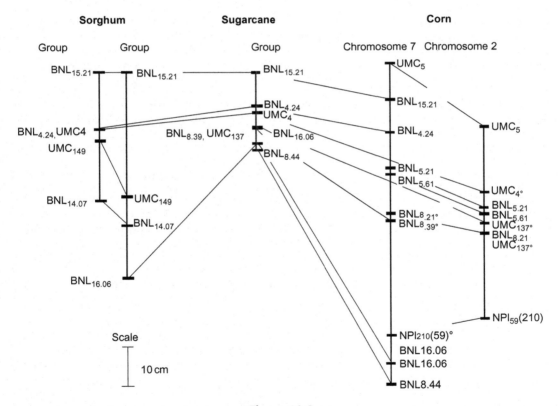

Figure 12.3

Genetic map of *loci* comparison and ordering and recombination rate among sugarcane, maize, and sorghum synteny. Two linkage groups are found for the chromosome segments of maize and sugarcane, gathered in only one sorghum chromosome. *Source: Grivet et al. (1994).*

microsatellites (SSR), and markers derived from SCARs, CAPs, etc. and, more recently, one of the most promising, SNPs. The discovery that simple sequence repeats — SSR — also occur in ESTs brought the opportunity to produce microsatellite markers in a simple and direct way, i.e., through the electronic search of ESTs databases. Scott et al. (2000) used this new methodology and generated 10 SSR markers derived from an ESTs database in grape containing 5000 ESTs. The database generated by the SUCEST project, funded by the Foundation for Research Support of São Paulo (FAPESP), is a source of microsatellite markers that enables the identification of this type of marker for the genetic improvement of sugarcane. Da Silva (2001) identified 20 SSR markers in the database of SUCEST and applied these markers in commercial varieties and wild species of *Saccharum*. Techniques based on the identification of genomic mutations, such as TILLING and ECOTTILLING, have also been applied to sugarcane.

Transcriptome techniques, however, are more promising because they generate a larger set of functional genetic information. In literature, there are many techniques used to study the gene expression of sugarcane, including the synthesis of cDNA from the mRNA tape. Some of them are: Northern blot, cDNA-AFLP, real-time PCR, subtractive hybridization, full-length libraries, differential display, and differential expression. The most powerful techniques applied to the study of expression are the microarray and oligoarray, as they generate large amounts of information in a short time and at a relatively low cost.

Proteomics techniques have made it possible to identify several proteins as well as to study the protein–protein, protein–DNA, protein–RNA, and protein–metabolic interaction processes. These studies of post-transcriptional and/or post-translational regulation have aroused great interest and become immediately applicable tools.

Metabolomics offered a new perspective for compounds previously considered as waste, with potential for improvement and use by biotechnology as is the case of dextran, xanthan gum, sorbitol, glycerol, beeswax refined pie, antifungals, etc. Soon, genetically modified plants will be like biofactories.

Biotechnological and molecular development promoted the rapid development of the sugar and alcohol sector. This section highlights the products obtained by rapid hydrolysis of sugarcane bagasse for animal feed, paper production, pharmaceuticals and products with applications in the chemical and pharmaceutical industry such as furfural, which is used in the synthesis of organic compounds.

Researchers from the IRD and INRA developed a natural method to convert crushed sugarcane into raw material for paper through the action of an enzyme produced by the *Pycnoporus cinnabarinus* fungus. This microorganism uses ethanol as substrate and produces a considerable amount of laccase, an enzyme delignification, and prepares the

Table 12.1: Parameters of yield and unwanted byproducts in broth fermentation by different strains of yeast.

Parameters*	Yeast			
	Pe-2	Vr-1	Cat-1	Pan**
Yield (%)	91.0	90.5	91.2	88.1
Glycerol (%)	3.38	3.20	3.54	4.70
Trehalose (%)	9.5	10.6	10.3	6.0
Final feasibility (%)	94	95	97	61

*Average of six fermentation cycles.
**Baker's yeast.

Table 12.2: Parameters of evolution in broth fermentation promoted by molecular biology in the selection of species and strains of genetic-elite standard.

Parameters	1977	2005
Fermentation yield	75–80%	90–92%
Distillation yield	95	>99%
Level of bacteria in wine	10^8 to 10^9/mL	10^5–10^6/mL
Recirculation of yeast	~70%	>90%
Fermentation time	18–22 h	6–10 h
Yeast content in wine	4.7–6%	8–17%

substrate for paper production without damaging the environment, unlike the methods traditionally used by the pulp and paper industry.

Also noteworthy is the fact that besides the cultivation of sugarcane, molecular biology also focuses on the selection of yeast for fermentation processes and on alcohol production. The main advances in this area are illustrated in Tables 12.1 and 12.2.

12.4 Genetic Engineering and Genetically Modified Varieties

GM plants are those whose genome has been modified by the introduction of exogenous DNA through different biotechnology-generated methods. This exogenous DNA can be derived from other individuals of the same species or a completely different species and may even be artificial, that is, synthesized in laboratory. The term genetically modified organism (GMO) is also often used to indicate, in general, any individual that has been genetically engineered using recombinant DNA techniques.

The most widely used techniques for obtaining plant transformants are the microparticle bombardment method and agroinfection by *Agrobacterium tumefaciens* (Figure 12.4). The agroinfection method faced serious implementation issues since, naturally, the bacterium did not colonize the culture. In 1998, Enriquez-Obregon et al. succeeded in using the

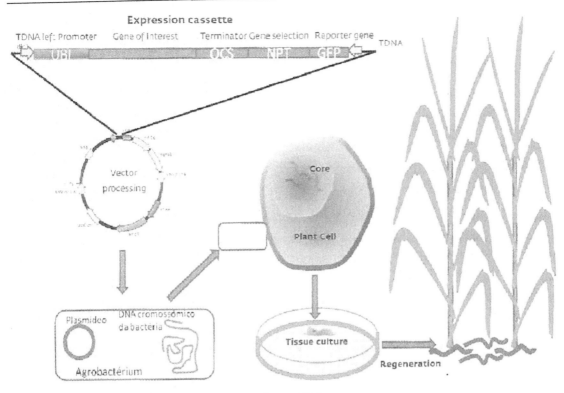

Figure 12.4

Construction of an expression cassette of an *Agrobacterium* transformation plasmid, followed by agroinfection and gene transfer to a plant cell, which undergoes differentiation in tissue culture, resulting in regenerated plants.

method. Currently, most researchers and laboratories use agroinfection and the method improved by Arencibia et al. (1998).

As in other horticultural species, the objective of the first changes in sugarcane was the inclusion of genes that confer antibiotic resistance and herbicide tolerance (Bower and Birch, 1992; Gallo-Meagher and Irvine, 1996). Among herbicide-tolerant genes embedded in sugarcane, special mention should be made of the glyphosate-tolerant gene.

12.4.1 Roundup Ready Varieties

The first transgenic herbicide-tolerant varieties were developed by Monsanto. These varieties are known as Roundup Ready, because they are tolerant to the herbicide glyphosate, whose trade name is Roundup. The herbicide glyphosate blocks the synthesis of aromatic amino acids in plants by binding to and inactivating the EPSPS enzyme, which is essential in the biochemical pathway of these amino acids synthesis.

The effect of glyphosate on susceptible plants is owed to the competition with the phosphoenolpyruvate (PEP) substrate by the binding site to the 5-enolpyruvyl-shikimate-3-phosphate synthase (EPSPS) enzyme, which converts SPP and shikimate-3-phosphate into 5 enolpiruvilchicamato-3-phosphate. This competition ends up by inhibiting the synthesis of the amino acids tryptophan, tyrosine, and phenylalanine (Siehl, 1997). Tolerance in GM plants is due to the introduction of the *cp4 epsps* gene, originally present in the CP4 strain of *Agrobacterium* spp. This gene encodes the CP4 EPSPS enzyme synthesis, whose action is not blocked by glyphosate. When expressed at normal levels, the CP4 EPSPS is able to provide the plant with 3-phosphoshikimate 1-carboxyvinyltransferase or also 5-enolpyruvylshikimate-3-phosphate synthase (EPSP synthase) in an amount sufficient to suppress the inhibition of synthesis of aromatic amino acids (Singh and Shaner, 1998). RR varieties are available in soybean, corn, cotton, and canola.

Whereas herbicide tolerant varieties reduce the amount of chemicals applied in agriculture, they should also be preferred by final consumers and those concerned with environmental conservation.

Varieties tolerant to other herbicides have been developed by other companies, such as varieties tolerant to sulfonylurea (DuPont), bromoxil (Calgene), glufosinate ammonium (Bayer CropScience), ALS inhibitors (Dow AgroSciences), and ALS inhibitors (Embrapa).

12.4.2 Insect-Resistant Varieties

One of the most efficient biological control agents is the bacterium *Bacillus thuringiensis* (Bt). This bacterium, which is native to the soil, has been used to control crop-defoliating caterpillars since 1980, when it was discovered that it produces a protein (delta-endotoxin) toxic to Lepidoptera, i.e., the caterpillars of butterflies. When the caterpillars ingest the spores of this bacterium, these germinate in the digestive tract, producing a delta-endotoxin crystal, which in basic medium binds to receptors of cell membranes of the digestive tract, changing its osmotic regulation and ultimately killing the insect. The digestive tract of most mammals, including humans, has an acidic pH, which rapidly degrades Bt protein, if ingested, and does not have the receptors delta-endotoxin would bind to. Thus, the delta-endotoxin is harmless to people and other vertebrates. The insecticide Bt has been considered one of the safest pesticides to humans and the environment, so that its application is permitted in organic cultivation.

The first sugarcane plants resistant to insects were obtained by inserting the gene truncated CryIA (A) b *Bacillus thuringiensis*, but with low protein expression and partial larvicidal response (Arencibia et al., 1997). Better results were obtained by Braga et al. (2001, 2003), who used a new construction of the expression cassette and obtained several GM crops resistant to sugarcane borer in two varieties under field conditions. The gene constructs used in other species such as maize and cotton were also tested in sugarcane. One example is the cassette which expresses the proteinase inhibitor II of potato and the cassette gene from *Galanthus nivalis* lectin, produced by Nutt et al. (1999), Nutt (2005) and Chen et al. (2004), which confers resistance to the larvae of the root borers in transgenic sugarcane. Transgenic sugarcane plants were also obtained by inserting genes that expressed soybean trypsin inhibitors (Falco and Silva, 2003).

One concern with Bt varieties is the development of resistance by insects and the consequent loss of efficiency of Bt protein in its control. Since the plant is continuously producing the Bt protein in its tissue, the chances to develop insect resistance may be higher. To circumvent this potential problem, producers who adopt the technology are required to implement a resistance management program (RMP), which establishes the planting of escape tracks for insects to reproduce in conventional plants that do not produce Bt. Additionally, producers are encouraged or sometimes asked to implement the rotation of GM crops with conventional ones.

Some of the genes that confer resistance to caterpillars and borers and were introduced in different agronomic species are: 1Ac Cry, Cry 1F, 3A Cry, Cry 3Bb1, as well as VIP genes, which are also cloned from *B. thuringiensis.* Currently, there are many investigations underway attempting to assess the efficiency of different Cry genes for various sugarcane pests.

12.4.3 Disease-Resistant Varieties

Since diseases are caused by various microorganisms, the development of resistance to each of them depends on recognition of specific genes. For fungal diseases, the development of transgenes is still limited by the lack of identification of *R*s genes, which recognize the elicitors of pathogen infection. For viral diseases, several transgenic events are described as resistant to the sugarcane mosaic virus (SCMV) (Joyce et al. 1998; Ingelbrecht et al., 1999), the Fiji disease virus (FDV) (McQualter et al., 2001) and the sugarcane yellowing virus (ScYLV) (Rangel et al., 2003). Most strategies use genic silencing techniques to obtain transgenic plants resistant to viruses. The techniques widely used with promising results were antisense and RNA of interference (RNAi), which enable interfering in the genic expression even in the case of polyploidy sugarcane.

12.4.4 Non-Flowering Varieties

Flowering in sugarcane is undesirable because it causes a significant proportion of photoassimilates to be directed to the production of flowers and seeds rather than to the accumulation of carbohydrates. Figueiredo (2003) and Ulian (2006), using an antisense

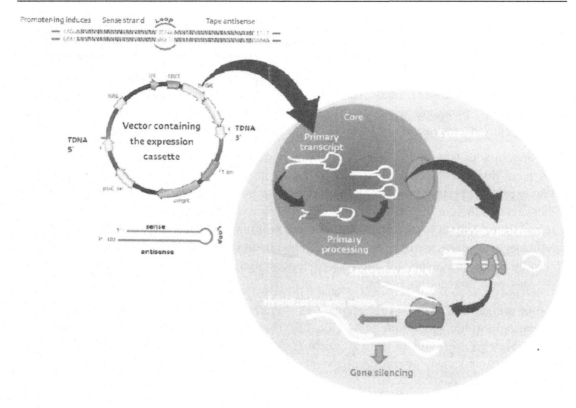

Figure 12.5
Genetic transformation of plants using the gene silencing technique, currently one of the greatest prospects for reducing the expression of undesirable traits. The insertion of the gene sequence itself in the antisense direction, separated by some random bases followed by the sense direction of the gene promotes annealing of the transcript and processing by enzymes such as Droscha, RISC, and Dicer, generating a fragment complementary to the mRNA, which, when subject to hybridization inhibits the translation of this protein.

construction of a gene in the MADbox group, which controls flowering, obtained GM plants incapable of flowering, which is desirable for sugarcane producers (Figure 12.5).

12.4.5 Varieties with High Sucrose Content

An example of how to obtain plants with desired traits besides those previously mentioned was reported by Wu and Birch (2007). They obtained the heterologous expression of the sucrose isomerase gene from a bacterium, and this gene was directed to the vacuole of transgenic sugarcane plants. The transgenic plants were able to store twice the amount of total sugars in mature stalks, due to the accumulation of isomaltulose, an isomer of sucrose.

Suppression and overexpression of genes involved in carbohydrate metabolism is the subject of many studies that seek to increase sucrose and other metabolites in sugarcane.

The large amount of carbon fixed in sugarcane has made it ideal for the production of new products such as biopolymers, proteins with specific properties and potential pharmaceutical applications, among others. The production of phydroxibenzoic, a biopolymer in sugarcane plants reached 7.3% and 1.5% of dry weight in leaf and stem tissues, respectively, as reported by McQualter et al. (2005). This metabolic is converted into glucose conjugates by uridyne diphosphate (UDP) glucosyltransferase and stored in the vacuole. Proteins with pharmaceutical properties were obtained by Wang et al. (2005), which produced the stimulating factor of granulocyte and macrophage colonies (GM-CSF) in sugarcane. Unfortunately, in the field the plants could accumulate only 0.02% of the total soluble protein as GM-CSF.

12.4.6 Drought-Resistant Varieties

The genetic transformation of sugarcane for obtaining varieties tolerant to different stresses, especially water, has always been the goal of many breeding programs. However, the identification of genes that regulate this response at the appropriate level is still incipient, since this characteristic is quantitative and governed by many genes. The use of techniques of differential gene expression enabled Rocchi et al. (2006) to identify genes responsive to water deficit. The authors identified some genes responsive to abscisic acid, which regulate the closure of stomata and a transcription factor homologous to rice DREB2. Overexpression of this gene family in other species such as *Arabidopsis thaliana* produced drought-resistant plants. In 2006, Zhang et al. obtained sugarcane plants with normal growth overexpressing the trehalose synthase gene of the edible maitake mushroom (*Grifola frondosa*). The transformants that accumulated trehalose were responsive to drought stress in field conditions. In 2007, Molinari et al., using a stress-induced promoter, obtained transformants expressing specifically the *Vigna aconitifolia* P5CS gene, which encodes the 1-pyrroline-5-carboxylate synthetase, an enzyme responsible for proline biosynthesis. The increase in proline production by transgenic plants was observed only under stress conditions, conferring them drought tolerance.

Table 12.3 illustrates the different genes being studied in sugarcane for the development of GM varieties.

The ease with which many sugarcane varieties can be processed and the identification of the sequences of thousands of genes that this plant expresses opens up new possibilities for obtaining promising transgenic plants (Ulian, 2008). Despite evidence of the great potential of sugarcane genetic engineering, there has been no commercial release of transgenic sugarcane to date.

Table 12.3: Different genes being studied in GM sugarcane.

Identification	Name	Phenotype
cp4 epsps	5-enolpyruvylshikimate-3-phosphate synthase	Glyphosate-tolerant
Pat	Phosphinothricin N-acetyltransferase	Glufosinate ammonium-tolerant
bxn	Nitrilase	Bromoxynil-tolerant
cry1Ab	Delta endotoxin cry1Ab	Insect-resistant
cry1Ac	Delta endotoxin cry1Ac	Insect-resistant
vip3A	Delta endotoxin vip3A	Insect-resistant
Lec 2	Lecithin 2	Insect-resistant
Neo	Neomycin phosphotransferase	Antibiotic-resistant
DREB2	Dehydration-responsive element binding	Drought-resistant
P5CS	"1-pyrroline-5-carboxilate synthetase	Drought-resistant
TPS	Trehalose phosphate synthase	Drought-resistant
Si	Sucrose isomerase	High sucrose content
SCLFY	LEAFY/ Floriculata -MADbox	Flowering inhibition
GM-CSF	Granulocyte and macrophage colonies stimulating factor	Total soluble protein accumulation
SCMV	Sugarcane mosaic virus	Virus-resistant
FDV	Fiji disease virus	Virus-resistant
SCYLV	Sugarcane yellow leaf virus	Virus-resistant

Analysis of the risks of GM crops to human health, animal health and the environment in Brazil are the responsibility of the National Technical Biosafety Committee (CTNBio), a body created in 1995 within the structure of the Ministry of Science and Technology. The function of this committee is to advise the federal government on the country's biosecurity policies. Formed by 54 members from the scientific community and government, CTNBio acts appropriately and independently, *inter alia*, in the assessment of research and marketing requests involving GMOs.

Bibliography

Arencibia, A., Vazquez, R.I., Prieto, D., Tellez, P., Carmona, E.R., Coego, A., et al., 1997. Transgenic sugarcane plants resistant to stem borer attack. Mol. Breed. 3 (4), 247−255.

Arencibia, A.D., Carmona, E.R., Téllez, P., Chan, M.-T., Yu, S.M., Trujillo, L.E., et al., 1998. An efficient protocol for sugarcane (Saccharum spp. L.) transformation mediated by Agrobacterium tumefaciens. Transgenic. Res. 7, 213−222.

Bower, R., Birch, R.G., 1992. Transgenic sugarcane plants via microprojectile bombardment. Plant J. 2 (3), 409−416.

Braga, D.P.V., Arrigoni, E.D.B., Burnquist, W.L., Silva Filho, M.C., Ulian, E.C., 2001. A new approach for control of *Diatraea saccharalis* (Lepidoptera: Crambidae) through the expression of an insecticidal Cryla (b) protein in transgenic sugarcane. International Society of Sugar Cane Technologists. Proceedings of the XXIV Congress, Brisbane, Australia, 17−21 September 2001. vol. 2, pp. 331−336.

Braga, D.P.V., Arrigoni, E.D.B., Silva-Filho, M.C., Ulian, E.C., 2003. Expression of the Cry1Ab protein in genetically modified sugarcane for the control of *Diatraea saccharalis* (Lepidoptera: Crambidae). J. New Seeds. 5, 209−222.

Bremer, G., 1961. Problems in breeding and cytology of sugar cane. Euphytica. 10, 59−78.

Chen, P.H., Lin, M.J., Xue, Z.P., Chen, R.K.A., 2004. Study on genetic transformation of GNA gene in sugarcane. Acta Agric. Univ. Jiangxiensis. 26 (5), 740–743, 748.

Da Silva, J.A.G., 2001. Preliminary analysis of microsatellite markers derived from sugarcane expressed sequence tags (ESTs). Genet. Mol. Biol. 24 (1–4), 155–159.

D'Hont, A., Glaszmann, J.C., 2001. Sugarcane genome analysis with molecular markers, a first decade of research. Proc. Int. Soc. Sugarcane Technol. 24, 556–559.

D'Hont, A., Grivet, L., Feldmann, P., Rao, S., Berding, N., Glaszmann, J.C., 1996. Characterization of the double genome structure of modern sugarcane cultivars (*Saccharum* spp.) by molecular cytogenetics. Mol. Gen. Genet. 250, 405–413.

Ewing, B., Hillier, L., Wendl, M.C., Green, P., 1998. Base-calling of automated sequencer traces using phred. I. Accuracy assessment. Genome Res. 8, 175–185.

Falco, M.C., Silva-Filho, M.C., 2003. Expression of soybean proteinase inhibitors in transgenic sugarcane plants: effects on natural defense against *Diatrae saccharalis*. Plant Physiol. Biochem. 41, 761–766.

Falco, M.C., Neto, A.T., Mendes, B.J., Arias, F.J.Z., 1996. Isolation and cultivation of sugarcane protoplasts. R. Bras. Fisiol. Veg. 8 (3), 175–179.

Figueiredo, L.H.M., 2003. Caracterização do gene LEAFY de *Saccharum* ssp. e análise filogenética entre diferentes espécies vegetais utilizando as sequências da família LEAFY/Floricaula. Master's Dissertation, University of São Paulo, Ribeirão Preto, Brazil. 137p.

Gallo-Meagher, M., Irvine, J.E., 1996. Herbicide resistant transgenic sugarcane plants containing the bar gene. Crop Sci. 36 (5), 1367–1374.

George, E.F., 1996. Plant propagation and micropropagation: plant propagation by tissue culture. Exegetics, Edington. II, pp. 37–66.

Grivet, L., D'hont, A., Dufour, P., Hamon, P., Roques, D., Glaszmann, J.C., 1994. Comparative genome mapping of sugar cane with other species within the Andropogoneae tribe. Heredity 73, 500–508.

Ingelbrecht, I.L., Irvine, J.E., Mirkov, T.E., 1999. Posttranscriptional gene silencing in transgenic sugarcane. Dissection of homology-dependent virus resistance in a monocot that has a complex polyploid genome. Plant Physiol. 119 (4), 1187–1198.

Joyce, P.A., McQualter, R.B., Bernad, M.J., Smith, G.R., 1998. Engineering for Resistance to SCMV in Sugarcane Acta Hort. 461, 385–391.

Lu, Y., D'Hont, A., Paulet, F., Grivet, L., Arnaud, M., Glaszmann, J.C., 1994. Molecular diversity and genome structure in modern sugarcane varieties. Euphytica 78, 217–226.

McQualter, R.B., Harding, R.M., Dale, J.L., Smith, G.R., 2001. Virus derived transgenes confer resistance to Fiji disease in transgenic sugarcane plants. International Society of Sugar Cane Technologists. Proceedings of the XXIV Congress, Brisbane, Australia, 17–21 September. vol. 2, pp. 584–585.

McQualter, R.B., Chong, F.B., Meyer, K., Dyk, D.E., van O'Shea, M.G., Walton, N.J., et al., 2005. Initial evaluation of sugarcane as a production platform for p-hydroxybenzoic acid. Plant Biotechnol. J. 3 (1), 29–41.

Molinari, H.B.C., Marur, C.J., Daros, E., Campos, M.K.F., Carvalho, J.F.R.P., Bespalhok Filho, J.C., et al., 2007. Evaluation of the stress-inducible production of proline in transgenic sugarcane (*Saccharum* spp.): osmotic adjustment, chlorophyll fluorescence and oxidative stress. Physiol. Plant. 130 (2), 218.

Nutt, K.A., 2005. Characterization of Proteinase Inhibitors from Canegrubs for Possible Application to Genetically Engineer Pest-Derived Resistance into Sugarcane. PhD: Queensland University of Technology, Brisbane, Australia.

Nutt, K.A., Allsopp, P.G., McGhie, T.K., Shepherd, K.M., Joyce, P.A., Taylor, G.O., et al. 1999. Transgenic sugarcane with increased resistance to canegrubs Proceedings of the 1999 Conference of the Australian Society of Sugar Cane Technologists, Townsville, Queensland, Australia, 27–30 April 1999. pp. 171–176.

Rangel, P., Gomez, L., Victoria, J.I., Angel, F., 2003. Transgenic plants of CC 84-75 resistant to the virus associated with the sugarcane yellow leaf syndrome Proc. Int. Soc. Sugar Cane Technol., Mol. Biol. Workshop, Montpellier, p. 4 30.

Rocchi, P., Jugpal, P., So, A., Sinneman, S., Ettinger, S., Fazli, L., et al., 2006. Small interference RNA targeting heat-shock protein 27 inhibits the grows of prostatic cell lines and indices apoptosis via caspase-3 activation in vitro. BJU Int. 98, 1082–1089.

Siehl, D.L., 1997. Inhibitors of EPSP synthase, glutamine synthetase and histidine synthesis. In: Roe, R.M. (Ed.), Herbicide Toxicity: Toxicology, Biochemistry and Molecular Biology. IOS Press, Amsterdam, Netherlands, pp. 37–67.

Singh, B.K., Shaner, D.L., 1998. Rapid determination of glyphosate injury to plants and identification of glyphosateresistant plants. Weed Technol. 12 (3), 527–530.

Ulian, E.C., 2006. Genetic manipulation of sugarcane. In: Mauritius van der Merwe, M.J., Groenewald, J.H., Botha, F.C. 2003. International Society of Sugarcane Technologists 5th ISSCT Molecular Biology Workshop, MSIRI, Reduit. Isolation and evaluation of a developmentally regulated sugarcane promoter Proc. South African Sugar Cane Technol. 77. 146–169.

Ulian, E.C., 2008. *Desenvolvimento de variedades geneticamente modificadas visando tolerância ao estresse hídrico.* Workshop on Sugarcane Genetic Improvement and Biotechnology. Session 8. Project on Public Policy Research Program. 8p.

Vasil, I.K., Vasil, V., White, D.W.R., Berg, H.R., 1979. In: Scott, T.K. (Ed.), Plant Regulation and World Agriculture. Plenum Pub. Corp., New York, pp. 63–84.

Wang, M.C., Bohmann, D., Jasper, H., 2005. JNK extends life span and limits growth by antagonizing cellular and organism-wide responses to insulin signaling. Cell. 121 (1), 115–125 (Export to RIS).

Wu, L., Birch, R.G., 2007. Doubled sugar content in sugarcane plants modified to produce a sucrose isomer. Plant. Biotechnol. J. 5, 109–117, The first engineered plant species accumulating high concentration isomaltulose without interfering growth and no decrease in sucrose storage by introducing sucrose isomerase gene with vacuolar compartmentation in sugarcane.

Zhang, S.-Z., Yang, B.-P., Feng, C.-L., Chen, R.-K., Luo, J.-P., Cai, W.-W., et al., 2006. Expression of the *Grifola frondosa* Trehalose Synthase Gene and Improvement of Drought-Tolerance in Sugarcane (*Saccharum officinarum* L.). J. Integr. Plant Biology. 48 (4), 453–459.

CHAPTER 13

Quality Control in the Sugar and Ethanol Industries

Celso Caldas[1] and Fernando Santos[2]
[1]*Central Analítica LTDA, Maceio, AL, Brazil* [2]*Federal University of Viçosa, Vicosa, MG, Brazil; University Federal do Rio Grande do Sul, Rio Grande do Sul, Brazil*

In recent years, the ethanol sector has become prominent in the international arena due to worldwide concerns about energy, environment, and social security. In this context, Brazil, the world leader in the production of sugarcane, has gained international prominence in the media, encouraging new foreign capital investments and leading farmers to increasingly expand their production.

In the face of this optimistic scenario, the number of industries that produce sugar and ethanol in Brazil grows every year. According to the Ministry of Agriculture, Livestock and Food Supply, Brazilian industries have at least three characteristics that differentiate them from the industries of other countries. The first feature presented in this context is the large scale of its production of sugarcane. This is possible due to the enormous territorial size of the country, as well as to the edaphic and climatic conditions of Brazil and to the crop management processes already implemented. The second characteristic is related to the commercial products that are manufactured, among them sugar and ethanol, as well as its byproducts, such as *cachaça* and Panela, and the co-generation of electricity by some industries, especially in the State of São Paulo.

Regarding the differences between Brazilian and foreign processing plants, one factor that stands out here is the geographic distribution of units within the country, where the diversity of rainfall regimes allows for the installation of plants in a broad territorial space and for the year-round production of sugarcane.

Any processing plant has in its structure a specific sector for quality control, usually tied to a lab, whose main purposes are:

- quality control of raw materials;
- monitoring of production processes, especially to evaluate and quantify industrial losses; and
- quality control of final products.

With regard to sugar and ethanol processing, those are precisely the roles of the quality control sector. To monitor such quality in manufacturing, it is imperative to count on good lab physical structures, accurate and reliable equipment, competent and well-trained technical staff and an efficient analysis plan.

Once interaction between these basic requirements mentioned above is ensured, it is also essential to use appropriate methodologies which are simple, fast, and accurate, and also to take advantage of all the relevant control management tools available to provide their technicians with reliable figures attesting to the quality of raw materials and their products and with numbers indicative of the performance of plants, always measured in terms of yield and efficiency.

Good quality control begins with the definition of the parameters to be used for the assessment of sugarcane. The analyses established for the remuneration systems for sugarcane by the Consecanas (Councils of Sugarcane, Sugar and Alcohol Producers) cannot be understood to be the only items in quality control for those raw materials. Many others must be considered and, where possible, implemented in the plants. However, these are more complex, laborious analyses which require more specific equipment, besides the need for better-qualified staff. Thus, besides the analyses of pol, brix, purity, reducing sugars, and fiber, which are mandatorily performed on the cane received by the processing plant (regardless of whether it is internally produced or acquired from suppliers), other analyses (in which are verified, for instance, maturation stage, level of infestation by spittlebugs or sugarcane borer, mineral and vegetal impurities, and mainly burning period) must become routine operations.

Regarding the quality parameters for the assessment of raw materials in accordance with the recommendations of the Consecanas, we emphasize the sampling systems in the transportation trucks. The use of tilted probes certainly provides a much more representative sample than the samples obtained by horizontal probes.

Conversely, in analytical terms, one can attest, without fear of error, that the determinations of pol and the estimation of reducing sugars by linear regression equations based on the purity of the juice are bound to be abandoned. This is caused by that fact that, due to the increasingly frequent use of the liquid chromatography technique for the analysis of juice, it is now possible to identify more carbohydrates in the sample. This happens because chromatography is a more sensitive technique which suffers less interference in the analyses of pol, thus being more efficient. In this evaluation, one should also mention the Near Infrared Spectroscopy (NIRS) technique. These tools, although faster and aided by sophisticated mathematical quantification tools, such as Fourier's Transform, are no more than comparative equipment, that is, they work from correlations obtained with primary methods such as polarimetry. Therefore, for the NIRS, all interference from primary methods is taken along through the readings required for mathematical and statistical calculations. This technique is more accurate, as are the primary methods used, and, in this case, we once again mention liquid chromatography as the ideal primary method technique for the determination of sugar contents.

In this case, where are the flaws of the methods established by the Consecanas? Mainly in the determination of the percentage of pol in sugarcane, which can be translated as a percentage of apparent sucrose. Note that the word "apparent" already indicates the inaptitude of this technique, due to the presence, in the juice, of other optically active substances besides (obviously) sucrose. The interference of these other compounds becomes elevated when the analyzed sugarcane is green or has had high burning times. In the first case this is due to concentrations of glucose and fructose, which are higher than on mature canes. These compounds, respectively, dextrorotatory and levorotatory, interfere considerably in the polarization of light caused by sucrose, which is also dextrorotatory. In the case of stalks with burn times of up to 36 h, beyond the interference of these sugars, there is an increase in the deviation of polarized light to the right, due to the dextran polymer, formed by the action of the bacterium *Leuconostoc mesenteroides*. In these cases, the sucrose content starts to be overestimated and this undermines the entire assessment of the quality of raw materials, both for the remuneration system and for the assessment performed by the plant, given that it accounts for more sucrose entry into the industry than what actually comes in, leading to higher costs.

Among the control processes which are not established in the official sugarcane remuneration systems mentioned above, we highlight maturation, extraneous matter and burning time indices or stages. As the name indicates, maturation stage is the reduction of vegetative growth and sucrose accumulation in the stalk; that is, more sucrose and less reducing sugars is the main determinant in defining the moment of cutting sugar cane. This determination is made by means of the ratio between Brix at the tip and Brix at the base of the stalk, resulting in the so-called maturation index (MI).

$$MI = \frac{Brix_{tip\ of\ the\ stalk}}{Brix_{base\ of\ the\ stalk}}$$

The determination chosen is refractometric Brix, for three reasons: (1) because it can determine Brix contents while the cane is planted, (2) because Brix content indicates the amount of all dissolved solids in the juice and (3) the possibility of analyzing the tip and base of the stalk and know its maturation stage, since sugarcane grows from the bottom up. In this case, when the Brix at the top of the stalk resembles the Brix of its base, one can verify that it is mature.

The extraneous matter which is found in sugarcane and classified as vegetable and mineral constitutes an important parameter of quality of raw materials, mainly due to negative factors that these residues cause in the processing of sugar and ethanol (Figure 13.1).

Figure 13.1
Extraneous matter delivered to the processing plants. (A) mineral impurity, (B) vegetable impurity.

Among the vegetable impurities are those related to the constitution of the stalk itself (leaves and straw), which lead to high levels of phenolic compounds, starch, aconitic acid, minerals and others. As for the mineral impurities (dirt, sand), these cause wear, bushing, clogging, and incrustations, in addition to shortening the lifespan of the equipment. In most processing plants, the control of vegetable and mineral impurities is accomplished with a high degree of reliability, considering that these parameters are used to award bonuses to loader operators and to sugarcane suppliers, as well as to penalize outsourced transportation systems. For the correct analysis of mineral impurities, it is necessary to analyze the organic material of each type of soil in sugarcane farms as well as in the clean sugarcane stalks, commonly called "whites".

In terms of quality of the cane, the burning time is perhaps the most compromising factor of its quality, due to spontaneous sucrose inversion reactions, with the formation of acids and also polymers such as dextran, that occur as a consequence of it. Currently, it is a common procedure to establish a goal of 36 h for burning time, besides other actions such as an increased use of mechanized harvest, which reduces burning time. The delay of harvest after burning and cutting causes the soil micro-organisms found in the stalks to find a natural "culture medium", propitious to their proliferation and to the development of decomposition reactions.

Once the procedures and actions for better knowing and improving the quality of the analyzed sugarcane are defined, attention is then turned to quality control, with a view to supervising the productive processes of sugar and ethanol. In this respect, the laboratory, as executor of the quality programs, has its focus directed to the quantification of losses of sugar and ethanol, without neglecting, obviously, to assist in the work of the processing plant, reporting on the characteristics of materials being processed, as well as of their by-products.

Achieving this goal requires the establishment of a perfect analytical plan to be executed by the laboratory. In many scenarios, a quality control program of a particular industrial unit is deployed at another without any modification. This is not a correct procedure and should be avoided, because one must always remember that the laboratory is a service provider to the production process, so the definition of the analytical plan to be executed will depend on process conditions, which are specific, and therefore the plan will be inherent to each plant.

The definition of the analytical plan starts with the definition of the materials that will be sampled and analyzed. After defining these materials, the plan then is given sequence with the establishment of the sampling frequency, which will depend on how much the quality or composition of the product to be analyzed varies. The more variable the process and the quality of its materials are, the higher the frequency of sampling and its subsequent analysis will be. The establishment of this sampling frequency should happen in such a way that does not overload the laboratory, which could compromise the quality and continuity of analysis. It must, however, meet the need for information that will guide the production processes. Therefore, the first step in defining the frequency of sampling is to know perfectly well what the purpose of the information to be generated by the laboratory will be.

Once the materials to be sampled and analyzed are defined, the next step is to establish the types of samples. Samples can be instantaneous, composite or continuous. Instantaneous samples are defined as those that are collected in the process and immediately analyzed. They are appropriate to stages which vary significantly, or even to equipment that needs to be constantly evaluated, for those can represent points of loss or cases in which the work of the plant needs to be adjusted to these results.

As for composite samples, as their name indicates, they are drawn from samples collected in an instantaneous way or by a continuous process, the latter being called continuous samples, to facilitate the differentiation between them. The greatest care with the composite samples, such as in the case of continuous samples, is with their preservation, since they cannot lose their initial characteristics. These types of samples are widely accepted, especially for materials which do not deteriorate easily.

In many situations, continuous sampling is more appropriate, including raw materials, intermediate and final products. However, statistics themselves indicate that these samples may mask the variability of the measurements, thus hindering corrections in the process and equipment adjustment. Therefore, it is necessary to establish a criterion for defining the products to be sampled by a continuous process.

It is recommended that samples be collected continuously when the materials sampled represent the raw materials, process losses, or final products. Thus, in the manufacturing process of sugar and ethanol, juice samples for sugar and ethanol (from must and molasses) should be collected continuously, since they are raw materials for sugar and ethanol factories, respectively. Another action to be suggested is the continuous collection of samples of shredded cane, as it is in fact the raw material used for the production of sugar and/or ethanol. However, considering that the analyses of this raw material indicates not only the amount of sugarcane, but also the performance of the shredding process, samples must be collected instantaneously and analyzed immediately, for, depending on the results, it is possible to extemporaneously repair the blades and hammers which are used on the preparation process for sugarcane.

The final products, sugar and ethanol, can also be collected continuously. However, for the determination of sugar humidity content and ethanolic levels in the ethanol produced, instantaneous samples can be adopted, because in possession of that data one can evaluate the performance of the sugar dryer and of the ethanol distillation column, respectively.

In relation to materials representing losses in the process, samples of rinsing water, cane bagasse, filter cake, and vinasse should also be collected continuously. One case deserves comment: the filter cake. Although this material actually represents the loss of sugar in the process, a continuous collection and analysis every 4 h, for example, could mask the performance of the filter, thus delaying adjustments. Therefore, we have a material whose instantaneous samples, as well as the continuous ones, are recommended. Instantaneous samples are indicated to investigate the performance of the filters, and the continuous ones to evaluate the losses of the process. These two procedures will guarantee the performance evaluation of the filtration process for sugarcane juice.

Finally, the continuous samples should be collected by devices designed for the reality of each plant, those often being activated by solenoid-type valves, pistons or electrical devices. In many cases, the difficulty in designing and putting this into practice lies in the physical condition of the materials to be sampled. Thus, the adaptation of continuous samplers for a perfect performance will depend on various adjustments, which should be made during the operation of such equipment.

As has already been mentioned, one important aspect of the sampling process (regardless of the type of samples) refers to the conservation procedure for those. The care with instantaneous samples lies in the transportation of material from the collection site to the laboratory. Depending on the layout of the plants, specific containers must be used, avoiding contamination and changes in composition, for example, loss of components by volatilization or increased humidity (by absorption of water from rain or dew. The loss of humidity to the environment can also be observed when air humidity levels are considerably low. Composite samples, either compiled from instantaneous samples or from continuous ones, need greater attention, since their analysis takes longer, and consequently the likelihood of chemical reactions that may alter their compositions is higher. When it comes to diluted liquid samples, such as sugarcane juice and must, mercury-based preservatives (saturated ethanolic solution of mercuric chloride, for example) are very efficient, though dangerous to health, because mercury has a cumulative effect in the body.

The best option in this case is to use an antibiotic that is specific to the ethanol industry, after performing a few tests to determine the optimal concentration of the said antibiotic to be used. Cooling is another conservation procedure often mentioned in the literature and used in plants, but samples should not be frozen as there may be changes in the original composition during thawing.

Solid samples, such as bagasse, filter cake, etc., are difficult to preserve, since you cannot add any product to those, in which case the homogenization of these products in the sample would be relatively difficult. In these cases, it is customary to use a container with a false bottom, where cotton soaked in chloroform or ammonia is placed so that its vapors come into contact with the samples, conserving them. Other methods are mentioned in the literature, but the aforementioned ones guarantee the preservation of samples.

After the definition of all the items discussed above, it is time for the development of the sampling manual, which should be specific to each unit, as they will take into account the specificities of each plant. This manual, an important management tool, must be elaborated by the laboratory's responsible technician. It is also recommended that the manual be mandatorily used by the technical staff of the laboratory and updated every season (process changes and the elimination or addition of new analyses are normal procedures). As a rule, the following items should be part of the manual:

- identification of collection sites;
- description of the collection sites;
- types of samples and sampling frequency;
- procedure for sample collection; and
- types and frequency of analyses.

The collection sites should be clearly defined and established in locations that allow for representative samples. That achieved, these sites should be carefully identified with signs. There is no specific standard for the production of identification signs for the sites, the definition of which being decided on a per company basis. However, the material of this signs should be durable and they should be colored in a way that allows for good visualization of the collection site. As for size, there is no pattern, but a 40 × 30-cm plate can be established. Information on the signs should be as concise and relevant as possible, so that their view is not somehow compromised (Figure 13.2).

The identification and numbering of sampling sites will facilitate both the development of the manual and training at the beginning of the harvest. The numbering should preferably accompany the flowchart of the factory. However, upon the appearance of a new collection site, the numbering should be sequential from the one which already exists, avoiding the destruction of the signs.

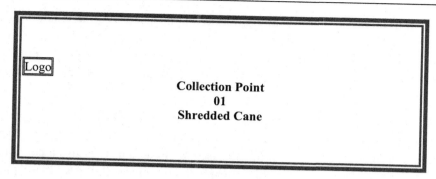

Figure 13.2
Model of the nameplates of the collection sites.

After the definition of the collection sites, they must have an objective description, for analysts to easily locate them throughout the plant.

Building on the example of identifying collection site #01 — shredded cane — the location could be described as follows: in the pile of shredded cane, before the return of the *cush-cush*.

This part of the sampling manual must include, for each material, the type of sample (instantaneous, composite, or continuous) and frequency of sampling, if samples are instantaneous or compound. Regarding the frequency of continuous sampling, it makes sense to mention them, even if they are obtained by samplers driven by pistons or solenoid valves in predetermined intervals. For the sake of illustration, it is worth mentioning that many samples are collected continuously through dripping or fillets falling in containers.

Descriptions of sampling collection procedures are certainly part of the most important item of the sampling manual. These descriptions, besides being clear, must be detailed enough to leave no doubt as to how the analyst should proceed. It is worth mentioning that, for any laboratory, having excellent facilities, being equipped with modern and precise equipment and having an experienced and well-trained technical staff will be to no avail whatsoever if mistakes are made in their sampling procedures. This would result in unreliable results which are inconsistent with the reality of the process and, thus, the laboratory would not be fulfilling its role within the production process.

Still using the example of shredded cane with its collection site defined and identified and types and frequency of sampling established, the procedure can be thus written up: climbing up the bridge on the conveyor of shredded cane and, with the aid of a "fork" type sampler, take samples across depth and width of the cane spread (lateral and middle) for a period of approximately 15 min, collecting the sample of sugarcane in a 15-L bucket or in plastic bags, which should, ultimately, amount to a total of about 3 kg of shredded cane, which will be dedicated to analysis.

It is recommended that after the definition of all these items, which constitute the analytical plan to be developed by the laboratory, a general framework be posted where visible in the lab, so that analysts can have easy access to all information regarding the process of sampling and analysis that constitutes quality control in the plant.

Finally, it is the laboratory's duty to perform the relevant analyses in the final products of the plant. This cannot be considered to be the main task of the laboratory, since all the others are also of immense relevance. However, one must consider that the commercialization of products is the end-goal of any industry, being the factor which ensures its economic viability. In this respect, the quality of final products is a prerequisite for acceptance in competitive markets.

Analyses of these materials must happen during the production processes and also in the storage sites for sugar and at the ethanol tanks. The frequencies of sample collection should be established according to the nature of the products, and may even be different for different types of analyses. For example, when you are making white sugar, determination of color should be more frequent than the analysis of ash. In sugar processing plants, the types of analysis must be established according to sales contracts, whereas for ethanol the determinations performed must be in accordance with the ordinances from the Brazilian National Petroleum Agency (ANP), as to what regards ethanol fuels. When the ethanol is of the export type, as is the case with sugar, its quality must be determined as a function of the quality parameters established in the commercialization contracts.

For any analysis performed in a laboratory, it is necessary to define the analytical method that leads to quantitative or qualitative knowledge of all (or some) of the elements and molecules that constitute the sample. When choosing this analytical method, it is important to highlight the instrumental methods which are currently prominent in analytical chemistry, be it qualitative or quantitative. As a rule, the procedures must be practical and provide reliable results, always with elevated levels of exactness and precision, and without neglecting factors such as time and cost.

Other points must be taken into consideration in the choosing of methods, such as: the required amount of samples, the need for pre-treatment, the need for performing duplicates or triplicates, etc.

In a macro-perspective classification, we can subdivide analytical methods into classic (also called traditional) and instrumental. In any of the cases, the methods are based, for practical purposes, into three fundaments:

- development of a chemical reaction and quantification of the consumption of reagents in the formation of the products;
- electrical measurements; and
- optical measurements.

In some analyses, the methods are based on a combination of optical/electrical measurements and chemical reaction, such as, for instance, in potentiometric titrations.

The traditional methods are gravimetry, titration (by neutralization, redox, precipitation and compleximetry) and volumetry. These methods are based, respectively, in the formation of precipitate of little solubility in a solution, in the quantification of a reagent or product in a reaction and, finally, in the measurement of the volume of a gas. The electrical methods are those that measure electric current, tension, or resistance levels, in relation to the concentration of certain compounds in solution. Optical methods measure the absorption and the emission of radiant energy at given wavelengths.

Regardless of the method being traditional or selective, one factor that must be studied and understood in detail is the interference that certain substances cause in the execution of the said method and that can prevent a direct measure of the concentration of the elements and compounds being analyzed. Considering, thus, the interference in chemical analyses, methods can be classified as specific or selective. Specific methods are classified to be those which can measure concentrations of substances regardless of the presence of any other compounds or elements. These are the ideal methods, but they are not always routinely found in laboratories. Selective methods are those which can be utilized in chemical analyses of certain substances, even with interfering agents.

In the sugar and ethanol producing units, the methods used in the labs are grounded on manuals and compendiums already consolidated in the technical and academic areas, and the choice of those is therefore not a very difficult one. It is worth remembering, however, that this choice is one decision made within a large complex structure of considerations to be made, including, for instance, decisions regarding chemical reagents, analytical equipment, competent and accredited staff, and validation of results.

Bibliography

Caldas, C., 1998. Manual de análises selecionadas para indústrias sucroalcooleiras. Sindicato da Indústria do Açúcar e do Álcool no Estado de Alagoas, Maceió, 422p.

Caldas, C., 2005. Teorias básicas das análises sucroalcooleiras. Central Analítica, Maceió, 172p.

COPERSUCAR, 2001. Centro de Tecnologia de Cana (CTC). Manual de controle químico da fabricação de açúcar, Piracicaba, 261p.

Fernandes, A.C., 2003. Cálculos na agroindústria de cana-de-açúcar. second ed. STAB — Sociedade dos Técnicos Açucareiros e Alcooleiros do Brasil, Piracicaba, 240p.

Leite, R.A., 2000. Compostos fenólicos de colmo, bainha, folha e palmito de cana-de-açúcar. Dissertação (Mestrado), Universidade Estadual de Campinas, Campinas, 135p.

Payne, J.H., 1989. Operações unitárias na produção de açúcar de cana. Nobel, STAB, São Paulo, 245p.

Ripoli, T.C.C., Ripoli, M.L.C., 2004. Biomassa de cana-de-açúcar: colheita, energia e transporte. Piracicaba, 302p.

Spri, Color components in sugar refinery processes, 1985. Tech. Rep. (19), 14.

Stupiello, J.P., 2000. Relação açúcares redutores/cinzas. STAB. 19 (2), 10.

CHAPTER 14

The Sugar Production Process

Claudio Soares Cavalcante[1] and Fernando Medeiros de Albuquerque[2]
[1]CSC Engenharia, Campinas, São Paulo, SP, Brazil [2]F. Medeiros Consultoria, Recife, PE, Brazil

Introduction

Before discussing the process of producing sugar itself, some considerations will be made about the influences that the process suffers when due importance is not given to the quality of raw material, the processing of the juice, and the design of the plant and its equipment. These factors may have little or much influence, affecting the execution of the process, since these are steps in a process in which sucrose is not manufactured, but recovered.

Describing these difficulties is to lay down all the reasons that lead to greater or lesser efficiency in the recovery of sucrose. It is important to name them now, because it would not be possible to understand later how all this can lead to lack of control and losses to such an important sector in sugar production, whatever the type of sugar to be produced. The factors cited, although not part of the pan boiling process itself, define its degree of efficiency and quality, for they shape the operating conditions the pan boiling process will be subject to.

14.1 Quality of the Raw Material

As previously mentioned, there are many difficulties in conducting the pan boiling process. One such difficulty is due to the quality of raw material.

Factors such as: diversification of varieties, types of varieties (early, middle, late), maturation stage, cultivation practices, climatic conditions, longevity of the sugarcane crop, soil type, quality and quantity of fertilizer, quantity and quality of the vinasse applied, among other agronomic factors not mentioned here, determine the quality of the raw material that will be processed. Furthermore, the varieties received at the factory for processing are never the same, they are never in their entirety those which should have been harvested and never have the same burn, cutting and transport time to arrive at the factory. There are a number of factors that often hamper all agricultural operations and it is not possible to perform a perfect job that brings the best raw material from the sugarcane field to the industry. It is clear that today, with mechanized harvesting of raw sugarcane, many of these problems no longer exist, but others occur with greater interference in the process conditions. The quality of the pan boiling process and performance of the boilers depend greatly on the quality of sugarcane delivered to the factory and this, among all factors, is what contributes most to a better or worse performance of the boiling phase.

Raw materials that may suffer some form of sucrose inversion losses, of course, will negatively affect the performance of the boiling pan and, consequently, the boiling process as a whole, due to the presence of other non-crystallizable sugars.

14.2 Treatment of the Sugarcane Juice

It is a phase that defines the maximum recovery of sucrose and sound quality of the final product, therefore all operations of this stage of the process should be carried out with extreme correction, from the beginning to the end.

The treatment starts with the discharge equipment, transportation, and preparation of the cane, without forgetting the extraction of juice, when the highest levels of contamination take place. Such equipment must always be kept clean and well taken care of, since all the raw material goes through them and the raw material receives various types of preparation and exposure of sucrose. In order to keep the sucrose free from microorganisms, continuous and effective cleaning of the machinery must be carried out, to avoid the contact of the molecules of sucrose with the microorganisms that deplete it, transforming it into other non-crystallizable sugars.

The juice, before being preheated, undergoing sulfitation, having its pH adjusted, heated, and decanted, goes through various equipment which must be maintained with the greatest precision cleaning and care possible, not to increase the proliferation and growth of micro-organisms forming polysaccharides that undermine the whole process of clarification, boiling, and consequently, the final sugar. The care to prevent the proliferation of bacterial contamination is of fundamental importance for conservation and recovery of the sucrose available in the raw material.

The chemical treatment before decanting is intended to cause a precipitation reaction, forming several products which, in most part, should be eliminated from the process, rendering the sucrose free for its crystallization reaction.

The juice treatment procedures consist of submitting it to a series of steps that involve physical (straining, heating, flash step) and chemical transformations (reactions promoted by the addition of chemicals), aiming at maximum elimination of non-sugars, colloids, turbidity and color, favoring greatest sedimentation.

Of all these steps, the heating has a very important function, so the heating pans must be purchased and evaluated according to their inner surface. Its design must be made according to the speed of the juice inside it (between 1.4 and 2.0 m/s) for ensuring good internal thermal exchange and to prevent the formation of deposits. It is also important to remember that high speed, while diminishing the deposits, also results in great load loss. Non-condensable gases affect the heat transfer, because they hinder the condensation of steam, lowering its temperature and decreasing heat transfer capacity.

The objectives of heating the juice are: to eliminate microorganisms by sterilization, to complete chemical reactions with the alkaline agent, to coagulate it, to flocculate the insoluble impurities, and to remove the gases efficiently, which is obtained by the flash step, before it enters the clarification chamber. If the flash step does not occur, the gas bubbles that are attached to the flakes slow down sedimentation.

The clarification is the stage where all undesirable impurities are removed for the proper functioning of the later stages. In order to achieve the best results, it is essential that all previous steps are rightfully carried out and the flows are constant and stable. Thus, the precipitation and removal of all that is not sucrose — and is prejudicial to the heating process and the final product — are more efficient. At this stage, we can not forget the filtering, an important process for recovering the sucrose also contained in the mud, which is removed as an impurity in decanting.

This process is no longer hampered by the discontinuity of the processes and lack of stability due to the lack of available technology for automation equipment. Today's technology enables constant flow rates and temperatures, and stable pH, giving the treatment gains in efficiency and quality.

Even in the refining process, the quality of the juice treatment is of great importance, sugars from deficient juice treatment are raw materials of poor quality for the refineries, resulting in low efficiency, high costs in treatment liquor, and reduction in the efficiency of the equipment, in addition to losses of production and productivity.

14.3 Project Design and Equipments

Speaking about design and equipment is very important because new plants will not save room and height. A few years ago, projects began to incorporate other concepts. In the sugarcane reception sector, equipment and spaces intended for weighing, sampling, unloading, feeding and preparation of cane have been redesigned with different sizes and heights, improving and facilitating cleaning and reducing losses. In the extraction sector, the height of the ground floors became much higher, facilitating the procedures for cleaning and disinfecting, and the monitoring of losses.

It is very common in new industrial plants that the height of the mills and the distances between the factory floors are higher, enabling anyone to walk underneath the mills. A large space can also be observed between one mill and another, facilitating cleaning and allowing better maintenance, which brings major advantages, even to prevent leaks, facilitate transshipment, etc., which are major factors in the recovery of sucrose by the factory.

In both sugar and ethanol plants the old design concepts have been replaced by these new ones. The area for juice treatment, for example, which is common to both, has been modernized. This area is designed so that operations occur on the same floor, avoiding many stairs, landings, and especially many workers. The new plants break old paradigms, because there are already projects in which fermentation occurs in the same area of juice treatment, allowing a number of operational advantages, because all equipment is on the same floor, and also allowing workers to multitask.

In sugar refineries, the projects may even follow the same floor height used in the area where juice treatment and the boilers are located.

This height is essential for the functionality of the heating area. The distance between the boiling pans and the crystallizers, the crystallizers and the centrifugals, and the centrifugals and the ground floor are extremely important details. They may seem minor, but they influence the whole operation, mainly the monitoring and control of losses. The new standards for the heating area of a sugar factory present several advantages. One of them is that the heat transfer from one evaporator to another becomes easier and faster, once the evaporation time for the boiling pans is reduced, avoiding waste of time between one heating and another. It also facilitates the visualization and control of leaks, overflows, valve maneuvers, the cleaning of the pipes, etc. In crystallization, a very important advantage can be cited: the process of heat transfer and centrifuging, because it is possible to lower the massecuite from the last crystallizer and centrifuge on the first, improving the yield of crystals in the massecuite due to the fact that they obey a proper resting time, increasing the size of crystals and depleting the liquor of sucrose also contained in honey. Centrifuging also presents advantages as, for example, avoiding sanitary conveyors which are close to the ground for the transport of sugar, which causes losses that are impossible to calculate, and the contamination of the final product. All these details are important. The manufacture, specially that of sugar, must be treated in the highest standard of the organized food industry, which works with raw perishable, subject to the proliferation of microorganisms (bacteria) harmful to the whole process and the health of consumers.

14.3.1 Evaporation

Evaporation is very important in the pan boiling process, for it is the evaporation which defines a good part of the operations, depending on the concentration of the juice when processed into syrup. In the evaporation area, the first step in the process of recovery of sugar from the juice takes place. The usual practice is to concentrate the clarified juice to about 65% brix, which requires the removal of approximately 75% of water in the juice. The need for economy of steam requires the adoption of the principle of multiple effects, i.e., an adequate facility must have the capacity to evaporate water and provide steam for heating the juice and the operation of the boiling pans. The section also provides the evaporation of condensed water to feed the boilers. The set of evaporators consist of an equipment which, when well-designed and operated, contribute indisputably to the proper heat balance of the factory, increasing or reducing the efficiency of various other equipment, including the boilers. Good concentration syrup is a product of extreme importance: when its concentration is low, there are many negative effects, causing increased consumption of steam and heating time, and loss of the capacity of the equipment.

Well designed and operated evaporators must devote a lot of attention to the withdrawal of non-condensable gases such as O_2, N_2, CO_2, SO_2, CH_3, CH_2, OH, NH_4OH, ammonia organic acids (in aqueous environments, ammonia is in the form of NH_2OH). The removal of these gases is made through plugs placed on the outside of the rollers, in the position farthest away from the entrance of steam.

The high temperature in the evaporator causes thermal decomposition of amides and sucrose reducing sugar bicarbonates, and as a result, CO_2 and ammonia are produced, mainly. Moreover, the pressure drops associated with the flow of juice contribute to the release of gases dissolved in the juice. The presence of these gases, known as non-condensable, in the vapor used for heating should be considered as an important control factor. The insufficient withdrawal of non-condensable gases results in a significant decrease of the temperature difference and a less intense heat transfer, which means less evaporation of the juice, lower temperature of the juice (for heating pans) and heating time increased, in the boiling pans.

14.3.2 Heating–Crystallization

The separation of sucrose from the impurities associated to the sucrose solutions is the ultimate objective of sugar production. This objective is achieved through the crystallization of sucrose, which is subsequent to the separation of crystals by means of centrifugal force, once other impurities would have been removed during clarification. The presence of non-sugars in the honey exerts adverse influence on the crystallization process. The separation of the non-sucrose sugars, through crystallization, requires several repeated crystallizations, as many as are economically viable.

14.3.3 Supersaturation

In order to enable the formation and growth of sugar crystals, the essential condition is supersaturation. Generally, the crystallization of low-purity materials requires a higher supersaturation than for sugar solutions of high purity. It is important to review the relationship of solubility, saturation, supersaturation, but also the effects of non-sugars in these relations.

14.3.4 Solubility

The solubility of sugar in the water varies with the temperature of the solution, which is expressed by the amount of pure sugar dissolved by weight. The solubility of sugar in the water also changes according to the purity of the solution, which in its turn, depends on the nature of the non-sugars.

14.3.5 Saturation

When a solution contains the total amount of sugar which it is able to dissolve, it is defined as saturated.

14.3.6 Supersaturation Index

If there is no crystallization in the saturated solution, it is necessary that a state of supersaturation be created. No sugar may be dissolved in a saturated solution. In this case, it was established that, by eliminating water by evaporation, a more concentrated solution would be reached, without exceeding the saturation level, before starting the crystallization process. In any solution, after evaporation of part of its water, it is clear that it will contain higher amount of sugar per unit than it previously did. Consequently, the amount of water removed by evaporation of a saturated solution provides greater or lesser degree of supersaturation. Clearly the same result can be obtained by cooling the solution.

Supersaturation index is the ratio between the amount of dissolved solids per unit of water in the supersaturated solution and that contained in the saturated solution of the same purity and temperature.

Example: Let us consider a saturated solution of sugar at 80°C. Using the saturation curve, it is possible to verify that this saturated solution contains 363 parts of sugar for each 100 parts of water. If this solution is concentrated at a temperature of 80°C until the contents reach 436 parts of sugar for each 100 parts of water, the supersaturation achieved is as follows:

$$\text{Supersaturation Index: } 436/363 = 1.20$$

Methods for measuring saturation

The degree of supersaturation can be measured only by indirect means. The most common methods are:

- Brix refractometer
- Heightened boiling point
- Electric conductivity
- Consistency

14.3.7 Crystallization

Crystallization is considered not only the initial phenomenon of the formation of sugar grains, but also the set of operations which are carried out to make the grain into the desirable commercial size. However, it is necessary to consider all the operation in the vacuum pans, in the several ways of producing the sugar grains, of different systems of massecuite, etc. The vacuum pans are the heart of the sugar manufacturing plants. The recovery of sugar by the factory, the quality of the sugar to be produced, the sugar plant's ability to take in more or less milling, the thermal balance, all depend on the handling of the vacuum pans.

A well-managed factory must handle the operation of the vacuum pans aiming at producing high-quality massecuite.

Crystallization is the initial formation of the grain and it must be conducted correctly, for it is the basis of a good process. It is known that the sugar solution, when subject to water evaporation, reaches a suitable saturation point, producing sugar grains. Thus, the process of crystallization becomes easy and simple, with all the physical phenomena that occur during the pan boiling process.

The systems of crystallization can be summarized in three distinctive methods:

- Waiting—spontaneous crystallization;
- Shock—induced crystallization;
- Seeding—implanted crystallization (complete).

Waiting method

The waiting method is the oldest method in the sugar industry and it is no longer used. It is inefficient and presents numerous weaknesses:

- difficulty in controlling the number of grains formed;
- due to the extremely high concentration, the formation of clusters is inevitable; and
- this method is recommended for crystallizing with high purity.

Shock induced crystallization

Shock induced crystallization consists of an operation very similar to the one mentioned above, except for the fact that it produces the nucleation of the grains when the solution reaches a level of concentration, through the injection of crystals which crystallize in the same system. This method has the advantage of allowing the control of the speed of the nucleation and of enabling crystallization to occur at the stable target zone.

Complete seeding method

The complete seeding method is the best method for the establishment of the grain, in this case, it does not form on the boiling pan, once the supersaturation is never allowed to exceed the stable target working zone. The first requisite is an automatized vacuum pan with trustworthy instruments, which allow for an accurate reading of the correct supersaturation at any time.

Efficiency of crystallization

Efficiency of crystallization is the percentage of sucrose converted into crystals.

Factor which influence crystallization:

- impurities present in the honey or syrup (percentage and type of non-sugar)
- volume of the nuclei;
- granulometry of the sugar;
- automated control of the graining and the boiling processes;
- breaking of crystals in the continuous centrifuges.

Crystallization rates

Crystallization rate is the speed with which sucrose in the mother liquor is transferred to the crystals. This rate depends crucially on the degree of supersaturation and purity of the mother liquor, and the viscosity of the massecuite, since the temperature has great influence on the viscosity.

14.3.8 Optimal Concentration of the Grains

In order to ensure the maximum depletion of honeys in minimum time and to simultaneously achieve the maximum purity in low-purity sugar, experience suggests that the two primary conditions regarding the quantity of crystals should be harmonized within the practical limits imposed by the equipment installed. These conditions are specified below:

(a) maximum area of the crystal per volume unit of massecuite; and
(b) maximum porosity of the layers of crystals in centrifuges.

These properties are of opposing influence and should be carefully used in relation to local conditions.

Maximum surface of crystals

Two factors contribute to this goal:

Maximum concentration of crystals per unit of final massecuite.

The surface of the crystals is directly proportional to their concentration or to their actual volume whatever their size. To obtain the maximum area of the crystals, their concentration must be the highest possible one. But this concentration is limited by the physical need to maintain the consistency of massecuite workable.

Final size of the crystals

The size of the crystals has considerable influence in shaping the surface of absorption per unit of massecuite, whatever the concentration of the crystal may be. The experience of many years of commercial practice has allowed us to determine the normal size of the final crystals of the low-purity massecuite within the range of 0.25−0.35 mm.

Maximum porosity of the crystal layers in the centrifugal

Two factors contribute to the porosity of the crystals:

Magnitude of the crystal

This specification is undesirable for the purpose of depletion of a massecuite with a maximum area of crystals. Since the separation of crystals from the mother liquor is needed, what happens is that the flow of honey finds resistance in the crystal layer in the centrifugal. Therefore, the bigger the crystal, the higher the porosity of the layer, and thus the size of the crystal is an important factor.

The speed and degree of flow of the mother liquor (honey) layer of the crystal in centrifugals of low purity are determined by the following factors: the centrifugal force, the mother liquor viscosity, sugar wall thickness and pore size.

The viscosity of the mother liquor is determined by the following factors: controllable temperature, purity, density and the uncontrollable factor of the material used.

Uniformity of crystal size

In crystals of different sizes, the porosity of the wall is reduced, which slows the flow of honey and results in a deficient centrifugal process.

By controlling the vacuum pans and crystallizers, we can achieve the uniformity of the crystals and hence perform a better centrifugal process. As for the quantity of seed to be used in massecuite, it is advisable to consider the influence of the factors listed above and the local conditions.

14.3.9 Uniformity of the Crystals

A uniform grain means better sucrose absorption power, sugar free of impurities and of better quality.

It is almost impossible that the sugar crystals have uniform size, not only because the seed particles originally ground are not uniform, but also due to the fact that the crystals grow at different rates. What you can do is keep the amount of these crystals to a minimum. Good circulation is one of the factors of greatest importance to reduce the amount of undesirable crystals (Figures 14.1—14.3).

14.3.10 Irregular Crystals

It is good to note that the presence of irregular crystals is harmful to the production of a good-quality sugar.

Figure 14.1
Good-quality commercial sugar.

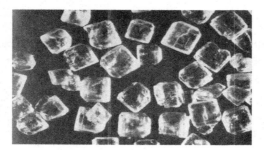

Figure 14.2
Excellent-quality commercial sugar.

The irregular crystals are influenced by controlled and uncontrolled factors.

Controlled irregular crystals are those whose formation can be avoided by good operation of the vacuum pans. Non-controlled irregular crystals are those from various types of deterioration of sugarcane.

Clusters

The clusters are formed more easily in the high-purity massecuites; in low-purity ones, they never form. The clusters which are formed can never be destroyed. They are formed in the upper layer of the supersaturation zone (stable target), just before the formation of fake grain. Before the formation of fake grain, cluster may be formed (Figure 14.4).

14.3.11 Twin Crystals

Some impurities are what define the characteristics of a crystal. In the cane sugar plant, crystallization happens at high concentrations with a high content of reducing

Figure 14.3
Excellent-quality commercial sugar.

Figure 14.4
Commercial sugar with high cluster content.

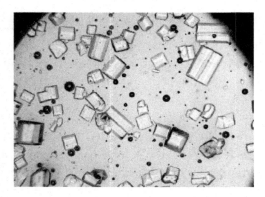

Figure 14.5
Commercial sugar with high levels of twinning.

sugars and there is a tendency for crystals to form twins. The formation of twin crystals comes mainly from the composition of non-sugars, rather than from the operational factor. The collage of the crystals is made by keeping the honey between them, and this causes considerable increase of ash and color change in the final product (Figure 14.5).

14.3.12 Fake Grain

The conditions that contribute to the formation of fake grain are:

- high evaporation, leading to high supersaturation;
- insufficient circulation;
- honey with high turbidity;
- grain in the feeding honey;
- false entry of air.

The formation of the large size grain with little nucleus, which has minimum area available for crystallization, is slow, as is the rate of absorption of sugar, compared to the concentration by evaporation. (Figures 14.6 and 14.7).

14.3.13 Acicular Crystals

Acicular crystals are the ones formed and developed from the processing of raw materials with high levels of dextran, gums, pectin, etc. These substances are concentrated in specific faces of the sucrose crystal by inhibiting the normal deposit of sucrose (Figures 14.8−14.10).

Figure 14.6
Massecuite with the presence of poor-quality fake crystals.

Figure 14.7
Massecuite with the presence of poor-quality fake crystals.

14.4 Operation of the Vacuum Pans

The introduction of automation of vacuum pans and the selection of the best tools for controlling the process is a goal to be achieved. However, we do not recommend investing in poorly designed equipment. A well-designed vacuum pan has great influence on the cooking time, quality and depletion of sugar honeys.

Some criteria for design of a pan with beaters which provides efficient operation in the preparation of quality massecuite are:

- good circulation;
- appropriate S/V ratio;
- tube height;
- maximum height of the massecuite during cooking;
- diameter of the central tube;
- diameter of the body; and
- volume of the nuclei.

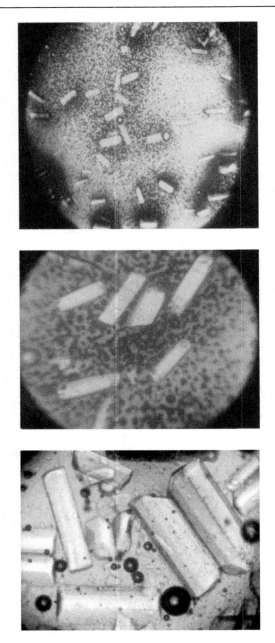

Figures 14.8–14.10
Poor quality massecuite coming from the processing of raw materials with high content of polysaccharides.

Nowadays, there are good technical reasons for the changes in the selection of the vacuum pans and for the replacement of the pans with beaters used in the traditional preparation of massecuite for continuous pans. The system of cooking in continuous pans is more effective in terms of usage of thermal energy, depletion of mother liquor and uniformity of crystals. The continuous pan system theoretically does not spend less water and less steam. In practice, it avoids peaks of steam consumption, which affect levels of tapping of the evaporation system, which in turn affect the consumption of steam and cause the the brix of the syrup to vary.

The continuous pan system is indispensable for automation, for it allows for reliable control of the absolute pressure rates, in order to avoid changes in temperature and, as a consequence, the minimum variation of the brix in the pan. It also maximizes the depletion of mother liquor.

The major advantages of continuous pans are:

- increase in actual hours/pot;
- they require no mechanical circulators;
- in practice, the continuous pan system has about 35—40% more capacity
- than a pan with beaters of the same volume, and
- high degree of movement of massecuite (good circulation ensures high
- productivity and good-quality sugar depletion of mother liquor).

14.4.1 Preparation of the Vacuum Pan and Materials

For any crystallization method adopted, a previous preparation of the vacuum pan and the material to be used in crystallization is required, because the crystallization rate of sucrose is influenced by many factors and not only by the method used. These factors are listed below.

- Steady and well-dimensioned vacuum in the pot.
- Pressure (absolute pressure or vacuum). In the preparation of massecuite, automatic control is highly beneficial to avoid changes in supersaturation due to temperature variation and loss of movement, resulting from the pressure oscillation. The increasing pressure fluctuation will fuse the grain in the primary phase of the massecuite, considering that the pressure increase will raise the temperature of cooking and therefore reduce the supersaturation. Decreasing pressure fluctuation, increasing vacuum brings the temperature down and this, in turn, increases the supersaturation. In this case, either new grains would arise or a cluster would be formed. It is even possible that both occur simultaneously.
- Supply of steam in the roller in the same proportion of water to evaporate.
- Uniform concentration of the massecuite to allow maximum evaporation.

- Good movement of the massecuite into the pan, to achieve maximum evaporation, while the grain in constant movement reaches the portions of honey in the state of saturation.
- It is very important to have sufficient amount of crystals or grain nuclei, so that greater amounts of sucrose are deposited.
- Low viscosity of the massecuite.

Considering these factors, one should take the following measures:

- the pan that will be used for crystallization must be cleaned thoroughly before being supplied the material to crystallize, because large grain residues in the crystallization affect the uniformity of the newly formed grain;
- it is essential to control water and steam injection properly; and
- the pouring of the solution in the pan must be done carefully to avoid the initial violent boiling, which will cause drag.

14.4.2 Volume of the Nuclei

It is certainly appropriate that this volume is as little as possible, because the nuclei are introduced in the form of syrup, or honey-rich magma with high purity, and only at the end of cooking can the purity be reduced by the addition of low-purity honeys.

The concentration of the liquid to crystallize should be monitored, observing the steam pressure in rollers, the pre-determined vacuum and the temperature of the liquid. These parameters having been set for solutions of equal quality and purity, one should always produce the same crystallization conditions.

14.4.3 Graining of the Sugar

It is extremely important to manufacture very well-formed sugar crystals, simple and perfect, without being twinned or clustered. With appropriate crystallization technology, one can say that, to form single uniform crystals, without twinning or clusters, it is necessary to work in the smallest concentration possible. Therefore, it is important to work with the seed (nuclei) obtained from a system that did not transfer any twinning or clustering, with good agitation so that the crystals do not stick, and with good control of concentration (supersaturation) of the mother liquor.

Once the seed is introduced, one must remember that a very small grain seed has planted. Thus, the grain offers little contact surface and require good initial conditions to begin to develop without producing clusters.

It is recommended at this time to slow down the pan. In order to do so, the steam pressure should be reduced in the rollers (2 lb/pol^2), never completely closing the valve once; it is very important that the massecuite keeps moving.

Figure 14.11
Good-quality graining.

In the case of pans with mechanical circulators, one can further reduce the steam ($1\ \text{lb/in}^2$). The vacuum and temperature of the liquid during crystallization should be adjusted according to quality, purity and viscosity of the liquid to crystallize. It is estimated, as a general rule, that the ideal adjustment of the vacuum be set at $25\ \text{in/cm}^2$.

The pan must be fitted with an intake of water at room temperature. The diameter of the line can vary according to the capacity of the pan. In the initial moments after the introduction of the seed, with reduced movement of the pan, one should start to inject water, with the goal of keeping the mother liquor in a degree of saturation such that the newly introduced grain can develop without formation of new grain.

This operation must continue until it is observed that the grain has absorbed as much free sucrose solution as possible. Reaching this point, the operating conditions of the pan should be restored and the addition should start normally (Figure 14.11).

14.4.4 Material to Be Crystallized

Crystallization can be conducted within a wide range of purities. There are technicians who prefer to crystallize at low purities. Others prefer high purities. In the suspensions of crystals in low purity, the chance of clusters being formed is small. It is therefore very well known and recommended to make the crystallization (graining) in an environment of lower purity (lower than that of the syrup), with the purity ranging from 72.0 to 74.0.

However, very low purity causes weakening of the grain, grains that have a shaky development and feel slimy. The initial process, for having a small grain with little surface contact, the low purity of mother liquor and a grain containing impurities, leads to the slow growth of the grain, causing great loss of time.

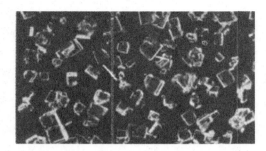

Figure 14.12
Commercial sugar derived from the crystallization of high-purity material.

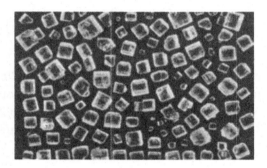

Figure 14.13
Commercial sugar derived from the crystallization of low-purity material.

High purity for crystallization produces the effect of very rapid and abnormal growth of some of the grains over others, producing uneven crystallization. Also, a new crystallization may also occur, originating from a minimum error on the part of the person operating the machinery (Figures 14.12 and 14.13).

14.4.5 Degree of Supply

After crystallization, a continuous and well-regulated supply of the material free of crystals (syrup or honey) should be maintained. The intensity of this natural supply shall vary according to many influencing and local conditions. Overall, it was established that the level of supply should be adjusted to the degree of crystal growth, so it has a friendly merger between the volume of crystals and the total volume of the mass, which should be maintained until the end of the process.

14.4.6 Density of Supply

The vacuum pans lose part of their capacity, when they are supplied with low density solutions. Experience shows that it is practical and advantageous to supply pans with materials of high density and free of grain. This procedure undoubtedly does not harm the process and provides good fuel economy, by increasing the capacity of the pans.

Supply of the pans carried out at a brix of 65–70% results in a significant decrease in the capacity of the pans.

14.4.7 Final Height of the Massecuite

It is prudent to ensure that the level of massecuite in the pan does not exceed 1.50 m above the top mirror, to avoid the problem of circulation due to low heat transfer and hydrostatic pressure in the last stage of cooking.

14.4.8 Final Concentration of the Massecuite for Discharge

The final concentration of massecuite is vitally important in low-purity cooking. The massecuite should be concentrated until the water content in the mother liquor is minimal, so as to obtain a completely depleted final molasses. This is achieved when the cooking process is finished, maintaining a satisfactory movement of the massecuite in the pan during the last 40–60 min. With the introduction of water, this movement in the pans at higher temperatures is reflected in a decrease in supersaturation. In this case, as a precaution so as not to form secondary grain when the massecuite lower in the crystallizer, it is advisable not to start the cooling of the massecuite immediately after discharge into the crystallizer.

14.4.9 System of Massecuites

In order to achieve the desired depletion, one must control the different purities of the massecuite. The depletion is influenced by several factors, some are controlled and others are not. The controlled ones are those that can be achieved through proper operation. The uncontrolled, from the standpoint of manufacturing, are derived from the quality of raw material to be processed.

The optimum conditions for obtaining quality massecuite are:

1. The massecuite should be cooked until the maximum concentration within the practical limits of work capacity of crystallizers and centrifugals.
2. The purity of low grade massecuite should be sufficient to give a maximum content of crystals.
3. The concentration of the mother liquor should have the adequate density to produce molasses with saturation temperature near 55°C, after the massecuite treatment in the crystallizer.
4. The total surface area of crystals of the massecuite in relation to the volume of mother liquor should be maximal, limited to the conditions of fluidity of the mass.

5. The uniformity and size of crystals must provide porosity that allows good separation between crystals and molasses in the centrifuging operation.

6. The viscosity of the honey and saturation temperature should be maximum, to allow for achieving good separation in the centrifuge between crystals and mother liquor.

7. One of the most important features of the pan boiling operation is the movement and temperature condition.

8. A well-designed pan boiling process should, in theory, try to achieve a level of 60% of sucrose present in the form of crystals at the end. This will obviously depend on the brix and the purity.

9. It is important that the instrumentation used for seeding the pan can be used also in the conditions after the formation of crystals. Conductivity and fluidity are recommended, as the Brix refractometer and high boiling point do not take into account the effect of the crystals. Depending on the instrumentation adopted, the supersaturation indicated in the instrument should be assessed in accordance with local conditions. In order to do this, experience is needed (Figures 14.14 and 14.15).

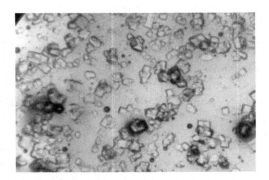

Figure 14.14
Poor-quality massecuite from operational failure.

Figure 14.15
Good-quality massecuite.

14.4.10 Seeding

The amount of seed that must be introduced into the pan, to induce crystallization, depends on the following factors:

- Use of sugar in the preparation of good-quality seed. Sugar selected to prepare the seed should not contain clusters. These clusters will appear in the final product, because they will not be destroyed.
- Manufacturing of the material destined to seeding from large and well-formed grains.
- Thinness of the seed: the thinner, the more nucleuses are introduced by an equal weight.
- Size of the crystals of low-purity massecuite.
- Experts are unanimous as to the formation of clusters; this is possible from pre-clustered particles that form during nucleation (graining). Therefore, it is recommended that the seed is subjected, before use, to a strong agitation to break the clusters contained therein.
- Avoid storing the seed without continuous shaking. The seed should be introduced in the mixture as a supersaturated liquid, preventing the penetration of air (Figure 14.16).

Calculation of the seed

Example:
PZA MC = 60.00
PZA MF = 38.00
PZA Sugar = 100.00
SEED = 0.008 mm.
SUGAR = 0.3 mm.
% Solids in the sugar.

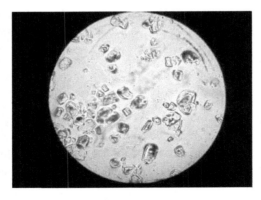

Figure 14.16
Poor-quality seed.

$$\text{R.C.} = \frac{60 - 38}{100 - 38} \times 100 = 35.48$$

Number of cookings

$4 \times 500 = 2000$

$2000 \times 100 \times 1.522 = 304,400$ kg

$304,400 \times 0.3548 = 108,001.12$ kg sugar

kg (sugar)/kg (seed) = $(TC)3/(TS)3$

$108,001.12$/kg (seed) = $(0.3)3/(0.008)3$

kg (seed) = $108,001.12 \times 0.0083/0.33$

kg (SEED) = 2.048 kg

Development of the crystal nuclei

D = $108,001.12/2.048 = 52,734.94$.

In practice, the quantity of seed will be adjusted in function of the depletion and the desired size of the commercial sugar grain.

14.4.11 Magma

The purity of the magma must be the highest possible (purity of "C" sugar), because it favorably affects the quality of sugar, due to the reduction of ash and filterability. Low purity of the magma means that large amounts of final honey are circulated, increasing the volume of massecuite.

The magma used as nuclei for the massecuite "A" should be of high purity. When high-purity magma with uniform crystals and minimum level honey are used, good-quality sugar will be produced and low honey recirculation will occur.

Conditioning of the magma

An effective treatment of the magma nuclei B/C with warm water — in order to eliminate the fine crystals from the breaking in the continuous centrifuges — is of vital importance. These crystals must be dissolved somehow before they start supplying it with syrup or honey. If this is not done, the small crystals present will create what, in ordinary language of the refineries, has been called "dust" or false grain, greatly impairing the quality and depletion of sugar produced.

Figure 14.17
Good-quality massecuite with proper magma nuclei treatment.

Figure 14.18
Massecuite made without proper treatment for the removal of magma fine.

The "C" sugar, from the low-purity massecuite, contains high ash content and high color. When used as magma, the core of this sugar contains impurities, so its quality is impaired (Figures 14.17 and 14.18).

Thinning of the magma

Thinning is a wide spread technique adopted to improve the purity of the magma, because it favorably affects the quality of sugar. However, it has the disadvantage of increasing breakdown of the crystals due to the reduction of the surface layer of honey that surrounds the crystal.

14.4.12 Thinning Honey

The thinning honey, due to its high content of impurity, must be reintroduced to the low-purity pan boiling processes (crystallization and supplying of the massecuite C).

Recasting of the magma ("C" Sugar).

This is also adopted to improve the quality of sugar, because the recrystallization purifies the crystals.

14.4.13 Double Massecuite System

This system is considered appropriate in the case of low-purity syrup.

The system of two massecuites can, in some cases be successfully applied. It has several disadvantages such as the following: the quality of sugar is not good, because, due to reintroduction of honeys, it becomes mandatory if one wants to have a good depletion of final molasses, generally higher due to the limited flexibility of the system.

14.4.14 Triple Massecuite System with Double Magma

This system, when well controlled, is very good and ideal to produce high-quality sugar with great industrial yield.

Advantages of the system

- It produces high-quality sugar, because the massecuite "A" is prepared only with syrup.
- Minimum reintroduction of honey, because the massecuites are made with low viscosity, reducing cooking time and reflecting favorably on the recovery of sugar.
- It reduces the operations of crystallization, since only the massecuite "C" is crystallized with seed.
- Best depletion, since the massecuite "B" becomes an intermediate depletion massecuite.

14.4.15 Number of Pan Boiling Processes to Be Produced for Each Crystallization

Our experience indicates that the optimal number of cooks in pan boiling processes to produce low-purity processes (massecuite C), from each crystallization, is four. This calculation is made based on the proportion of one to four. However, if the pans have different volumes, the appropriate corrections must be made in order to keep the proportions correct.

We recognize that the theory recommends the highest number of boiling as possible, however, in practice, in most cases, when we try to achieve the greatest number of nuclei, the result is little uniformity in the massecuites, lack of grains and therefore, low depletion.

14.4.16 Depletion—Maximum Recovery of Sucrose

The decrease in purity between the syrup and the final honey can reach 50 points in the triple massecuite system.

Important points to consider

- Reduction/ash ratio must be greater than 1:5;
- impurity/water ratio—low value demonstrates poorly elaborated boiling process; and
- the concentration of pan boiling process should be conducted to obtain the minimum impurity/water ratio of 3:5.

Depletion of the honey

Depletion of the honey:

- non-sugars—most salts increase the solubility of sucrose;
- reducers decrease the solubility of sucrose;
- better depletion provides better energy saving; and
- better adequacy of the sugarcane juice treatment system.

Factors that determine a better depletion

- juice treatment;
- syrup treatment;
- crystallization efficiency; and
- separation efficiency.

14.4.17 Complementary Crystallization by Cooling

Treatment of massecuite in crystallizers depends on the following factors:

- yield and size of crystals;
- amount of impurities in the massecuite (reducing sugars, ash, etc.);
- brix and viscosity of the massecuite;
- type of crystallizer;
- type of centrifugal; and
- state of depletion of the massecuite honey reached in the pan.

The crystallizers are designed to complete the crystallization started in the vacuum pans.

To get a good depletion, it is important to achieve minimum retention of massecuite in the crystallizers. In order to further improve the efficiency of the complementary crystallization by cooling, the installation of a serial circulation system in the horizontal crystallizers or the use of vertical ones with suitable capacity is necessary.

The cooling rate depends on the type of massecuite, the nature of impurities and design of the crystallizer. The minimum temperature of crystallization depends on the system and desired saturation temperature. The cooling should be conducted so that the massecuite is maintained in a continuous motion, in order to diminish the internal differences of temperature and saturation, thus avoiding the risk of forming new nuclei. Either way the agitation should be slow, for the gradual reduction of temperature and consequent decrease in the solubility of the mother liquor, continuing the additional crystallization.

Another option available is the re-heating of the massecuite. It is worth noting though, the availability of reheating of massecuite.

After the crystallization rate reaches a practical limit on the crystallizer, it is necessary to effectively separate the crystals and the mother liquor by centrifuging. Viscosity is one of the factors that influence the efficiency of the centrifugals. Therefore, to reduce the viscosity before centrifugation, we recommend reheating the massecuite until the saturation temperature. The heating of the massecuite should not exceed 50°C above the saturation temperature of the mother liquor.

Bibliography

Baikow, V.E., 1967. Manufacture and Refining of Raw Cane Sugar. Amsterdan, 453p.

Caldas, C., 1998. Manual de Análises Selecionadas: Para Indústrias Sucroalcooleiras, Maceió, 424p.

Castro, S.B., Andrade, S.A.C., 2007. Tecnologia do Açúcar, Recife, 420p.

Clarke, M.A., 1996. Sugar y Azúcar, Artigo técnico.

Falcão, R.N., Castro, S.B., 1985. Tecnologia do Açúcar, Recife, 185p.

Fernandes, A.C., 2003. Cálculos na Agroindústria da Cana-de-açúcar, Piracicaba, 240p.

Fundora, G.F., Coletâneas Artigos Técnicos.

Gillett, E.L.G., 1948. Sugar Crystallization, Artigo técnico.

Honing, P., 1969. Princípios de Tecnologia Azucarera. Continental, México (Tomo I 645p. Tomo II 448p. Tomo III 582p.).

Hugot, E., 1969. Manual da Engenharia Açucareira, vol. 2. Editora: Mestre Jou, São Paulo.

Payne, J., 1989. Operações Unitárias na Produção de Açúcar de Cana. Nobel: STAB, São Paulo, 245p.

Pose, R., 1976. Tecnologia Del Azucar Sistema General de Cocimiento, Montevideo, 115p.

Procknor, C., Coletâneas Artigos Técnicos.

Spencer, M., 1967. Manual Del Azúcar de caña. Móntaner y Simón, Barcelona, 940p.

Stupiello, J.P., Coletâneas Artigos Técnicos.

Zarpelon, F., Coletâneas Artigos Técnicos.

CHAPTER 15

Ethanol Fermentation

João Nunes de Vasconcelos

Federal University of Alagoas, Maceió, AL, Brazil

15.1 General Information Regarding Ethanol

Ethanol is commonly determined to be the chemical component ethanol or ethyl alcohol, with molecular presentation CH_3CH_2OH (52.15% carbon, 37.73% oxygen, and 13.12% hydrogen). It solidifies at 130°C, in mass form. It boils at 78.35°C, under 760 mmHg of pressure. Its density at 20°C is 0.7893 g/mL and, at 15°C is 0.7943 g/mL. It is a clear, colorless liquid, with aromatic odor, neutral reaction, and ardent flavor. It burns with a blue flame of little luminescence but is very hot, and in its pure state it combusts spontaneously by exposure to sun beams. Because of its water absorption rate, it is a very good agent in preventing putrefaction, since it dehydrates tissues submerged in it.

Anhydrous alcohol is miscible with gasoline in any proportions. Its addition to gasoline improves the quality of that fossil fuel. The octane rating of gasoline varies from 60 to 65, not allowing for very high compression rates. For that reason, it is common to resort to anti-knocking agents, such as tetra-ethyl lead and methyl tert-butyl ether (MTBE), with a view to preventing gasoline vapors and compressed air in the cylinders of combustion engines from detonating spontaneously before the desired compression level is reached. If such undesired detonations occur, the engine will have worse fuel-to-power conversion ratios and will produce vibrations. Ethanol has a high octane rating (150) and successfully substitutes the toxic anti-knocking additives used.

Ethanol has been used as a fuel for a long time. The famous quotation often attributed to Graham Bell illustrates the matter well: "Ethanol is a wonderfully clean fuel." Henry Ford stated, one century ago: "Ethanol is the fuel of the future." At the time, Ford's most popular model, the Ford T, was bi-fueled, since it ran both on alcohol and on gasoline. During World War I (1914–1918), ethanol was used as fuel in spark-ignited combustion engines.

Brazil has been using some form of ethanol since the 1920s, when the Serra Grande Plant in Alagaoas developed the so-called carburating agent for Otto Cycle engines, which was a mixture of 79.5% anhydrous ethanol, 20% ether and 0.5% refined castor oil. However, only in 1931 did the federal government officialy enact ethanol as an engine fuel and determine the mandatory 5% ethanol presence in gasoline, by mandate of Decree 19,717. There was a shortage of gasoline during World War II, which was replaced with ethanol or syngas. With the end of the war, gasoline started being imported again and ethanol lost space, despite the fact that it continued to be mixed into gasoline. In 1966, by mandate of Decree 59,190, it was determined that the percentage of mixture into gasoline could not exceed 25% of anhydrous ethanol.

The production of carburant ethanol presents a number of advantages: environmental gains, because it can substitute gasoline in combustion engine vehicles; social gains, through the decentralized generation of jobs in Brazil and in other countries (in Brazil, sugarcane production occurs in most states); reduction of the dependency on fossil fuels imported from other countries; and fixation of populations in the countryside, reducing rural exodus levels.

15.2 The Ethanol Industry in Brazil

The history of ethanol in Brazil dates back to the time of the captaincies, when sugarcane brandy was produced from the residues of sugar production. Up until the end of the 19th century, when the production of ethyl alcohol from molasse started being implemented in the sugar industry (a fact which boosted the capacity for its production) the only type of ethanol utilized in Brazil was that used in distilled drinks, particularly *cachaça*.

Proálcool (The Brazilian Program of Incentive to the Production of Ethanol) was created in 1975 by mandate of Decree 76,593, from November 14, 1975, and regulated in 1975 and 1979 – Decree 83,700, from July 5, 1979, as a response of the Brazilian Government to the oil crisis of 1973. A second reinforcement to the policy came after the second crisis of 1979. Currently, Proálcool is one of the most relevant programs for the incentive of biomass liquid fuels in the world, superseded only by the United States, after 2005, in gross volume of ethanol produced (Table 15.1).

As a source of inspiration for the world and for the United States, Brazil, even though it has been superseded in production volume, was, still is and will be the most important ethanol-producing country, because its capacity for increase in production without compromise to food-producing areas is still far superior to that of the Americans. Also, considering the volume of consumption of liquid fuels in the two countries, Brazil's production of ethanol is proportionally much more representative than that of the United States.

With the implementation of Proálcool, there was a sharp increase in the capacity for production of ethanol in the Brazilian sugar–ethanol industry, together with the modernization and expansion of the existing attached distilleries, implementation of autonomous distilleries, creation of hundreds of thousands of direct and indirect jobs, and the manufacturing of vehicles running on 100% alcohol-fueled engines. At the peak of the program, Brazil had a fleet of 4.5 million vehicles running on 100% ethanol engines.

Table 15.1: Evolution of the production of ethanol in Brazil and in the
United States, between 1997 and 2007.

Year	Brazil (billions of liters)	USA (billions of liters)
1997	15.49	5.89
1998	14.12	6.45
1999	12.98	6.16
2000	10.61	6.47
2001	11.50	6.96
2002	12.62	8.43
2003	14.73	10.90
2004	15.10	13.38
2005	16.00	16.14
2006	15.90	18.40
2007*	20.00	25.00

*Estimate.
Source: Ministry of Agriculture, Livestock and Food Supply (2007).

In 1986, 96% of the light vehicles manufactured in the country were running on 100% ethanol-fueled engines (698,564 vehicles).

Before Proálcool, the Brazilian production of ethanol was approximately 600 million liters per harvest (1974/1975 harvest). After the implementation of the program, these numbers rapidly evolved, and in 1985/86 the harvest had already reached numbers of the order of 12 billion liters. These numbers remained stable until the 1995/96 harvest, as can be verified in Figure 15.1. In the 1997/98 harvest, numbers peaked at 15.4 billion liters, after which point they started dropping, reaching 10.6 billion liters in the 2000/01 harvest. Afterwards, production increased and numbers started picking up once more. Currently, production grows at elevated rates. In the 2007/08 harvest, 22,462,132 m^3 of ethanol were produced. These highs and lows in productions were due to the crises in Proálcool. However, the program took new dimensions after 2003, with the release of new bi-fuel vehicles; these vehicles run on mixtures, in any proportion, of hydrous ethanol and gasoline (the gasoline containing 24% anhydrous ethanol).

The production of sugarcane evolved from around 74 million tons in the 1974/75 harvest to 496 million in the 2007/08 harvest (Figure 15.1), and still shows a growth trend. In the same figure, one can observe that the production of sugar presented similar growth trends, arising from a production of approximately 6.7 million tons in the 1974/75 harvest to 31.3 million in the 2007/08 harvest.

To process this monumental amount of sugarcane in Brazil, the country already had, in January 2008, 370 industrial units in 20 states of the nation. Out of those, 117 were autonomous distilleries (they only produced alcohol), 240 were attached distilleries (implemented next to sugar mills) and 15 were sugar mills (they only produce sugar).

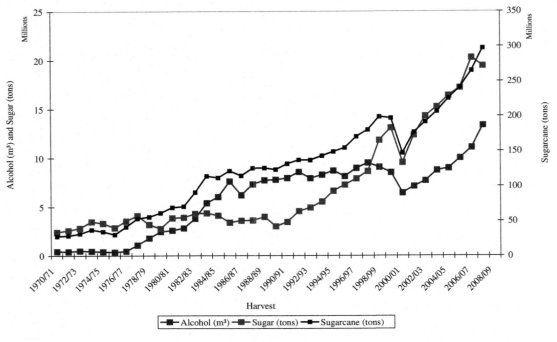

Figure 15.1

Evolution of sugarcane, sugar, and ethanol production in Brazil throughout the 1970/71 to 2007/08 harvests. *Sources: Elaborated by the author from data from the Institute of Sugar and Alcohol (IAA), Association of Independent Sugar and Alcohol Producers of Alagoas (Assucal), the São Paulo Sugar-Alcohol Agroindustry Federation (Unica) and from the Ministry of Agriculture, Livestock and Food Supply (MAPA).*

The production of ethanol from cellulose has been studied for decades, but there is not, as yet, economic feasibility for the activity. In the case of Brazil (considering the constant oscillations in the prices of ethanol and the growing demand for electric power, with constant pricing and growing demand), it is speculated that the economic feasibility of producing ethanol from the bagasse of sugarcane, which is expected to be a great technological breakthrough, will not be very easy to achieve, if one considers the current parameters.

In terms of productivity, there are cases in the literature of corn producing 8 tons/ha, while sugarcane can produce up to 85 tons of raw material for the same area. In terms of ethanol/hectare, the production for sugarcane is 7000 per liter, while corn will reach 3900. Therefore, the productivity of sugarcane ethanol is 1.8-times higher than that of corn for the same planting area. The ratio between renewable energy and fossil energy consumed mentioned in the literature was 8.9 for sugarcane ethanol in Brazil in 2005 and 1.3−1.8 for corn ethanol in the United States. In energy terms, corn ethanol is five-times more favorable to ethanol from sugarcane. These are difficult numbers to overcome.

15.3 Historical Background of Ethanol Fermentation

The word fermentation has its origin in the Latin word "fervere", which means boiling, or effervescence, and has been used since time immemorial. It came from the observation of the abundant and chaotic release of carbon dioxide during the fermentation of grape must for the production of wine, which gave the impression that the fermenting area was boiling.

Since prehistoric times, man has been benefitting from fermentation, unaware as he was of its causes. He produced cheese, milk, curd, and vinegar. Bread, found in Egyptian pyramids built more than 6000 years ago, involves fermentation by yeasts. Wine was known to the ancient Greeks, and the Bible mentions its existence. The Egyptians knew how to prepare bread and alcoholic beverages from cereal and fruit 4000 years ago. Egyptian documents from the fourth dynasty, from around 2500 BC, describe the production of beer, and an Assyrian manuscript from 2000 BC includes beer in the list of products included in Noah's Ark. When Christopher Columbus disembarked in America, he found Indians who drank beer prepared from cereal grains.

Despite having benefitted from these microorganisms for a long time, humanity was unaware of their existence until van Leeuwenhoek (1632−1723), with his rudimentary microscope, was capable of observing yeasts for the first time.

Only in 1815 did Gay-Lussac establish the equation of ethanol fermentation, written as follows

$$\underset{\text{(glucose)}}{C_6H_{12}O_6} \longrightarrow \underset{\text{(carbon dioxide)}}{2CO_2} + \underset{\text{(ethanol)}}{2CH_3 - CH_2OH}$$

In 1857, Louis Pasteur demonstrated that ethanol fermentation was conducted by yeasts, that those were living cells, and that fermentation occurred in the absence of oxygen (anaerobiosis) When those were excluded from the process, no fermentation would occur, which led to the conclusion that this was a physiological phenomenon directly connected to cell activity.

In 1887, Buchner studied the conservation of protoplasmic extracts from yeasts for injection in animals, utilizing sucrose, and accidentally discovered that ethanol fermentation occurred in the absence of free cells, when the sucrose was rapidly fermented by the yeast extract. From Buchner's studies, especially during the first half of the 20th century, one was able to elicit the enzymatic reactions responsible for the conversion of sugar into ethanol and CO_2 by yeasts.

15.4 Production Processes

Alcohol can be obtained through three different ways: distillation of alcoholic liquids, synthetic distillation and fermentation distillation.

15.4.1 Distillation of Alcoholic Liquids

Alcoholic liquids (beers, ciders, winemaking residues, sugarcane brandy, etc.) are not commonly used sources for the economic production of ethanol. Those are only used in exceptional conditions: (a) when there is a surplus in production, working as a market regulators; (b) when large amounts of fermented beverages show infection problems; and (c) when a special variety of ethanol needs to be obtained, for example to obtain grape ethanol for the production of Porto Wine. Wine is only used for ethanol production in regions where the product is cheap and in large harvest (surplus) years. This production process is not economically significant in Brazil.

15.4.2 Synthetic Distillation

Synthetic distillation is the production of ethanol from unsaturated hydrocarbons, such as ethene and ethine, oil gases and coal. In the literature, there are mentions of countries in which there are large reservoirs of oil and petrochemical technology is advanced, and therefore in which synthesis is an economically viable way of obtaining ethanol. However, such a claim is quite inconsistent, since the production of ethanol by synthesis in the world is of around 2.6 billion liters, for an installed productive capacity of 3.3 billion liters (Table 15.2). Production through fermentation, only in the United States and Brazil, reached values superior to 40 billion liters in 2007 and has been increasing rapidly and systematically year after year, not only in those countries, but also in others, which are adopting incentive programs similar to those implemented in Brazil.

15.4.3 Ethanol Fermentation (Fermentation of Carbon Hydrates)

This is the most commonly used process in Brazil and, in general, in the other countries. Brazil is a country of continental dimensions, with a wide diversity of climates and soils; as

Table 15.2: Synthetic ethanol in the world.

Year	Productive Capacity (1000 m^3)	Production (1000 m^3)	Consumption Fuel Chemical Industry/Solvents Beverages	International Trade
1997	3.259	2.245	− 2.245 −	296
1998	3.259	2.276	− 2.276 −	301
1999	3.259	2.306	− 2.306 −	305
2000	3.259	2.339	− 2.339 −	309
2001	3.290	2.379	− 2.379 −	315
2004	3.353	2.515	− 2.515 −	348
2007	3.359	2.628	− 2.628 −	374

Source: DATAGRO (2007).

a consequence, sugarcane is produced year-round in 20 states of the federation. The North and Northeast harvests begin when the Centre-Western and South harvests are near their end. This is important, because it allows the country to produce sugarcane during the whole year in Brazil.

15.5 Raw Materials

15.5.1 Introduction

The raw materials used for the production of ethanol through fermentation are of agricultural origin (renewable resources), dependent on photosynthesis. Not all crops, however, are economically viable sources of raw material for ethanol production. To be considered a viable raw material for the production of ethanol, a candidate must contain glucose, fructose, sucrose, starch, cellulose, etc.

The adoption of a certain raw material for the production of ethanol depends on a series of factors: availability and ease of transportation; production costs; industrial output in ethanol terms; having the adequate substrate (in economically viable concentrations) for the agent microorganism that performs ethanol fermentation; industrial cost of the transformation into ethanol; being easily obtainable; not demanding costly pre-treatment; not contributing to complicate the separation processes of products from the fermented substrate; and being economically advantageous and of easy stocking.

In Brazil, the raw materials used for the production of ethanol are sugarcane and molasse, a byproduct of the production of sugar. Because they are the only commercially viable raw materials presently available, only those two will be subject to more careful consideration.

15.5.2 Classification

The raw materials available for the ethanol industry are classified, for learning purposes, into: (a) sugar-rich raw materials; (b) starch-rich and fecula-rich raw materials; and (c) cellulose raw materials.

Sugar-rich raw materials

Sugar-rich raw materials are subdivided into:

(a) *Directly fermentable*: those which contain in their composition substances which do not need any sort of transformation to be absorbed and transformed into ethanol by the agent microorganism. Included in this category we have trioses, tetroses, and hexoses. Examples are: glyceric aldehyde ($C_3H_6O_3$), erytrose ($C_4H_8O_4$), glucose ($C_6H_{12}O_6$), and fructose ($C_6H_{12}O_6$). Those which only contain those sugars are not used in the industrial process of ethanol production, due to their cost of production (e.g., fruit).

(b) *Indirectly fermentable or non-directly fermentable*: contain sugar with gross formulas $C_{12}H_{22}O_{11}$ (sucrose, maltose, and lactose) and raffinose ($C_{18}H_{32}O_{16}$). The characteristic of these raw materials is that their sugars must be broken (through hydrolysis) before being absorbed by the agent microorganism of the transformation into ethanol. When the raw material contains sucrose, as in the case of sugarcane and molasses, hydrolysis can be performed via an acid (diluted acids) or enzyme (invertase). The yeast which produces ethanol also contains invertase, which performs hydrolysis of the sucrose, transforming it into an equimolecular compound of glucose (dextrose) and fructose (levulose). If hydrolysis is acid, the resulting sugars will be the same. Examples of raw materials with potential use in the production of ethanol are: beet (sucrose and raffinose), malt (maltose), and milk serum (lactose).

(c) *Mixed*: the most commonly used for the production of ethanol in Brazil, mixed sources are those raw materials which possess sugars classified (as above) into directly or indirectly fermentable. The most relevant ones are sugarcane and the molasses from sugar mills, which contain sucrose (in larger amounts), glucose, and fructose (in lesser percentages).

Starch-rich and fecula-rich raw materials

These comprise in their composition starch, a polysaccharide of general formula $(C_6H_{10}O_5)n$, used as a reservoir of energy by green plants and found in high levels in cereal grains, roots and spuds. Examples of starch-rich raw materials are: corn, grain sorghum, rice, barley, wheat, oat, rye, etc.; and fecula-rich: manioc, *cará*, sweet-potato, potato, arrowroot, sunflower roots, babaçu coconut, etc.

For the ethanol fermentation of these raw materials to occur, it is necessary to perform previous hydrolysis, often called saccharification, which transforms starch or the fecula into fermentable sugars. Saccharification can be chemical or enzymatic. In the case of ethanol fermentation, the yeasts do not possess the amylases, which are enzymes responsible for the hydrolysis of starch. For this reason, it is necessary to perform a previous hydrolysis of starch, breaking it into smaller, soluble sugars.

Cellulosic raw materials

They comprise, in their compositions, cellulose and hemicellulose. They also need, similarly to starch-rich and fecula-rich raw materials, previous hydrolysis (saccharification) to be transformed into sugars that can be fermented by the agent microorganism of ethanol fermentation.

Saccharification can also be performed biochemically or chemically. Cellulose-rich raw materials include: sawmill residues, wood, and its derivates, and agricultural residues such as hay, bamboo, corn kernels, and sugarcane bagasse.

15.5.3 Sugarcane

Sugarcane was introduced to Brazil in 1532, brought from Madeira by Martim Afonso de Souza. It was initially planted in the São Vicente Captaincy, where the first sugar mill was implemented in Brazil. Its name was São Jorge dos Erasmos. The second sugar mill, built years later at the Pernambuco Captaincy, was named Nossa Senhora da Ajuda (both religious names). From these two captaincies, the culture extended to Bahia, Sergipe, Alagoas, Espírito Santo, and Rio de Janeiro. The biggest producer of sugarcane was, for more than four centuries, the Northeast *Zona da Mata* (area near the coast), especially the state of Pernambuco. Subsequently, production expanded to the Southeast region and the São Paulo state became the biggest producer (a status it continues to hold today), being currently responsible for 59.85% of the country's production of sugarcane (data from the 2007/08 harvest). In the 1950s, São Paulo surpassed Pernambuco in production while Alagoas (also in the Northeast) increased its production, also superseding Pernambuco in numbers. Alagoas increased its production by planting in the *tabuleiros* (coastal formation that extends throughout most of Brazil's coast). Sugarcane is currently planted in 20 Brazilian states, with prominence in São Paulo, Paraná, Minas Gerais, Goiás, Mato Grosso, Mato Grosso do Sul, Alagoas, and Pernambuco.

The composition of sugarcane depends on the variety chosen, the state of maturation, climate conditions, fertilization of the soil, height of topping, fertilization-irrigation with vinasse, sanity of the culture, time between cutting and processing (deterioration), physical, chemical, and micro-biological properties of the soil, age, and others. Table 15.3 presents the composition of the syrup of mature sugarcane that is ready to be industrially processed. Sugarcane contains around 75% water, 25% organic materials, and 0.5% mineral materials. High incidence of reducing sugars indicates that the sugarcane is "green", while low levels indicate "mature" sugarcane. The syrup contains about 82% water and 18% soluble solids (Brix), glucose $\cong 0.4\%$, fructose $\cong 0.2\%$, and sucrose $\cong 14\%$.

15.5.4 Molasse

Molasse, a byproduct of the production of sugar, is a dense, viscous liquid of dark brown tint, rich in sugars, and containing a small percentage of water. It has different names according to the region, such as exhausted honey, poor honey, final honey, residual honey, or simply molasse. Its density ranges from 1.4 to 1.5 g/mL and its productivity is 40 kg/ton of sugarcane. Ethanol output is 280−230 L/ton of molasse.

Its composition depends on variety, age, state of sanity, maturation, planting system, fertilization, and crop treatment of the sugarcane, as well as climate conditions, sugar manufacturing processes, if the sugarcane was harvested crude or burned, weather and storage conditions, etc.

Table 15.3: Average composition of the syrup of sugarcane.

Component	Percentage
Brix	19.5
Water	81.0
Sucrose	16.0
Glucose	0.30
Fructose	0.10
Total sugar amount	18.00
Reducing substances (Non-fermentable)	0.02
Nitrogen-rich matter	0.03
Sulfuric acidity	0.50
pH	5.50
Ashes	0.40
P_2O_5	0.02
K_2O	0.15
CaO	0.02
MgO	0.02
SiO_2	0.02
Vitamins	Variable

Source: Araújo (1982).

Molasse cannot be processed directly, and has to be thinned in good quality water and/or sugarcane syrup in the preparation of the must. Despite being rich in sugars, it is deficient in other nutritional sources, which makes some form of complementation with nitrogen and phosphorous salts necessary.

Table 15.4 presents the composition of molasse obtained in the state of Alagoas after the production of granulated and demerara sugar.

15.6 Agent Microorganisms in Ethanol Fermentation

15.6.1 Introduction

The most commonly studied microorganism for the production of ethanol is the yeast *Saccharomyces cerevisiae*, followed by bacteria *Zymomonas mobilis*. On an industrial scale, in Brazil, the predominant yeast is *S. cerevisiae*. In several industrial units it is still common to use breadmaking yeast, pressed or dried and granulated, in the beginning of harvest. In other units, at the end of the harvest, the ethanol-producing yeast is isolated from the rest and the agent microorganism is conserved in a nutrition culture until the beginning of the next harvest, where it will be re-used. There is an increasing number of distilleries which use isolated yeasts selected from their own fermentation processes.

Yeasts are heterotrophic, unicellular, chlorophyll-less microorganisms which perform both anaerobic and aerobic metabolism. They are affected by the "Pasteur Effect," that is, they

Table 15.4: Composition of molasse obtained in the state of Alagoas after the production of granulated and demerara sugar.

Element	Molasse Originated From	
	Granulated Sugar	**Demerara Sugar**
C (%)	23.66[a] ± 1.21[b]	22.26 ± 0.79
CaO (%)	1.36 ± 0.12	1.35 ± 0.10
MgO (%)	1.03 ± 0.10	0.99 ± 0.07
N (%)	0.49 ± 0.02	0.49 ± 0.03
K_2O (%)	3.51 ± 0.21	3.80 ± 0.13
P_2O_5 (%)	0.07 ± 0.01	0.15 ± 0.02
Cu (ppm)	16.85 ± 7.7	5.60 ± 1.60
Zn (ppm)	19.45 ± 4.01	11.96 ± 0.78
Fe (ppm)	225.16 ± 55.74	274.44 ± 24.84
Mn (ppm)	19.61 ± 3.53	38.22 ± 4.93
Brix (%)	78.61 ± 0.81	81.33 ± 0.88
Pol (%)	36.58 ± 1.13	33.58 ± 0.69
AR (%)	16.20 ± 0.49	19.20v ± 0.90
ART (%)	54.73 ± 1.14	54.65 ± 0.69
Purity (%)	46.54 ± 1.29	41.41 ± 1.01

[a]Average.
[b]standard deviation.
Source: Vasconcelos (1983).

fermentate in anaerobiosis, with low biomass formation and high amounts of ethanol; the opposite will occur in the presence of oxygen. They are abundant in nature, in the ground, in powder substances and in fruit in general, and are carriable by wind or insects.

Depending on where their cultures are established, yeasts present variable dimensions, with shapes that can range from ovoid to spherical or ellipsoidal. They may occur isolated, in pair, or, occasionally, in small chain or cluster formations. They reproduce vegetatively by spudding, form ovoid or spherical spores and fermentate vigorously.

S. cerevisiae is widely disseminated in nature, but the species is more often associated with industrial fermentation, particularly those for the production of alcoholic beverages and carburant ethanol. The species is also employed in the production of breadmaking yeast. Despite the existence of other microorganisms with potential for the production of ethanol via fermentation, yeasts are still the most important agents, and practically the only ones used industrially.

15.6.2 Characteristics of Industry-Relevant Yeasts

The improvement in the production of ethanol is a process that necessarily entails the adequate selection of the agent microorganism for ethanol fermentation. It also involves the selection of yeasts with high fermentation speeds and dominance and long-lasting lifespans during the harvest, good fermentation capacity, elevated sugar-to-ethanol conversion rates, small output of glycerol, low foam levels, tolerance to high concentrations of subtract and ethanol, resistance to acidity and high temperatures, genetic stability, flocculance (when one aims at avoiding the need for centrifuges), good fermentation efficiency (high output of ethanol), high productivity and elevated cell growth speeds, elevated ethanol output, and substrate consumption speeds.

With higher fermentation speeds, contamination risks decrease, and volume capacity needs for ethanol fermentation are also reduced. Resistance to high concentrations of ethanol is a big advantage from an industrial point of view, since vapor consumption during distillation is lower and there are lower levels of vinasse output and its consequent losses in ethanol. Furthermore, it is possible to fermentate musts with higher concentrations of sugar, allowing for lower consumption of water for thinning. One also has to consider that these elevated ethanol levels must be reached without compromise to the metabolism of the agent microorganism of ethanol fermentation.

Yeasts with high ethanol-to-sugar conversion efficiency rates are one of the most important demands from the industrial point of view. Another important characteristic is resistance to acidity, since sulfuric acid is added to the treatment of yeast to control bacterial infections.

Resistance to temperature is also an important factor, especially in the Northeast of the country, where room temperatures during the sugarcane harvest are high, and superior to those in the Center-Southern regions.

15.7 Preparation and Adjustment of Musts

A sugary liquid which is ready to be fermented is called *must*. There are three types of must: syrup must, molasse must (molasse plus water), and mixed (molasse plus syrup and, occasionally, also water, depending on the molasse/water ratios).

The concentration of sugars in the must must not trigger inhibition issues in the metabolism of the agent microorganism (repression of catabolism of the agent microorganisms by the substrate and/or by the product).

If the must is of the syrup type, due to imbibition in the process of extraction that happens in the molasses mill, the result syrup is used directly as must. In many cases, no other adjustments are made but heating and decantation. In this case, Brix levels are approximately 12° (12 g/100 g). If the must is of the mixed type, depending on the syrup/molasse ratio, there will be a need to add water to correct for Brix levels. In the case of the molasse must type, which is simply the mixture of molasse and water, the levels of sugar are adjusted to 18−20° Brix.

Thinning control for musts is necessary due to the following factors: (a) musts with low sugar concentration lead to higher fermentation speeds, but with lower alcoholic levels in the fermented culture (vine); they favor cell growth, increase consumption of vapor and water and allow for higher-volume fermentation tanks; infections become more likely, because of the lower antiseptic power of ethanol; cleanliness issues are reduced; they require higher volumes of water for thinning, which leads to a need for bigger fermentation tanks; they demand more power from distillation equipment, consume more vapor in the process, and also produce higher levels of vinasse; and (b) musts with high concentration levels lead to incomplete fermentation that is time-consuming and leads to loss of sugars, which entails lower efficiency rates in the distillation process.

Whichever raw material is used for the production of ethanol, it must necessarily be prepared previously in a way that adapts it to the demands of the yeasts. The closer these raw materials are to the conditions which are optimum for the yeasts, the higher the efficiency of the fermentation process will be. Despite anaerobiosis, the yeasts grow during fermentation. They also reproduce, and exercise all the activities related to the maintenance of their lives. These activities will occur as long as the minimum conditions required by the yeasts are satisfied when one prepares the must for fermentation.

15.8 Conducting of the Fermentation Process

15.8.1 Introduction

There are several ways of conducting the ethanol fermentation process, classified in non-continuous (batch feeding) and continuous, as well as those which reuse the inoculum.

In batch feeding, the filling of fermentation tank, inoculation and discharge operations for the fermented must are performed intermittently. In other words, the inoculum (yeast) is placed in the fermentation tank, the must is added and, at the end of the fermentation, the fermented must (wine) is sent to the separation section. Because in the industrial production of ethanol the distillation sector works continuously, several fermentation tanks are necessary, so that one can have, at regular intervals, fermentation tanks in the beginning, middle and end of the fermentation process. This ensures a continuous supply, through a portable vat, of wine for the distillation tanks.

When the conducting of the process happens in a continuous way, the fermentator (reacting agent) is continuously fed with the must containing the nutrients. The fermented must is also continuously removed from the tank at the same rate of feeding of the input must.

15.8.2 Non-Continuous Processes

The non-continuous fermentation processes (batch feeding) can be conducted in different ways. The filling of fermentation tank, fermentation, discharge of the fermented output, and cleaning operations are performed intermittently. Compared with the continuous processes, the non-continuous processes offer lower productivity levels, as the idle time that happens in the batch-feeding process is non-existent in continuous fermentation operations. In the processes that re-use the inoculum after the end of the fermentation, the microorganism is separated by decantation or centrifuge, treated for its purification and reactivation and, subsequently, reused in a new fermentation cycle. These processes are economically advantageous, due to the reusing of the microorganisms. Compared with the processes which do not reuse the inoculum, the consumption of substrate is reduced to allow for cell multiplication, since this consumption aims at the replacement of losses inherent to the process and/or due to occasional accidents.

15.8.3 Process with One Inoculum for Each Fermentation (Simple or Conventional Batch Feeding)

In this type of fermentation process, each fermentation tank receives an inoculum that has been recently prepared, is pure and has been activated; all of which being necessary requisites for a satisfactory handling of the fermentation process. After the end of the fermentation, the fermented environment, containing the agent microorganisms of ethanol fermentation, is sent to the separation section. The fermentating recipient is washed, receiving a new (pure and vigorous) inoculum, which corresponds to 20−25% of the full usable capacity of the fermentation tank, for the next fermentation cycle. Despite offering excellent working conditions, this process has the inconvenience of consuming large amounts of substrate, besides needing a single propagation unit (multiplier) of the agent microorganism for fermentation, which leads to low global efficiency levels in relation to the consumption of substrate.

It demands higher levels of attention and care during conduction, and requires more labor. The products involved lead to outputs with high levels of purity, due to the lower risk of contamination and the physiological purity of the agent microorganism, which constitutes the inoculum. The fermentation in each tank is independent from the others, therefore being more pure.

This process, also known as "pure culture," is the oldest one. It offers good working conditions, presenting a much lower risk of contamination. However, its economic cost/benefit ratio is lower, especially as to what regards the production of ethanol. This type of fermentation process is used in laboratory tests; there is no record of this type of conduction for the fermentation process being used commercially in Brazil for the production of ethanol.

15.8.4 Batch Feeding by Cuts

In this type of fermentation process, fermentation tank 1 receives adequate amounts of inoculum and is fed by the fermentation environment until it completes its expected volume of work. When concentration levels in the fermentation environment reach around 50% of the concentration of the feeding must, half of the usable volume of tank 1 is transferred to tank 2, and both are fed until they complete their respective volumes of work. It is said that there has been a "cut" from one fermentation tank to the other. In tank 1, fermentation is left to be completed, and the fermented environment is sent forward to distillation. Tank 2, on the same criteria elected for tank 1, is cut to tank 3, and the process repeats.

The successive cuts can lead to decreased efficiency in fermentation and productivity, especially when it works with unsterilized musts, as is the case with industrial ethanol fermentation. This process is very commonly used in the ethanol industry at the beginning of each harvest, with a view to quickly multiplying the fermentating agents before starting the recovery process for the inoculum through centrifuge work. In this process, there will always be consumption of a given amount of substrate for cellular multiplication instead of ethanol production. In the case of undesirable contamination, this would rapidly spread throughout the entire process, making it more difficult to control.

15.8.5 Decantation Process

In this case, after the end of the fermentation process, the material is left to rest, and after a certain length of time (4—6 h), the upper part of the material in the fermentation tank (usually 80% of the volume being worked) is removed from the tank, and the rest is adequately treated and reused as inoculum for a new fermentation cycle. It is a less efficient process than the Melle—Boinot process and is used in micro-distilleries for the production of ethanol and in small-sized cachaça distilleries for the production of sugarcane brandy. This procedure is adopted in small industrial units, due to the elevated cost of continuous centrifugal separators, which are only economically viable in large-size units (Figure 15.2).

The decanted yeast is treated with sulfuric acid, preferably with stirring, at pH levels ranging between 2.2 and 3.0, after which it remains still for 1—3 h before the beginning of a new fermentation cycle. In this process, there are high levels of yeast loss through wine and higher fermentation times. Fermentation efficiency and productivity are lower, with higher levels of vinasse output. The consumption of nutrients and sulfuric acid is higher, besides the difficulty in working with sugar-rich fermentation environments. This type of fermentation process allows for the production of wine with low ethanol levels. The treatment of the yeast is made in a non-homogeneous way and there is a reduction in the nominal capacity of the distillation equipment. In the case of *cachaça* mills, especially small-sized ones, there is normally no treatment with sulfuric acid.

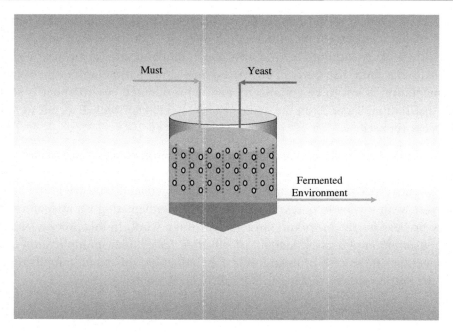

Figure 15.2
Representative schemata of ethanol fermentation by decantation.

15.8.6 The Melle—Boinot Process

Among those processes which reuse the inoculum is the Melle—Boinot, which is utilized in practically all distilleries in Brazil which work with non-continuous fermentation processes. It consists of using continuous centrifugal separators for the separation of yeasts from the wine before their being sent to the distillation tanks.

This process consists, in general lines, of the following operational procedures (Figure 15.3): after the end of the fermentation process, the fermentation environment is sent to a holding tank, called a lung-tank, which feeds the separation section, where two fractions are produced. The first fraction, called de-yeasted wine, goes ahead to another holding tank, called portable vat, which feeds the distillery. The second fraction, usually equivalent to 10—15% of the fermentation tank's volume capacity, is a concentrated suspension of yeasts, called yeast cream or yeast milk. This suspension is sent to pre-fermentation or treatment basins, where it is thinned with water and receives sulfuric acid until pH levels of between 2.2 and 3.0 are reached, with mechanic stirring, and remains resting for 1—3 h. Afterwards, it is used in a new fermentation cycle.

In this type of fermentation process, because the filling pattern is variable, it can be affirmed that industrial ethanol fermentation in Brazil, if made through the non-continuous

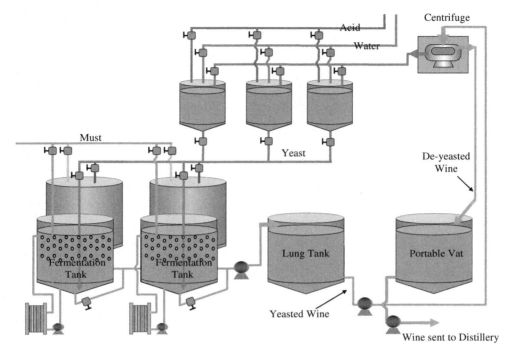

Figure 15.3
Schemata for the Melle−Boinot ethanol fermentation process.

process, is very similar to conventional batch feeding, with variable flow of must feeding. The volume of inoculum in the pre-fermentation tanks, after treatment, varies from 20% to 30% of the volume of the fermentation tanks, which can hold volumes of the order of thousands of cubic meters, depending on the working capacity of the distilleries.

Batch feeding is defined as the process in which the introduction of nutrients can be continuous or intermittent, with constant or variable feeding of the substrate to the fermentating tank, and in which there is no removal of portions of the fermentation environment except at the end. It is applied in the processes in which one of the nutrients, if fed in its totality in the beginning of the process, inhibits the process in some way. This inhibition can be physical or physiological. Those are intermediate processes between continuous and non-continuous processes.

These processes are economically advantageous, due to the reusing of the agent microorganisms of ethanol fermentation. Compared with the processes which do not reuse the inoculum, the consumption of substrate is greatly reduced to allow for cell multiplication, since this consumption aims at the replacement of losses inherent to the process and/or due to occasional accidents.

This type of fermentation process, compared with the others, presents lower levels in losses of yeast, fermentation time, consumption of nutrients, volume of vinasse, and costs with sulfuric acid. Furthermore, it presents higher levels of fermentation efficiency and productivity, allowing for the use of more concentrated fermentation environments (musts). The wine produced has a higher level of ethanol and there is no periodic renewal of the yeast in the fermentation tanks, unless when serious contamination problems arise.

Currently, around 70% of the production of ethanol in Brazil is still made through this fermentation process, which presents the following disadvantages: a need for high-volume fermentation tanks; variation in the conditions of the fermentation environment, as a consequence of the process being conducted non-continuously (in batches); it presents high periods of idleness; and the need for continuous centrifugal separation for the reusing of ethanol yeasts, which has high maintenance and installation costs.

15.8.7 Continuous Processes

In this type of fermentation process, the feeding of must (addition of substrate) and the removal from the fermented environment are processed continuously and at the same rate. Because of the absence of the idle periods which usually happen in the batch processes (filling, discharge and cleaning periods, etc.), these processes present high productivity rates when compared to the corresponding processes when made in batches.

Conventional fermentators, if equipped with adequate mechanical stirring, can be considered to be of the complete mixture (CSTF) type, whose schemata is presented in Figure 15.4. The processes are normally conducted in a series of reactors and with cell reutilization. Due to the homogeneity conditions in the reactor, the concentrations of the several inputs in each tank are the same as those which leave it. In this configuration, four reactors are placed in a series, and the first operates the feeding of the must. The other fermentators in the series are fed with partially fermented environments coming from the previous reactor, at the same entry flow as the first one. The feeding flow of substrate to the fermentator is constant and equal to that of removal, ensuring a constant volume for the fermentation environment.

Continuous fermentation, when compared to the non-continuous type, presents as advantages: higher installed volumetric capacity for the same production output; better ease of automation; better uniformity throughout the operation, which consequently leads to better-quality ethanol; less expenses with labor; less idle periods; higher uniformity in the product; microorganisms are submitted to the same environmental conditions in each bio-reactor; and because the fermentation process is usually conducted permanently, optimum environmental conditions for the microorganism can be achieved throughout the process.

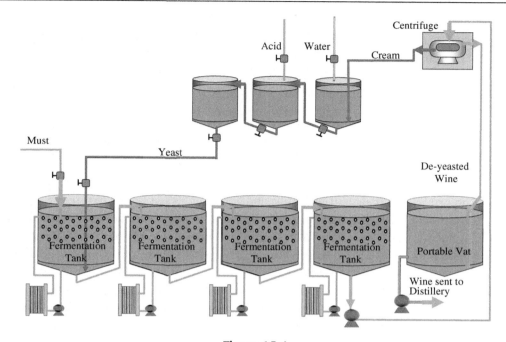

Figure 15.4

Representative schemata of the ethanol fermentation process with four reactors in series and with recycling of cells.

The following disadvantages can be mentioned: difficulty in maintaining steady states due to constant variations in the feeding flows and in the concentrations of TRS in the musts, especially if the distilleries are attached to sugar processing plants; possibility that the microorganisms will suffer genetic mutations; lack of knowledge of the kinetics involved in the process (dynamic aspects of the behavior of ethanol yeasts); and difficulty in maintaining asepsis in the conduction process for long periods.

15.8.8 Processes with Immobilized Cells

One of the most modern technologies currently available for the production of ethanol is the use of immobilized yeast cells in inert supports. This process eliminates the stage of separation of agent microorganisms during ethanol fermentation, allows for the possibility of working with high concentrations of cells in the reactor and presents all the advantages of the aforementioned processes. Immobilization refers to any technique which severely hinders the free circulation of cells, which can be achieved with cell aggregation or by their retention on top of or inside an inert support.

The material to be utilized as an inert support must present convenient mechanical properties: non-toxicity to the cells; high cell-retention capacity in the reactor; biochemical and chemical inertia; low sensitivity to possible mechanical pulls, be it of compression by weight, shearing tensions caused by fluid, or occasional internal gas pressures; high diffuseness of nutrients and products which are formed; elevated porosity; and pores which present as characteristics the easy diffusion of nutrients and products formed, and not allowing, at the same time, for the escaping of immobilized microorganisms; also, being simple, low cost and available in the market/region of interest.

The immobilization process must be simple, of low cost and easy reproducibility, avoiding denaturation conditions such as elevated temperature or pH levels, and the use of organic solvents.

Fermentation processes conducted with immobilized cells, when compared to those which use the conventional processes with free cells, present the following advantages: lower contamination risks; easy separation of the fermented environment from the biocatalysts, eliminating the need for centrifuges; possibility of repeatedly using the biocatalyst (batch or continuous processes) due to the higher stability levels of immobilized cells; increased output rates, since reproduction is limited in the immobilized cells, which causes a consequent reduction in the use of substrate for cell growth; better contamination control; elimination of the external recirculation of cells; also, as the system can operate on high levels of thinning without high levels of cell redistribution in the reactor, it allows for high output levels; due to the use of immobilized cells in the process, it is possible to work with high cell density in the fermentator, regardless of thinning rates, which also leads to increased productivity for a given volume capacity in the reactor; reduction of separation and recirculation costs; increased durability of the biocatalyst; and a widened optimum range for pH levels. This process has, on the other hand, the following disadvantages: resistance to the mass transfer of some stands; abrasion of the stands; and breaking of the stands through the evolution of carbon dioxide.

The use of continuous processes of ethanol fermentation on an industrial scale with flocculent yeasts has shown promise, and there are industry units already operating with this process commercially in Brazil. This type of fermentation process eliminates the centrifuges from the industrial process for ethanol production, which are costly to install and maintain. The advantages of the elimination of the separation stage remain when this process is compared to that in which free cells are used. However, when compared with the continuous fermentation process with immobilized yeasts, the process which uses flocculent yeasts presents the additional costs of equipment for sedimentation and recirculation of the agent microorganisms of ethanol fermentation.

The process of producing ethanol with immobilized yeasts differs from the process with flocculent yeasts in that, in this case, the bottom of the environment is formed by cell-support systems, and in the previous one by cell flocculance. In the case of the support system being sugarcane stalk or the corn kernel, at the end of the harvest the stalk-yeast or

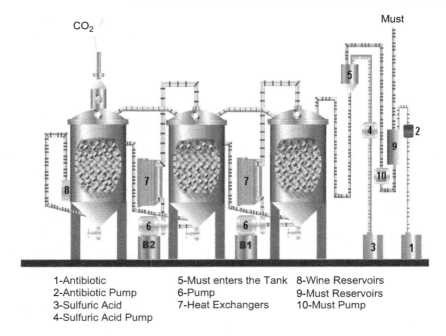

1-Antibiotic
2-Antibiotic Pump
3-Sulfuric Acid
4-Sulfuric Acid Pump

5-Must enters the Tank
6-Pump
7-Heat Exchangers

8-Wine Reservoirs
9-Must Reservoirs
10-Must Pump

Figure 15.5
Continuous ethanol fermentation with immobilized yeast in sugarcane stalks or corn kernels.

kernel-yeast systems can be used for animal feeding, since the material is of good quality. However, even if those materials are not used as animal rations, they can be discarded without harm to the environment, since they are natural and biodegradable. This process is currently being piloted in an industrial unit in the state of Alagoas, and counts on three fermentators in series with 810 L of working volume each (Figure 15.5).

Continuous processes, after these considerations, can be classified as: processes conducted with free cells, in which the microorganisms are in suspension in the fermentation environment; flocculent cell processes, where the cells are in suspension in the fermentation environment, forming cell flocculances; processes with immobilized cells in inert supports, forming fixed or fluid beds. In the last two cases, the objective is the elimination of cell recirculation, since the fermentating microorganisms are retained in the fermentator. With this procedure, an expensive stage in terms of implementation and maintenance is eliminated from the process.

15.9 Physical and Chemical Factors Which Affect Performance for Ethanol Fermentation

Several factors affect performance of ethanol fermentation, that is, the efficiency rate of the conversion of sugars into ethanol. Worthy of highlight are the following: concentration of sugars, nutrients, pH, stirring, oxygen, temperature, nutrient, antibiotics, agent microorganism, among others.

15.9.1 Concentration of Sugars

Knowing that Total Reducing Sugars (ART in Portuguese) and ethanol, in some concentrations, exert an inhibiting effect on the metabolism of yeasts, and that this effect is synergic, one cannot neglect the important of performing a well-controlled addition of sugars, not only to control or even minimize inhibiting effects, but also to increase efficiency and productivity in the fermentation process.

The adoption of a profile of adequate feeding levels reduces fermentation time and increases fermentation efficiency, not necessarily demanding investments and modifications in the engineering of the process. Also, depending on the filling profile adopted, the consumption of anti-foaming agents can be reduced. In this stage, the must is added to the fermentation tank according to a pre-established timetable, in a way to compatibilize the speed of addition of sugars (must) with their consumption by the agent microorganism of the fermentation process.

The concentration of sugars can affect both the growth of yeasts and the production of ethanol. If the goal is the production of biomass, it is convenient that the process be conducted with low concentration of sugars and aeration, to avoid that, even in the presence of oxygen, respiration is repressed in high sugar concentrations (CRABTREE effect).

In the case of ethanol fermentation, whose goal is the production of ethanol, the repressive effect of fructose and glucose on the respiratory chain is beneficial. High concentrations of sugars can cause elevated concentrations of ethanol in the fermented environment (wine) or incomplete fermentations, with formation of undesirable byproducts, which can severely reduce the efficiency of the fermentation process.

Adequate concentrations of sugar increase the speed of fermentation and productivity (higher outputs of ethanol with the same installed volumetric capacity and in the same time frame), improving the performance of the fermentation process, as it allows for smaller cell growth levels and smaller formation of glycerol for the same amounts of glucose/fructose metabolized.

Brix values in the must are controlled to remain between 18° and 20°, when working with molasse must, and variable, depending on the ratios in the syrup: molasse, between 12° and 18°, in the case of mixed must. If the must is of the syrup type, it will depend on the level of imbibition in the molasse mill, during the crushing step. In this case, Brix levels will be lower, reaching down to 12°.

Sugar excesses will lead to osmotic stress in the yeast, maybe leading to inhibition of metabolism. Even if such an effect did not occur, there would a higher output of ethanol, which would inhibit the metabolism of the yeast.

15.9.2 Concentration of Oxygen

Aeration tends to lead to lower ethanol outputs, because the yeast will often present the Pasteur effect, oxidating carbohydrates through respiration, allowing for higher rates of cell multiplication. Oxygen is necessary for cell multiplication; however, for the fermentation process to happen adequately, conditions should be anaerobic. Excessive oxygen levels may provoke other effects, such as oxidation of lipids in the mitochondrial membranes, with reflexes on the output of yeasts. Oxygen acts as a final acceptor of electrons in oxidative phospholyration and as an essential nutrient in the synthesis of lipids.

15.9.3 Stirring

Mechanical stirring is one of the most important operations in industrial ethanol fermentation, bringing a series of benefits, such as: lower temperature levels; prevention of the formation of "fermentation tank deposits"; lower fermentation time (shorter periods of residence in the tank); higher ethanol output; higher fermentation efficiency and efficiency in the process as a whole; height uniformity of the output; better performance of centrifuges; more representative samplings; better cell viability. Mechanical stirring, if adequately performed, is the best form of hygienization of the fermentation environment, both in batch feeding and continuous processes conducted with free cells. It promotes uniformity in the fermentation environment, keeping the yeasts in suspension, allowing for efficient contact between the agent microorganism of ethanol fermentation and its substrate.

One of the most important facts in the use of mechanical stirring in ethanol fermentation (as in any fermentation process) is that, if the operation is well-conducted, the environment becomes homogeneous and the sampling will be representative of the substrate, wherever one chooses to collect it in the tank.

15.9.4 Temperature

Yeasts are mesophile microorganisms, and the most adequate temperatures in industrial practices vary from 30°C to 35°C, with maximum values of 38−40°C, when there is no efficient control of temperature in the fermentation environment. As the process is exothermic, temperature is normally controlled through the use of plate heat exchangers, installed on the surroundings of the fermentators. Literature in the area indicates that increases in temperature increase ethanol's inhibiting effect, probably due to the fact that production speed is higher than diffusion-through-membrane speeds.

Elevated temperatures favor the appearance of bacteria and the evaporation of ethanol (in the case of open-lid fermentators). Temperature exerts influence on fermentation time (and productivity output in ethanol) and the appearance of undesired infections. The optimum temperature for yeasts ranges from 26°C to 35°C.

15.9.5 pH and Acidity

Yeasts are acidophilus microorganisms and work well in broad pH ranges. Best results, however, are obtained for pH levels ranging from 4.0 to 4.5. The levels of pH have a marked influence on industrial fermentation, due to their importance in bacterial contamination control, and to their effect on the growth of yeasts, fermentation rates and formation of byproducts. The fermentation of sucrose is more affected by pH than that of glucose, since invertase activity in yeasts is more affected by low pH levels than fermentation potential.

In the processes where the yeasts will be re-utilized, the inoculum is treated in pre-fermentators (treatment basins), where it receives sulfuric acid until pH levels are between 2.2 and 3.0, and remains resting for 1−3 h. Afterwards, it is used in a new fermentation cycle. This procedure allows for a reduced bacterial load. The combination of pH values of 2.5 in the inoculum and 5.5 in the must and 20−30% working volume in the must and 70−80% in the fermentators, together with the influence of other products produced during fermentation, leads to a fermentation environment with pH levels close to 4.0.

Sulfuric acid is added in sufficient amounts so as to favor the action of yeasts and hinder the development of bacteria. On the other hand, the addition of sulfuric acid favors the inversion of sucrose, which does not fermentate directly. The yeast has invertase; however, the use of sulfuric acid for the activation of this inversion makes fermentation easier and more active.

15.9.6 Concentration of Cells

High cell concentration leads to low fermentation times, while low concentration leads to higher fermentation times. High cell concentration leads to increased productivity and low fermentation times, regardless of the type of fermentation process employed.

15.9.7 Nutrients

The concentration of nutrients in the must is one of the most important factors for the good conduction of the ethanol fermentation process. If present in inadequate quantities, higher or lower than necessary, nutrients can lead to negative reflexes on the performance of ethanol fermentation, affecting cell multiplication and fermentation speed. Yeasts need an environment which contains sources of carbon, such as glucose and fructose; besides, the environment must be a source of vitamins, nitrogen, phosphorus, sulfur, potassium, magnesium, calcium, zinc, iron, copper, cobalt, iodine, and others in small amounts.

S. cerevisiae cells use phosphorous in the $H_2PO_4^-$ formula, which prevails in 4.5 pH levels. Sulfur can be assimilated from sulfate, sulfite or thiosulfate. The sulfur is added to the treatment of the inoculum (addition of sulfuric acid) and in the sulfitation of the syrup (addition of sulfur) in the manufacturing of granulated sugar. Nitrogen is absorbed by the yeasts in ammoniacal (NH_4^+), amidic (urea) or aminic forms. Adequate sources of nitrogen are important for the synthesis of amino acids and proteins and for the growth and physiology of yeasts.

It is worth highlighting that, as the must contains part of or the totality of several of these nutrients in adequate quantities, it is always convenient to determine the average composition of the musts so that complementation is as accurate as possible.

15.9.8 Antibiotics/Antiseptics

In Brazil, musts are of the syrup, mixed (syrup plus molasse), and molasse types. In none of these cases is the resulting must sterile. The addition of antibiotics is a generalized practice in the creation of an environment that is favorable to the development of yeasts and unfavorable to the development of undesirable microorganisms in the fermentation process. Each product acts differently, acting on a determined group of microorganisms. The evaluation of both the bacterial load and the efficacy of commercial antibiotics will be the determining factor in the choosing of the anti-bacterial agent.

15.10 Stages of Ethanol Fermentation

After the addition of the inoculum (yeast) to the must, ethanol fermentation starts, being divided, for classification purposes, into three different stages, which are called preliminary stage, main or turmoil stage, and final or complementary stage. It is important to highlight that it is difficult to precisely determine when those stages begin and end.

The preliminary stage, which begins with the initial contact between the yeast and the must, is characterized by the predominance of cell multiplication, with little releasing of carbon dioxide, little formation of foam, small elevation of temperature in the fermentation environment, and little formation of ethanol. This stage is short-lived and depends on the type of ethanol fermentation. It is the stage where the yeasts adapt to the environment.

The main stage is characterized by intense production of ethanol, carbon dioxide, and heat, being the most long-lasting of the three. In this stage there is formation of foam, increase in acidity levels and reduction of density of the environment, as a consequence of the conversion of sugars into ethanol. To control the rapid temperature increase in the fermentation environment, plate heat exchangers are often used in ethanol fermentation on industrial scales.

In the complementary stage, the releasing of CO_2 is reduced, with less stirring of the environment and decreased temperature, as a consequence of the end of the supply of fermentable sugars. At the end of this stage, the environment's surface presents without stirring and free of foam.

In industrial practice, when fermentation occurs in batches, it is common to add the treated yeast to the tank, and subsequently perform the programmed addition of must. In this case, the fermentation practically begins in the main stage, which, depending on the filling configuration, can be more uniform, with moderate elevations of temperature and foam.

15.11 Control of Ethanol Fermentation

During ethanol fermentation, several supervision procedures take place in order to keep track of several analytical and operational aspects, so that the process can occur with optimum performance and without bigger abnormalities. The main aspects supervised are mentioned below.

15.11.1 Fermentation Time

Fermentation time, in the ethanol industry, ranges from 6 to 9 h, when fermentation takes place in batches. If it is continuous, this time can be estimated to be between 8 and 10 h. Numbers grater than this can indicate abnormalities, which need to be corrected as promptly as possible, since one is usually working with fermentation environments of hundreds of cubic meters of volumetric capacity. Elevated times may indicate contamination, low viability of yeasts, low concentration of yeasts in the fermentation environment, excess of sugars in the must, among other factors.

15.11.2 Foam Presentation

When ethanol fermentation occurs within normality, with good performance, the bubbles formed are regular and with a certain glow, keeping the same pattern throughout the surface of the fermentation environment, and are easily broken by the pressure exerted by the carbon dioxide released during fermentation. When there is contamination, the bubbles become bigger due to the coalescence of smaller bubbles and, when those break (not without difficulty), they already are big in size and irregular in shape, with certain opacity, not maintaining the same pattern, as is the case with regular fermentation. Addition of anti-foaming agents is a generalized practice in industrial ethanol fermentation, as a form of controlling the amount of foam generated during fermentation.

15.11.3 Smell

When ethanol fermentation is normal, the smell is penetrating, pleasant and pungent, due to the raw materials utilized in the preparation of the must. If the smell is outside this pattern, this could indicate possible contaminations.

15.11.4 Residual Sugars

The concentrations of residual sugars at the end of the fermentation stage must be the lowest possible. The usual methods for determination of sugars are not selective, and quantify, by expressing glucose-based results, the compounds that reduce Cu^{+2} to Cu^{+1}. Due to this fact, wines coming from molasse musts present residual sugars in far superior rates to those of syrup musts. Mixed musts present intermediate values.

15.11.5 Temperature of Fermentation (or of the Fermentation Environment)

Literature mentions ideal numbers close to 30°C. In the industry, especially in the Northeastern states, regular temperature is close to 34°C (33−35°C) and will more often than not reach 38−40°C, because of problems in cooling systems. As the process is exothermic, temperature is normally controlled through the use of heat exchanging devices, usually plaques, installed on the surroundings of the fermentators. If the temperature elevates, in a normal setting, it will probably be due to dirt in the plate heat exchangers. In this case, after cleaning the plate heat exchangers, performed with good-quality, high-pressured water, the temperature will return to normal. If temperatures are low, especially at the beginning of harvest and depending on the region, it is recommended to warm the must before mixing it with the yeast (inoculation). In the case of the Northeastern states, where temperatures are more elevated during the harvest than those of the Center-Southern region, this procedure is not necessary.

15.11.6 Acidity and pH

The routine procedure is to evaluate initial and final acidity levels. Final acidity must not exceed two-times the initial value. There are recommendations that this value does not exceed 50% of the initial value. If that occurs, it is possible that there is some form of contamination. In the literature, there are mentions of values ranging between 2.5 and 3.0 g of sulfuric acid/L, a value which should not supersede 5.0 g/L. When working with molasse or mixed musts, the strong buffer effect of the must can make it difficult to evaluate this parameter.

15.11.7 Fermentation Efficiency

Fermentation efficiency is the parameter that indicates sugar-to-ethanol conversion efficiency and can be calculated directly considering the sugars added (process efficiency) or those effectively consumed (fermentation efficiency). This parameter is the reflex of all the previous stages, and if those were well-performed, this parameter will certainly yield satisfactory results. In these calculations, Gay–Lussac's equation is taken as the basis.

15.11.8 Productivity

Productivity represents the amount of ethanol produced by unit of time and volume. This parameter, together with that of fermentation efficiency, is a good indicator of the performance of ethanol fermentation, especially when the isolation and selection of ethanol-producing yeasts is performed.

15.11.9 Output

Output is the measurement of the quantity of ethanol produced by area of sugarcane harvested, by ton of sugar, ton of molasse, etc. It is a good indicator of the performance of the fermentation process.

15.11.10 Concentration of Sugars

This is one of the most important controls in ethanol fermentation. When the process is performed in batches, control is performed at 1-h intervals, at the beginning or the end of the fermentation process, through Brix measurements. The values should be consistent with an indicative curve for the operational conditions of the fermentation process. There is an increase in Brix during the filling stage of the fermentator, in a more or less accentuated fashion, depending on the filling profile. At the end of this stage, decrease will take on a descending hyperbolic pattern.

If the profile of sugar consumption, as performed by the Brix curve, is dissonant from the pattern (too fast or too slow), there may be an imbalance between the addition and the consumption of sugars by the yeasts. This could indicate inadequate (deteriorated) raw materials, an inadequate percentage of yeast or sugars, excessive or incomplete cooling, inadequate tank filling profile, yeast with low cellular viability, among other factors. It is worth highlighting that a well-conducted fermentation process is a reflex of all the parameters mentioned. These controls are performed at each fermentation cycle, when it is processed in batches.

If fermentation is continuous, these parameters are equally important, and are controlled in all fermentation tanks (which are here inter-connected). In this process, a few of the terminologies and evaluation processes are different. For instance, in the batch fermentation process, we have fermentation time. In the continuous fermentation process, we have residence time, that is, the amount of time necessary for the must to enter the first tank and leave the last one as fermented must (wine). In this case, we do not talk about filling profile, but about thinning rate, which is the ratio between the flow of feeding of the tank and its volume. Acidity, pH, temperature, and Brix of the fermentation environment and of the wine are controlled in a similar fashion to that of the batch process.

Bibliography

Aiba, S., Humphrey, A.E., Mills, N., 1973. Biochemical Engineering. Academic Press, New York, 434p.

Amorim, H.V., 1977. Introdução á bioquímica da fermentação alcoólica. IAA-PLANALSUCAR (COSUL), Araras, 90p.

Amorim, H.V., Basso, L.C., Alves, D.M.G., 1996. Processos de produção de álcool. Controle e Monitoramento. FERMENTEC/FEALQ/ESALQ-USP, 103p.

Angelis, D.F., 1987. Leveduras. In: Microbiologia da fermentação etanólica. Universidade Estadual "Júlio de Mesquita Filho", Instituto de Biociências, Rio Claro, pp. 41−62.

Araújo, J. A. de, 1982. Obtenção do etanol por fermentação alcoólica. IAA-Planalsucar (Coone), Rio Largo, 83p.

Bajpai, P., Margaritis, A., 1987. The effect of temperature and ph on ethanol production by free and immobilized cells of *Kluyveromyces marxianus*. Grown on Jerusalem artichokes extract. Biotechnol. Bioeng. 30 (2), 306−313.

Brown, S.W., Oliver, S.G., 1982. The effect of temperature on the ethanol tolerance of the yeast, *Saccharomyces uvarum*. Biotechnol. Lett. 4 (4), 269−274.

Buzás, Z.S., Dallman, K., Szajsni, B., 1989. Influence of pH on the growth and ethanol production on free and immobilized *Saccharomyces cerevisiae* cells. Biotechnol. Bioeng. 34 (6), 882−884.

Castro, A.C., Salerno, A.G., 1991. Fermentação contínua em reator tipo torre usando levedura floculenta. STAB. Açúcar, álcool e Subprodutos. 9 (4/5), 46−50.

Cooperativa de produtores de cana, açúcar e álcool do Estado de São Paulo LTDA, 1987. Fermentação. Piracicaba, 434p.

Datagro: Datagro, Boletim Informativo Quinzenal, 2007. n.12, p5, São Paulo, SP, Brasil.

Fernandes, A.J., 1984. Manual da cana-de-açúcar. Livroceres LTDA, Piracicaba, 196p.

Jones, R.C., Pamment, N., Greenfeld, P.F., 1981. Alcohol fermentation by yeast-the effect of environmental and other variables. Process Biochem. 16 (3), 42−49.

Kretzchmar, H. 1956. Levaduras y alcoholes y otros productos de la fermentación. Editorial Reverte, S. A. Barcelona, 735p.

Leão, R.M., 2002. Álcool: energia verde. Iqual Editora, São Paulo, 256p.

Lehninger, A.L., Nelson, D.L., Cox, M.M., 1993. Principles of biochemistry. second ed. Worth Publishers, New York, 1013p.

Lima, U.A., 1985. Evolução dos processos fermentativos no Brasil. In: Semana de Fermentação Alcoólica "Jayme Rocha de Almeida", 4, Piracicaba. Anais, pp. 1−6.

Lima, U.A., Basso, L.C., Amorim, H.V., 2001. Produção de etanol. In: Lima, U.A., Aquarone, E., Borzani, W., Schmidell, W. (Eds.), Biotecnologia Industrial. Processos fermentativos e enzimáticos, vol. 3. Editora Edgard Blucher LTDA, São Paulo, pp. 1−43.

Llames, H.P., 1956. Fabricacion del alcohol. Savat, Madrid, 735p.

Maiorella, B., Blanch, H.W., Wilke, C.R., 1983. By-product inhibition effects on ethanol fermentation by *Saccharomyces cerevisiae*. Biotechnol. Bioeng. 25, 103.

Mapa — Ministério da Agricultura, Pecuária e Abastecimento, 2007. Balanço Nacional da Cana-de-Açúcar e Agroenergia, 139p.

McElroy, W.D., 1988. Fisiologia e bioquímica da célula. Edgard Blucher LTDA, São Paulo, 144p.

Mutton, M.J.R., Mutton, M.A., 2005. Aguardente. In: Venturini, W.G. (Ed.), Tecnologia de bebidas: matéria-prima, processamento, BPF/APPCC, legislação, mercado. Editora Edgard Blucher LTDA, São Paulo, pp. 485—524.

Novaes, F.V., 1982. Matérias-primas para a produção de álcool. Curso de Aperfeiçoamento de Produção de álcool. IAA-PLANALSUCAR (COONE), Rio Largo-AL, 5p.

Pelczar Jr., M.J., Chan, E.C.S., Krieg, N.R., 1997. second ed Microbiologia. Conceitos e aplicações, vol.1. Makron Books do Brasil, São Paulo, 524p.

Ramalho Filho, R., Vasconcelos, J. N. de, 1992. Do PROáLCOOL á valorização da cana-de-açúcar em Alagoas. In: Maimon, D. (Ed.), Ecologia e desenvolvimento. APED, Rio de Janeiro, pp. 235—258.

União da Indústria Canavieira de São Paulo — UNICA. Disponível em: <www.portalunica.com.br> Acesso em várias datas.

Vasconcelos, J. N. de, 1985. Estudo sobre a composição química de melaços do Estado de Alagoas. STAB: Açúcar, álcool e Subprodutos. 3 (5), 45—51.

Vasconcelos, J. N. de, 1987. Influência da complementação do mosto em nutrientes nitrogenados e fosfatados sobre o processo de fermentação alcoólica. Brasil Açucareiro. 105 (4, 5 e 6), 41—48.

Vasconcelos, J. N. de, 1987. Operação e simulação do processo de fermentação alcoólica em batelada alimentada com vazão variável de alimentação. Dissertação (Mestrado), Escola de Química da Universidade Federal do Rio Janeiro, Rio de Janeiro. 229p.

Vasconcelos, J. N. de, 1988b. Fundamentos teóricos da fermentação contínua Rio Largo. IAA-PLANALSUCAR (COONE). 80p.

Vasconcelos, J. N. de, 1998. Fermentação alcoólica contínua com levedura imobilizada em colmos de cana-de-açúcar. Tese (Doutorado), Universidade Federal do Rio de Janeiro, Rio de Janeiro, 113p.

Vasconcelos, J. N. de, 1999. Influência do pH do mosto sobre o desempenho da fermentação alcoólica desenvolvida por levedura livre e imobilizada em colmos de cana-de-açúcar. STAB. Açúcar, álcool e Subprodutos. 18 (2), 52—56.

Vasconcelos, J. N. de, 2003. Agitação no processo de fermentação alcoólica. STAB. Açúcar, álcool e Subprodutos. 21 (4), 16.

Vasconcelos, J. N. de, 2006. Concentração inicial de células no inóculo e sua influência sobre a fermentação alcoólica desenvolvida por leveduras. STAB. Açúcar, álcool e Subprodutos. 24 (5), 22—24.

Vasconcelos, J. N. de, 2008. Influência do perfil de enchimento das dornas no crescimento celular, na produção de etanol e no consumo de ART.. STAB. Açúcar, álcool e Subprodutos. 27 (2), 20—23.

Vasconcelos, J.N. de, Lopes, C.E., de França, F.P., 1998. Yeast immobilization on cane stalks for fermentation. Int. Sugar J. 100 (1190), 73—75.

Vasconcelos, J.N. de, Lopes, C.E., de França, F.P., 2004. Continuous ethanol production using yeast immobilized on sugar-cane stalks. Brazil. J. Chem. Eng. 21 (3), 357—365.

Vasconcelos, J.N. de, Rossetto, R., Ivo, W.M.P.M., Santiago, A.D., Barbosa, G.V.S., 2008. Impulsionando a produção e a produtividade da indústria sucroalcooleira. In: Albuquerque, A.C.S., Silva, A.G. (Eds.), Agricultura tropical — Quatro décadas de inovações tecnológicas, institucionais e políticas: produção e produtividade., vol.1. Embrapa Informação Tecnológica, Brasília, pp. 717—733.

Vasconcelos, J. N. de. Alcoholic fermentation with immobilized yeast in sugarcane stalks. In: International Workshop of Sugarcane Co-Products. ISSCT-STAB, Maceió-AL, de 12 a 16/11/2006. Anais em CD.

CHAPTER 16

Distillation

Thales Velho Barreto[1] and Antônio Carlos Duarte Coelho[2]
[1]Velho Barreto Inc. and ECO, Technology and Industrial Equipment Inc, Brazil
[2]Federal University of Pernambuco, Recife, PE, Brazil

Introduction

Distillation is an unitary physical operation that aims at separating the components of a mixture, according to the relative volatility of the components (the most volatile moves to the vapor phase, while the less volatile remains preferably in the liquid phase). This operation is performed by means of partial boiling and condensation of the mixture if one wants to separate and/or purify a mixture.

The boiling temperature is the temperature at which a substance goes from liquid to gaseous state. The value of this temperature depends on environment pressure and is used to characterize pure substances when subjected to distillation at atmospheric pressure.

Since the Middle Ages, there have been reports of distillation processes using stills (Figure 16.1), mainly to obtain alcoholic beverages from a fermented must. This procedure also enabled chemists of the past, called alchemists, to separate and/or concentrate various substances: ethanol, acetic acid, citric acid, turpentine, etc. As time went by, distillation was divided into categories according to its application. Today the known types of distillation are atmospheric, vacuum, azeotropic, extractive, and fractionated.

Important sectors, such as the petroleum and petrochemical industries, use the distillation process, alone or combined with other unitary physical operations, to produce numerous compounds and mixtures with varying degrees of purity: natural gas, naphtha, LPG, diesel, kerosene, lubricants, HFO, asphalt, ethylene, butadiene, etc. (Figure 16.2).

16.1 Principles of Distillation: Liquid–Vapor Equilibrium

In distillation, the vapors produced are usually richer in more volatile components than the liquid, which allows for the separation of fractions enriched in the desired components. The liquid and vapor generally contain the same components but in different relative amounts.

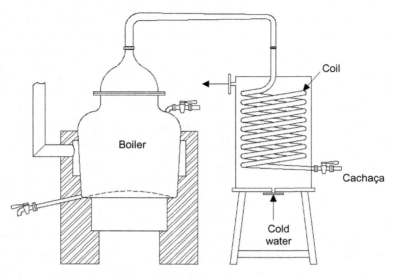

Figure 16.1
Molasse mill.

Figure 16.2
An image of a modern oil refinery with several distillation tanks.

The relative volatility is a measure of the differences in volatility between two components, that is, between their boiling points at mean sea level pressure.

The vapor—liquid equilibrium (VLE) curve of the most volatile component shows us the bubble points and dew points of a mixture under a certain fixed pressure. In ideal systems, the boiling

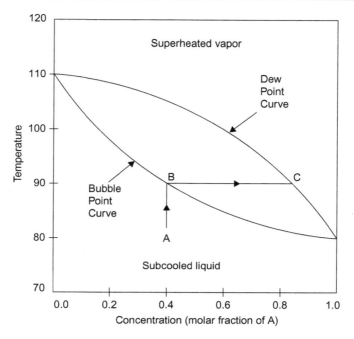

Figure 16.3
Bubble points and dew points of a mixture under fixed pressure.

point remains constant. The higher the relative volatility of two components, the greater the ease of separation is. The curve also indicates a binary mixture that, by having uniform characteristics of vapor−liquid equilibrium, is relatively easy to separate (Figure 16.3).

In a binary system under constant total pressure, analysis of a graph with y-axis being the mole fraction of gas and x-axis the mole fraction of the liquid phase can help assess the degree of difficulty in separating components by distillation.

The farther away from the diagonal the equilibrium curve is, the easier the separation by distillation will be. In Figure 16.4a, separation by distillation is easier than in the case of Figure 16.4b.

In non-ideal mixtures such as water−ethanol, there may be major differences between the volatility levels, depending on the concentration of the components in the mixture. For compositions between 5 and 50 wt% of ethanol, relative volatility is much greater than 1, whereas when approaching 95% in ethanol weight, the relative volatility is close to 1, reaching this value at the point of the curve which is called the azeotropic point.

When a liquid mixture produces vapor with the same composition, we have an azeotrope. The equilibrium curves intersect the diagonal lines (Figure 16.5).

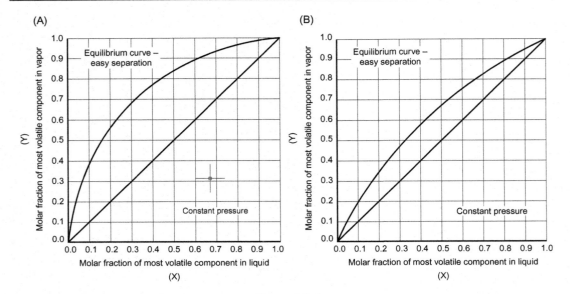

Figure 16.4
Vapor—liquid equilibrium (VLE) curves of binary mixtures at constant pressure.

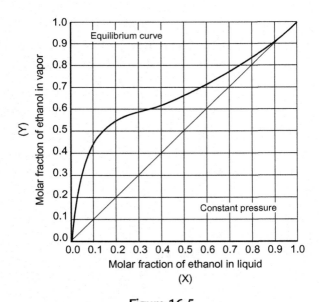

Figure 16.5
Vapor—liquid equilibrium (VLE) curve for binary mixtures, ethanol—water under constant pressure
of non-ideal systems with homogeneous azeotrope.

16.2 Distillation Columns

Industrially, distillation is made in equipment called columns, which may be of stages (plates or trays) or by contact (filling), which provide enrichment of the produced vapor and depletion of the distillate (compared to the more volatile component).

In 1925, McCabe and Thiele developed a graphical method for determining the optimal number of plates of a distillation column. Today, computer simulation methods are used.

In the ethanol industry in Brazil columns of plates or trays are used, which are installed in parallel at a distance (spacing). There are columns that have, over the feeding plate, a concentration zone (above) and a zone of depletion (below). The vapor, which is rich in the more volatile component, is cooled and condensed in condensers upon leaving the top of the column. At the bottom of the column, heating vapor comes in and the liquid (effluent) comes out, almost free of the most volatile component, as seen in Figure 16.6. In Figure 16.7, we have represented the main internal components of a column: bubblers, weirs, and dams. The plates are classified by the types of boilers: caps (Figure 16.8), orificed, or drilled holes (Figure 16.9), and valves (Figure 16.10).

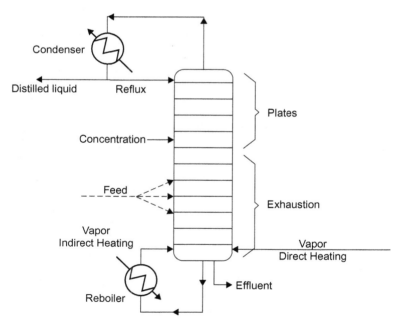

Figure 16.6
General outline of a column.

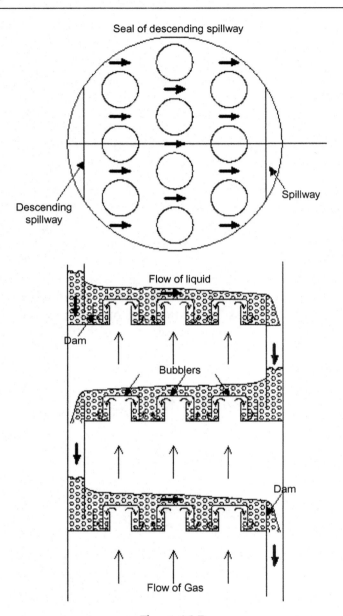

Figure 16.7
Main internal components.

Figure 16.8
Plate with caps.

Figure 16.9
Perforated plate.

Figure 16.10
Plate with valves.

The distillation columns are complemented by: heat exchangers (condensers, reboilers, regenerators, coolers) and flow, temperature and pressure control instruments. Each of them is described below.

16.2.1 Condensers

Condensers are responsible for condensing the enriched vapors at the top, promoting the reflux of liquids to the column and allowing the withdrawal of distillate. They use water as coolant.

16.2.2 Reboilers

Reboilers are indirect heaters. The process vapor heats the liquid at the bottom of the column, which is vaporized and returns to complete the heating of the system. The vapor condenses and the condensed water is used to feed the vapor generators (boilers).

16.2.3 Regenerators

Regenerators recover energy from the process. According to their functions, they will acquire special designations. As an example, we can mention the two types of heat exchangers which heat wine until boiling using thermal energy from fluids of the process — the pre-heater of wine (E), which harnesses the heat of phlegm from the top of the rectification column (B), and K heat exchangers, a set of four or more bodies which partially recovers the heat from vinasse (effluent from the distillation of wine).

16.2.4 Coolers

Coolers serve to cool the final product before it is stored.

16.2.5 Instruments for Flow, Temperature, and Pressure Control

Currently, there is a high level of instrumentation and automation in ethanol plants. Measurements are made of the flow of wine and of the final product. Measurements of temperatures and pressures in certain plates guide the operation and enable automation of the process. Figures 16.11 and 16.12 show ethanol distilleries.

16.3 Raw Materials

Ethanol, by fermentation, is produced from sugar-rich, starch-rich and cellulose-rich raw materials. In Brazil, the production of sugar is made from sugarcane, and processed in

Figure 16.11
500 m^3/day Apparatus — Rubi-Cooper Distillery, GO.

Figure 16.12
250 m^3/day Apparatus — Alcana, MG.

standalone or attached distilleries. A standalone distillery produces only ethanol and processes sugarcane, while an attached distillery (annexed to a sugar factory), besides the direct usage of sugarcane juice (mixed juice) and molasses, can also use the intermediate products of sugar production to compose the feeding must.

The second generation of ethanol, produced from cellulosic raw materials (mainly sugarcane hay, stalks and bagasse, byproducts of the sugarcane—ethanol industry), has been studied for years, but there is no industrial-scale production as yet.

The flowchart of ethanol production, with the main stages of the process, is shown in Figure 16.13.

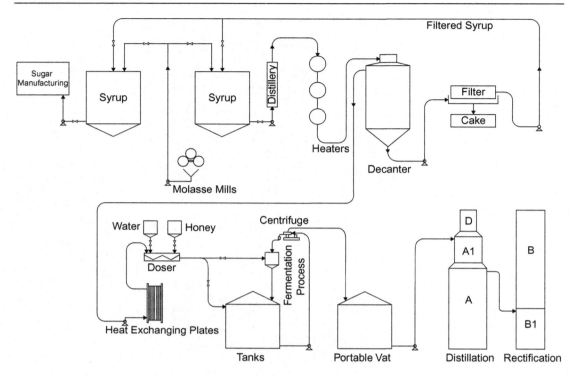

Figure 16.13
Flowchart for ethanol production.

16.4 Ethanol: Uses and Classification

Ethanol (C_2H_5OH), also called ethyl alcohol, is a substance produced by must fermentation and subsequent distillation. The final product, according to the grading and concentration of impurities, has several applications: bio-fuels for vehicles (hydrous and anhydrous ethanol), beverage production and raw material for pharmaceutical, cosmetics, and ethanol—chemical industries.

16.4.1 Fuel Ethanol

Hydrated ethanol

Hydrated ethanol has an ethanol concentration of 92.6—93.8 wt% (INPM, Portuguese acronym for percentage of ethanol by mass). It is used in cars running on ethanol (flex-fuel type).

Anhydrous ethanol

Anhydrous ethanol has an ethanol concentration of at least 99.3 wt% (INPM). It is mixed with gasoline at a maximum ratio of 26%. In a 3% proportion, it has replaced lead-tetraethyl, a highly polluting antiknock agent, a process in which Brazil was a world pioneer.

Complete specifications of the types of fuel ethanol are on the website of the National Petroleum Agency: www.anp.gov.br/doc/audiencia_publica/Minuta_de_Resolucao_AP_9_2005.pdf

16.5 Industrial Production of Ethanol in Brazil

16.5.1 Hydrous Fuel Ethanol

Hydrous fuel ethanol is obtained through two sets, distillation and rectification, and peripheral equipment (Figure 16.14). The first set consists of three overlapping columns (Table 16.1). Rectification is done traditionally through a column with two sections: exhaustion (B_1) and concentration (B). Its main features are shown in Table 16.2.

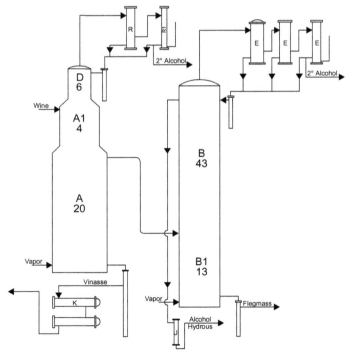

Figure 16.14
Flowchart of hydrous ethanol fuel.

Table 16.1: Main dimensions of the columns of the distillation sector.

Column	Number of Trays	Spacing (mm)
Distillation (A)	20	650
Purification of wine (A_1)	04	500
Concentration of heads (D)	06	350

**Table 16.2: Main dimensions of the columns of the
rectification sector.**

Column	Number of Trays	Spacing
B1	13	350
B	43	350

Values are benchmarks, varying according to the manufacturers. The diameters
of the columns are sized based on production.

Process description

Centrifuged wine with ethanol content in the range of 6.0−10.0% in volume will pass
through the wine pre-heater (E) in which is heated to 70°C in countercurrent with the vapor
on top of the rectification column (B) and then go to K heat exchangers for additional
heating to 92−93°C, with the partial heat from vinasse. After heating, the wine feeds plate
04 in column A1 so that the volatile products (lights or heads), mainly aldehydes and esters,
can go to column D, for concentration. The concentrated vapors leave through the top of
the column to condense at heat exchangers R and R1; part of these vapors return to the top
plate (reflux) and part are removed as second generation ethanol. The bottom product of
column D proceeds to the following rectification set.

The wine climbs down the plates through the weirs to the boiler (base of the column), leaving it
as vinasse, a byproduct which is almost free of ethanol, with less than 0.03% of ethanol volume,
and is a pollutant agent used in fertigation in sugarcane fields. Exhaust vapors (discharged vapors
from counterpressure turbines running at 1.0−1.5 kgf/cm^2 of pressure) or plant vapor (vapor
generated in the concentration of the juice in the first stage of the multiple-effect evaporators,
with 0.7 kgf/cm^2 pressure) feed the boiler of distillation column (A) through direct vapor injection
or through the reboiler (indirect) and climb up the plates through the bubblers, impoverishing (in
terms of ethanol concentration) the wine that goes down and enriching the vapors that rise up.
Above plate 20 of the distillation column, a phlegm comes out, with graduation corresponding to
the vapor from distilled wine, to feed the rectification sector between concentration − Column B
(above the feed) and exhaustion − Column B_1 (below the feed) sections. Column B_1 is heated by
exhaust or plant vapors in enough amounts to deplete the liquid which goes down the weirs and
reaches the boiler with less than 0.03% in volume of ethanol (flegmass).

In the concentration section (column B), the vapors are gradually enriched in ethanol, reach the top, go to condensers E, E_1, and E_2, returning to the top plate (reflux). Hydrous ethanol, the final product, is removed four plates below the top, to reduce contamination by volatile impurities. A heat exchanger (cooler J), using water as coolant, lowers the temperature of the ethanol to a maximum value of 33°C before storage.

The rectification system can only have the concentration section (column B). As in this case there is no depletion, the bottom liquid returns to column A_1.

In several plates at the rectification set, a mixture of higher alcohols known as fusel oil appears in higher concentration. According to the boiling points and solubility of these higher alcohols, such mixtures are classified as low oils and high oils, depending on the region inside columns B_1 and B where they are located. The main higher alcohols are: iso-amyl, iso-butyl, *n*-butyl, iso-propyl, and n-propyl.

In the production of fuel ethanol, fusel oil must be removed to avoid losses by flegmass (if there is a B_1 column) or vinasse (if there is no depletion of phlegm, which returns to column A_1) in this process. The byproduct must be removed, cooled, and stored, since there is a market for it in the chemical industry, especially if the concentration of iso-amyl ethanol is above 60%.

Control of the operation is performed through monitoring of the temperatures of certain plates and of the operating pressure of the columns (gauge pressure). The vapor consumption is a function of the ethanol content of the wine and of the distillation technology employed. In plants that have as main objectives to produce ethanol and to generate electricity for sale, the trend is the deployment of vacuum distillation, which allows for a consumption of vapor which is 50% lower than that in distillation processes performed under atmospheric pressure.

The consumption of cooling water depends on the efficiency of the condensers (project) and water quality. Table 16.3 gives some reference values.

Table 16.3: Operational data from the production of hydrous ethanol fuel.

Parameter	Unit	Value
Temperature — tray A_1	°C	106−110
Temperature — tray A_{20}	°C	98−102
Temperature — tray B_{1-1}	°C	104−108
Temperature — tray B_4	°C	88−92
Pressure — column A	mwc	4.5−5.5
Pressure — column B_1	mwc	3.5−4.5
Water consumption	m^3/m^3 of ethanol	45−60
Vapor consumption	kg/L of ethanol	2.0−3.2

16.5.2 Anhydrous Ethanol

Hydrous ethanol is an azeotropic mixture (same composition in vapor and liquid phases). To obtain anhydrous ethanol, one requires an operation called dehydration, whose types used are: azeotropic distillation, extractive distillation, and molecular sieve, which accounted respectively for 60%, 25%, and 15% of the production of ethanol in 2008.

Azeotropic distillation

In 1902, Young noted that it was possible to obtain anhydrous ethanol by distilling hydrous ethanol in the presence of benzene (Figure 16.15). The industrial process of azeotropic distillation is a consequence of the fact that benzene, ethanol and water, substances with boiling temperatures, respectively, of 80.0°C, 78.3°C and 100.0°C, form, in certain proportions, a mixture of azeotropic boiling point of 64.9°C, lower than any of the individual components. In the ternary mixture, the components are in the following proportions, by weight: benzene, 74.1% ethanol, 18.5%, and water 7.4%. These features allow the dragging of water into its

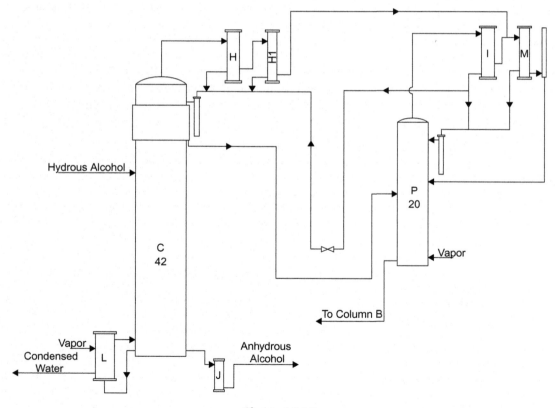

Figure 16.15
Flowchart for azeotropic distillation.

Table 16.4: Main dimensions of the columns of the dehydration sector.

Column	Number of Trays	Spacing
C	42	325
P	20	325

vapor phase through the ternary mixture of high volatility. Several other substances also have this property: hexane, *n*-hexane, trichloroethylene, cyclohexane, etc. However, for decades, benzol, trade name of benzene, was the drying agent used in virtually all anhydrous ethanol production facilities. In 2000, an ordinance of the Ministry of Labor banned its use because of its carcinogenic character. Currently, azeotropic dehydration makes use of cyclohexane and products of mixtures of hydrocarbons, in facilities designed for the dehydration of benzene and now adapted to the new conditions. The main dimensions of the columns of dehydration (C) and recovery of the drying agent (P) are shown in Table 16.4.

Process description

Hydrous ethanol, with graduation around 93° INPM, exits from the rectification column (B) and proceeds to plate 28 of the dehydrator (C), in which the addition of the dehydrating agent will promote the following distribution: in the upper area, the ternary ethanol, water and drying agent, which have lower boiling points; in the intermediate region, the binary ethanol and drying agent, with boiling points lower than that of ethanol; and in the lower area, the final product, anhydrous ethanol, of highest boiling point, as a bottom product. This fact makes it mandatory to use a reboiler (indirect heater L).

Vapors of the ternary mixture, leaving from the top of column C, are condensed in heat exchangers H and H1. The liquid mixture, thus condensed, goes to a decanter, in which it is separated into two layers: the upper, rich in dryer agent, goes back to plate 40 to form the ternary mixture; the inferior mixture, rich in water, goes to plate 12 in Column P. In this column, we have the concentration of the drying agent, which returns to column C. The product is a mixture with ethanol content around 50° INPM, which is reallocated into the rectification column. Table 16.5 presents reference values for the control of the operation.

Extractive distillation

In this process, a third substance, called extracting agent, is also added to the mixture of ethanol−water (Figure 16.16). The goal is the opposite of that in the azeotropic process. Being of low volatility, the extractor draws water into the bottom area of the column, while the anhydrous ethanol leaves through the top. In the past, glycerin was used in extractive distillation processes to obtain anhydrous ethanol. Currently, the extracting agent used is mono ethylene glycol − MEG. The system consists of two columns with the main features described in Table 16.6.

Table 16.5: Operational data regarding the production of ethanol fuel from hydrous ethanol.

Parameter	Unit	Value
Temperature — tray C_1	°C	82
Temperature — tray C_{14}	°C	68—72
Temperature — tray C_{33}	°C	64—66
Pressure — Column C	mwc	2.8—3.2
Temperature — tray A_1	°C	106—110
Temperature — tray P_{17}	°C	70—72
Pressure — column P	mwc	0.8—1.5
Dehydrat. agent consumption	L/m^3 anhydrous	0.6—1.0
Water consumption	m^3/m^3 of ethanol	40—50
Vapor consumption	kg/L of ethanol	1.5—1.6

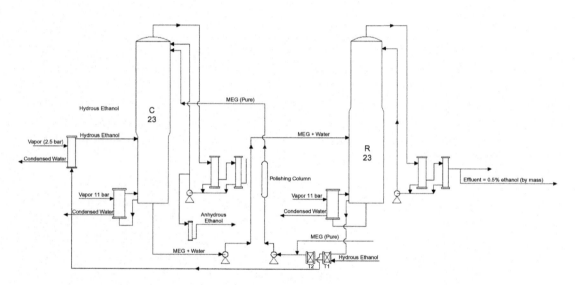

Figure 16.16
Flowchart of extractive distillation.

Table 16.6: Main dimensions of the columns of the dehydration sector.

Column	Number of Trays	Spacing (mm)
C	52	350/400/650
R	23	400/650

Process description

Hydrous ethanol is pumped into tray 19 of the dehydration column (C), being previously heated in a regenerative heat exchanger (T_1) by liquids from the boiler at column R (recovery of MEG), and then in a reboiler with vapor of 2.5 bar.

The extracting agent, MEG, is pumped to plate 46 to the dehydration column (C) and drags the water to the boiler, the mixture (water + MEG) becoming the bottom product. Anhydrous ethanol, now almost totally free of water, rises through the plates until the top and is condensed, so that part of it returns to the column (reflux) and part is cooled, proceeding, as a final product, to storage tanks. Heating of the column is performed by vapor of 11 bar of pressure through reboiler 2.

The mixture MEG + water from the dehydrating boiler (C) is pumped into the tray of column 11 to recover the extracting agent (R), which is heated with 11-bar vapor, from reboiler 3, and operates under vacuum. At the base of this column MEG is recovered and then returns to column C (plate 46), after going through a vase with ionic resin (polishing resin). This equipment prevents contamination of the anhydrous ethanol with 1.4 dioxane, which occurs in some facilities, due to the degraduation of MEG at high temperatures and low pH. Table 16.7 presents operational data regarding the process.

Molecular sieve

Molecular sieves are synthetic adsorbents composed of metal aluminosilicates (zeolites) (Figure 16.17). The ceramic structure of zeolites has a porosity of 3 Å and, by adsorption, retains water and allows for the passage of ethanol.

Process description

The hydrous ethanol is preheated in a regenerative system (TC1) with the anhydrous ethanol that comes out of the sieve. It is then vaporized in a reboiler (EVAP 1) with low pressure

Table 16.7: Operational data from the production of anhydrous ethanol with MEG.

Parameter	Unit	Value
Temperature — boiler C	°C	145–155
Temperature — boiler R	°C	160–170
Vapor heating of the columns	bar	11
Vapor heating of reboiler 1	bar	2.5
Pressure — Column C	kgf/cm^2	0.2–0.3
Vacuum — Column R	Pol Hg	18–20
Consumption of dehydration agent — MEG	L/m^3 anhydrous	0.15
Water consumption	m^3/m^3 ethanol	35–38
Vapor consumption — 11 bar	kg/L ethanol	0.45
Vapor consumption — 1.5 kgf /cm^2	kg/L ethanol	0.33

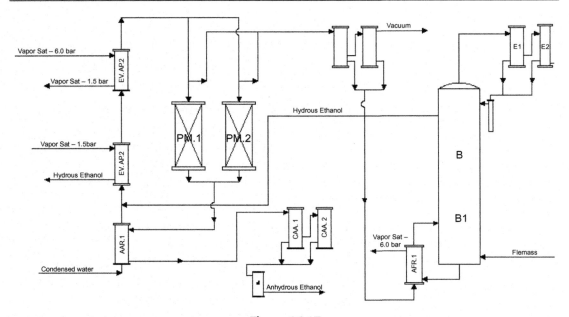

Figure 16.17
Flowchart of dehydration with molecular sieve.

vapor (1.7−2.5 bar) and superheated in a second reboiler (EVAP 2) with high pressure vapor (6 bar). In vapor form, it passes through the sieve (PM), which retains water and allows the passage of anhydrous ethanol. It will then condense in heat exchangers CR1 and CR2 and subsequently be cooled in (J), after which it proceeds to the storage tanks. As dehydration occurs, the sieve becomes saturated with water, which forces the regeneration of the zeolite, made automatically. For this reason, the facilities are composed of two vessels (PM1 and PM2): while one is operating the dehydration, the other is being regenerated. Regeneration is carried out in a vacuum and the liquid (phlegm), with graduation around 60° INPM, is sent to a rectification system (column B/B1), heated with steam at 1.7−2.5 bar, in which it is concentrated, and is then returned to the process as hydrous ethanol. The installation of the concentration system for the phlegm of regeneration, in a vacuum, will reduce the consumption of vapor (low pressure) through the utilization of energy in the rectification process. The operational data are shown in Table 16.8.

16.5.3 Special Alcohols

Impurities or byproducts

The fermented must shows a large amount of byproducts formed in the fermentation process. The quality of the ethanol depends on the concentrations of these impurities. The

Table 16.8: Operating data regarding dehydration with molecular sieves.

Parameter	Unit	Value
Temperature of ethanol upon entering the sieve	°C	130
Steam pressure — EVAP 1	bar	1.7–2.5
Vapor Pressure — EVAP 2	bar	6.0
Vapor pressure — B/B$_1$	bar	1.7–2.5
Vapor cons. — 7 bar pressure	kg/L ethanol	0.05
Vapor cons. — pressure from 1.7 to 2.5 bar	kg/L ethanol	0.7
Water consumption	m^3/m^3 ethanol	40–45

most common are: acetaldehyde, acrolein, crotonaldehyde, acetal, diacetyl, acetone, acetic acid, ethyl acetate, ethyl carbamate, methanol and higher alcohols: iso-amyl, iso-butyl, *n*-butyl, iso-propyl, and *n*-propyl.

It is possible to produce quality ethanol in poor installations under ill-conducted processes through the excessive withdrawal of volatile impurities in the final condensers at columns D and B and of fusel oil (higher alcohols) in the rectification column. Obviously, the cost of this procedure is very high. Efficient production of quality ethanol must have the smallest possible removal of impurities. For that end, the following favorable conditions should be observed: quality of raw material (sugarcane at the point of maturity and high purity), adequate carrying out of the fermentation process, appropriate technical resources in the distillation sector of the installations, adequate chromatography and analytical monitoring, and correct operation.

Anhydrous

Ethanol produced by azeotropic dehydration is intended for the biofuels market due to the natural contamination of the drying agent (hydrocarbon). The other two processes (molecular sieve and extractive distillation) allow for the acquisition of a final product that meets the technical specifications of various markets. There seems to be a clear tendency to utilize the molecular sieve in the new anhydrous ethanol plants, due to the non-contamination of the final products and its low consumption of vapor.

Hydrous

The production of special alcohols obeys the special physical and chemical specifications required by the client. Currently, the foreign market demands special alcohols with the designations: REN, Korea A (24), Korea B (40), and neutral (specifications in Appendices 16.1 and 16.2). Numbers 24 and 40 refer to the maximum amount of higher alcohols allowed. The acquisition of these products may require, in addition to the careful removal of low and high oils at rectification, additional procedures such as transfer and hydroselection to remove other impurities.

The two ethanols of the "Korea" type can be obtained through well-managed fermentation from quality raw materials, without using the additional resources aforementioned.

Transfer operation

The distillation of hydrous ethanol (transfer) concentrates the volatile impurities in a column with a recommended minimum number of 40 plates, which are withdrawn at the last condenser as ethanol of second rate. The purified ethanol leaves through the base. In a default installation of anhydrous ethanol, the dehydrating column (C) can be used for this operation.

Hydroselection operation

The process of hydroselection (a more effective process than the transfer) is the addition of hot water on top of the column, to change the solubility and volatility of the substances, causing them to split. From the top of the column, comes a concentrated mixture of volatile impurities, and at the base a purified hydro-alcoholic mixture of low graduation (10−12° INPM).

Neutral ethanol

The production of neutral ethanol from hydrous ethanol requires both processes, hydroselection and transfer, because it is a product that requires a high degree of purity (Figure 16.18). Due to its application, for instance, in the beverage and cosmetics industries, sensory evaluations are made besides instrumental ones. The heating of the columns should be performed through reboilers (indirect heating), to avoid contamination by vapors. Table 16.9 presents some characteristics of the equipment.

Process description The hydrated ethanol is pumped into the hydroselection column, entering ten plates below the top, at which point the water enters at 85−90°C, changing the volatility and solubility of various impurities, which are removed in the final condenser as a second-rate ethanol. The purified phlegm descends, at the plates, until it reaches the boiler with 10−12° INPM, whence it is pumped into the rectification column (B / B1). In this column, the following procedures take place:

- concentration of phlegm at the desired graduation in the final product, at the top;
- sending of partially purified hydrous ethanol to the demethylation column (transfer);
- robust removal of lower and higher oils (higher alcohols);
- elimination of impurities through the final condenser end, as the second ethanol; and
- exhaustion of phlegm in the form of effluent (flegmass).

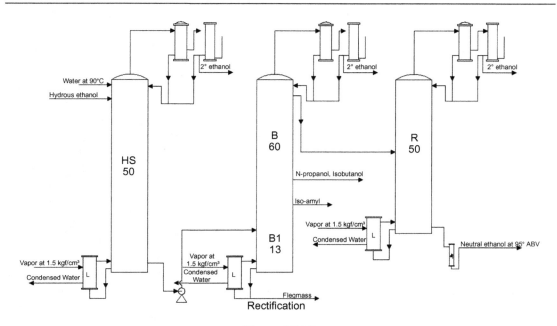

Figure 16.18
Flowchart for the production of neutral ethanol from hydrous ethanol.

Table 16.9: Main dimensions of the columns for production of neutral ethanol.

Column	Number of Trays	Spacing (mm)
Hydroselection (HS)	50	400
Depletion of phlegm (B_1)	13	350
Rectification (B)	60	400
Demethylation (R)	50	400

Table 16.10: Consumption of vapor and water from hydrous ethanol.

Parameter	Unit	Value
Vapor consumption 1.7–2.5 bar	kg/L	3.5
Water consumption	m^3/m^3 of ethanol	100

The hydrous ethanol enters the demethylation column for further purification. The impurities that were not removed until this stage of the process are removed in the final condenser, and the neutral ethanol, which leaves at the base of the column, is cooled and stored. Table 16.10 shows some data regarding this process.

Appendix 16.1 Specifications of Hydrous Ethanol for Export

Parameters	Units	Korean A (24)	Korean B (40)	REN
INPM Graduation at 20°C	% mass	93.0 min	92.6 min	93.8 min
Aldehydes (acetaldehyde)	mg/100 mL	6.5 max	10.0 max	1.0 max
Acidity (acetic acid)	mg/100 mL	3.0	3.0 max	1.0 max
Benzene, chloride, sulfur	—	Absent	Absent	Absent
Crotonaldehyde	—	Absent	Absent	Absent
Cyclohexane	—	Absent	Absent	Absent
Hydrocarbons	—	Absent	Absent	Absent
Dry extract	mg/100 mL	4.0 max	4.0 max	1.5 max
Esters (ethyl acetate)	mg/100 mL	9.0 max	15.0 max	1.0 max
I-propanol	mg/100 mL	2.0 max	2.0 max	1.0 max
Methanol	mg/100 mL	6.0 max	8.0 max	1.0 max
Permanganate test at 20°C	minutes	2 min	NS	10 min
Higher alcohols*	mg/100 mL	24.0 max	40.0 max.	3.0 max
Alkalinity	—	Negative	Negative	Negative
Appearance	—	**	**	**
Alpha color	—	NS	NS	5 max.
Acetal	mg/100 mL	10.0 max	10.0 max	5.0 max

NS: non-specified.

*Includes: *i*-propanol, *n*-propanol, *i*-butanol, *n*-butanol, and *i*-amylic.

**Clean and free of materials in suspension.

Appendix 16.2 Specifications of Neutral Hydrous Ethanol

Parameters	Units	Specifications
Graduation at 20°C	% volume	96.1 min
Esters, in ethyl acetate	mg/100 mL	0.5 max
Aldehydes, in acetaldehydes	mg/100 mL	0.5 max
Acidity, in acetic acid	mg/100 mL	1.0 max
Methanol	mg/100 mL	1.0 max
Higher alcohols	mg/100 mL	0.5 max
Benzene, chloride, sulfur	ppm	Absent
Acetal	—	Non-detectable
Acetone	—	Non-detectable
Crotonaldehyde	—	Absent
Hydrocarbons	—	Absent
Alkalinity	—	Negative
Permanganate test at 20°C	minutes	30 min
Appearance	—	*

*Clean and free of materials in suspension.

Bibliography

Barreto, T.V., Duarte-Coêlho, A.C., 2008. Otimização do dimensionamento de aparelhos para produção de álcool hidratado. Anais do 90 Congresso Nacional da STAB. STAB, Maceió.

Barreto, T.V., Duarte-Coêlho, A.C., Cruz, F.A.C.M., 2007. Consumo de água em destilarias de álcool. XXIV Simpósio da Agroindústria da Cana-de-Açúcar de Alagoas. STAB, Maceió.

Dantas, R.B., 1983. Fatores para consolidação da alcoolquímica; aspectos tecnológicos, econômicos e empresariais Anais do 2^0 Congresso Brasileiro de Alcoolquímica, Recife. IBP e ABIQUIM.

Foust, A.S., et al., 1982. Princípios das operações unitárias. Editora Guanabara Dois S.A, Rio de Janeiro.

Gaussent, P., 1984. Utilisation des chaleurs disponibles dans les unités d'ethanol. Salon Professionnel de la Maîtri se de l'Energie dans l'Industrie. ADEME, Paris.

Incropera, F.P., Dewitt, D.P., 2003. Rio de Janeiro: LTC Editora.

Kreith, F., Bohn, M.S., 2003. Princípios de Transferência de Calor. Pioneira Thomson Learning, Thomsom.

Kretzschmar, H., 1961. Levaduras y alcoholes. Editorial Reverté S.A, Barcelona.

Llames, H.P., 1956. Fabricacion del alcohol. Salvat Editores S.A, Barcelona.

Lyons, T.P., et al., 1995. The Alcohol Textbook. Nottingham University Press, Thrumpton.

Mariller, C., 1943. Distillation et rectification des liquides industriels. Ed. Dunod, Paris.

Mariller, C., 1951. Distilerie agricole et industrielle. Ed. Baillère et Fils, Paris.

Mauguin, P.H., Bonfilis, C., Nacfaire, H., 1994. Biocarburants en Europe développements, apllications, perspectives. Actes du 1er Forum Européen sur les Biocarburants. ADEME, Tours.

McCabe, W.L., Smith, J.C., 1973. Operaciones básicas de ingeniería química. Editorial Reverté S.A, Tokyo.

Menezes, T.J.B., 1980. Etanol, o combustível do Brasil. Editora Agronômica Ceres Ltda, São Paulo.

Nisenfeld, A.E.; Seemann, R.C., 1981. Distillation columns. Instrument Society of America.

NG Metalúrgica Ltda, 2008. NG Metalúrgica Geração. Boletim Informativo, Piracicaba, n. 20.

Perry, J.H., Chilton, C.H., 1980. Manual de engenharia química. Editora Guanabara Dois S.A, Rio de Janeiro.

Rasovsky, E.M., 1973. Álcool. Coleção Canavieira (12).

Shinskey, F.G., 1977. Distillation Control for Producitivity and Engergy Conservation. McGraw-Hill Kogakusha Ltda, New York.

Treybal, R.E., 1968. Massa Transfer Operations. McGraw-Hill Kogakusha Ltda, New York.

Winkle, M.V., 1967. Chemical Engineering Series: Distillation. McGraw-Hill Kogakusha Ltda, New York.

CHAPTER 17

Industrial Waste Recovery

**Sarita Cândida Rabelo[1], Aline Carvalho da Costa[2]
and Carlos Eduardo Vaz Rossel[1]**

[1]*Brazilian Bioethanol Science and Technology Laboratory—CTBE/CNPEM, Campinas, SP, Brazil*
[2]*School of Chemical Engineering, State University of Campinas, Campinas, SP, Brazil*

Introduction

Using the waste generated by the sugar and alcohol industry has become imperative both from the environmental point of view, and as a means to prevent wasting profitable material. The waste produced when processing sugarcane to produce sugar and alcohol, which is already being generated in significant amounts, should increase considerably during the next years, as this crop expands and new agro-industrial plants are implemented. This waste material can be used as raw material for new products. The flowchart in Figure 17.1 shows potential byproducts from this industry.

17.1 Bagasse and Straw Recovery

Sugarcane bagasse is the fraction of biomass resulting from the cleaning, preparation, and extraction of sugarcane juice. Several factors influence sugarcane bagasse composition:

- the use of fire or another method for straw removal before cutting;
- harvesting and loading methods resulting in greater or lesser dragging of dirt, sand and vegetable residue, i.e., manual, mechanical cutting, chopped cane, cutting to include the tip, etc.;
- the type of soil where sugarcane was planted (latosols, sandy soils, and other types of soils); and
- different procedures used for cleaning sugarcane.

Bagasse is heterogeneous in terms of size and particle format and regarding the three predominant components, the polymers: cellulose, hemicellulose, and lignin. Table 17.1 shows bagasse potential as a source of reducing sugars from the cellulose and hemicellulose present in the bagasse.

Sugarcane. DOI: http://dx.doi.org/10.1016/B978-0-12-802239-9.00017-7
365

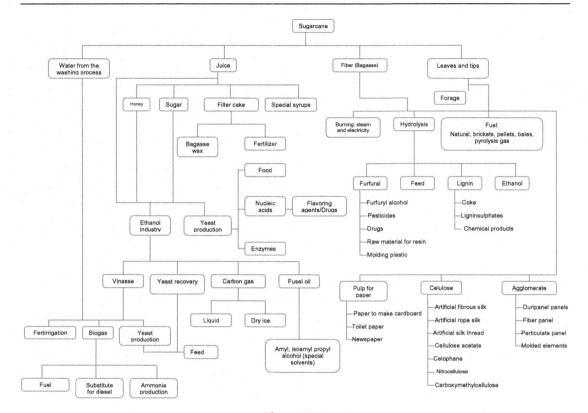

Figure 17.1
Potential products and byproducts of the sugar and alcohol agroindustry.

Table 17.1: Bagasse composition.

Composition (%) Dry Base	Bagasse
Glucose	19.50
Xylose	10.50
Arabinose	1.50
Galactose	0.55
Lignin	9.91
Organo solubles	2.70
Reducing sugars	1.85
Uronic acids	1.91
Ash	1.60
Moisture	50.00
Total hexoses	20.04
Total pentoses	12.00

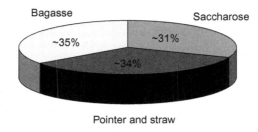

Figure 17.2
Energy provided by sugarcane.

Ethanol production currently uses saccharose as raw material, and that corresponds to using approximately one-third of the energy found in the sugarcane, as can be seen in Figure 17.2.

Bagasse is burned with very low energy efficiency to generate vapor and electric energy and the harvesting process causes straw to be wasted. With the growing use of automatic harvesting equipment and the gradual elimination of the practice of burning straw (Law 11241/02), it is hoped that sugarcane can be used more completely.

17.1.1 Biomass Hydrolysis

Both the bagasse and the straw are basically constituted of lignin, hemicellulose, and cellulose. Hydrolysis transforms hemicellulose and cellulose into sugars and these can be transformed into ethanol by fermentation. Cellulose hydrolysis produces glucose, which is easily fermented into ethanol by microorganisms, but hemicellulose hydrolysis produces mostly xylose. Few microorganisms ferment xylose into ethanol, therefore, several studies are attempting to find a viable way to use this fraction.

Although production costs for ethanol from lignin–cellulose material are still not competitive, several countries are investing in studies whose aim is to make this technology feasible. In Brazil, the "power cane" paradigm involves optimizing the harvesting process and the energy balance of the sugar mills, in order to increase the surplus biomass. This surplus can be transformed into ethanol, which translates into increased production without an increase in planted area. The expectation is that once technical obstacles are overcome and the process optimized, this will represent a 50% increase in ethanol production.

There are two hydrolysis routes for the lignin cellulose material: acid and enzymatic hydrolysis. Acid hydrolysis normally takes place at high temperatures and in pressurized reactors, and the most commonly used acids are sulfuric acid or hydrochloric acid. In this type of hydrolysis, if reaction conditions (temperature and acid concentration) are not well controlled, the final products (sugars and lignin) will be degraded, inhibiting metabolism of the fermentation which will take place later. Additionally, sugar yields are very low and the process requires expensive corrosion resistant reactors.

In enzymatic hydrolysis, an enzyme complex (cellulases), rather than acids, is used. This process takes place at low temperatures, without pressurization, and causes less pollution. Currently, one of the largest challenges that must be overcome is the cost of enzymes, although several research projects are being conducted, in several countries, aiming at reducing this cost.

Enzymatic hydrolysis processes, as well as some acid hydrolysis processes, require a pre-treatment to break the structure of biomass, since the presence of lignin and hemicellulose makes access to cellulose difficult. The main pre-treatment objectives are: reducing the degree of cellulose crystallinity, dissociating the lignin—cellulose complex, and increasing the surface area of the biomass, so that efficiency in the hydrolysis stage is increased.

Possible hydrolysis routes and fermentation of the resulting sugars are shown in Figure 17.3. In two-stage processes, first we have the hydrolysis (acid or enzymatic) and then comes sugar fermentation, when pentoses are fermented, followed by hexose fermentation, or the simultaneous fermentation of pentoses and hexoses.

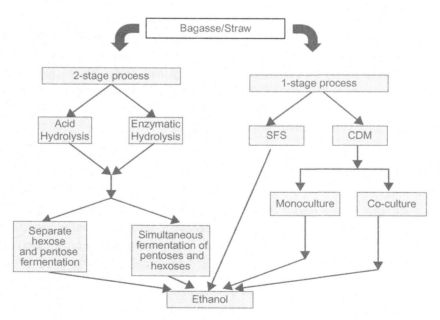

Figure 17.3
Hydrolysis and fermentation routes.

Table 17.2: Transformation potential of bagasse into ethanol (liters/ton of bagasse).

Scenario	Forecast Conversions	Ethanol Hexoses	Ethanol Pentoses	Total Ethanol
[1]	Hexoses: 60% fermentation: 89% pentoses: 70% fermentation: 0% distillation: 99.5%	69.1	0	69.1
[2]	Hexoses: 80% fermentation: 91% pentoses: 78.5% fermentation: 0% distillation: 99.75%	94.2	0	94.2
[3]	Hexoses: 80% fermentation: 91% pentoses: 85% fermentation: 50% distillation: 99.75%	94.2	37.9	132.2
[4]	Hexoses: 85% fermentation: 89% pentoses: 70% fermentation: 0% distillation: 99.5%	97	0	97
[5]	Hexoses: 95% fermentation: 91% pentoses: 85% fermentation: 50% distillation: 99.75%	111.4	37.9	149.3

[1] Pre-treatment and diluted acid hydrolysis with hexose recovery, employing current technology.

[2] Pre-treatment and diluted acid hydrolysis with hexose recovery and optimization of the hydrolysis reaction, to achieve the best values reported in the literature.

[3] Pre-treatment and diluted acid hydrolysis with hexose and pentose recovery, and reaction optimization to achieve the best values reported in the literature.

[4] Pre-treatment and enzymatic hydrolysis with hexose recovery, employing current technology.

[5] Pre-treatment and enzymatic hydrolysis with hexose and pentose recovery, employing optimized recovery.

In one-stage processes, hydrolysis and fermentation take place at the same time. In the Simultaneous Saccharification and Fermentation (SSF) process, cellulases of a cellulolytic microorganism (generally, a fungus of the *Trichoderma* genus) are used, along with an ethanol producing microorganism. In the Direct Conversion by the Microorganism (DCM) process, microorganisms produce the enzymes and effect fermentation. We can use single culture, in which a single microorganism hydrolyses the cellulose and ferments the sugars into ethanol, or co-cultures, in which one of the microorganism is responsible for the hydrolysis and the other for the fermentation. Each of these routes has advantages and disadvantages. They are being studied by several groups of scientists, but they have not reached a consensus yet.

The transformation potential of bagasse into ethanol is described in Table 17.2 for different scenarios.

17.1.2 Furfural

One of the possible processes for bagasse and sugarcane straw recovery is hydrolysis of the hemicellulose fraction and conversion of the resulting pentoses into furfural. The existing processes consist of treating the biomass in an autoclave, at high temperatures, using an acid catalyst. Average yield for this process is 70 kg of furfural per ton of bagasse.

Furfural can be used as an intermediary for furfuryl alcohol and molding resins, or as a solvent in petrochemical refining.

17.1.3 Using Bagasse for Cattle Feed

When incorporated into cattle feed, the digestibility of *unprocessed* bagasse is low, as the microbial flora in the rumen is not very effective in digesting the cellulose, hemicellulose, and lignin complex.

Procedures such as adding sodium hydroxide, lime, or exposure to water steam at high temperatures (150–280°C) have been developed to increase bagasse digestibility. They act by removing or segregating lignin and causing partial hydrolysis of hemicellulose into reducing sugars (pentoses). The rumen assimilates this treated bagasse better, and is able to metabolize the free pentoses and hexoses resulting from cellulose hydrolysis. Among these processes, the most widely used is Steam Explosion.

This steam treatment (discontinuous or continuous) is done in autoclaves, where bagasse is subjected to water steam at 15–25 bar for 5–20 min. Under these conditions, hemicellulose undergoes hydrolysis and releases organic acids that serve as catalysts. When subjected to hydrolysis, hemicellulose gives rise to pentoses (xylose and arabinose) and hexoses (glucose and galactose). Lignin melts and becomes segregated and cellulose becomes more exposed to the attack of rumen microorganisms.

After this auto hydrolysis stage, bagasse is subjected to instant expansion to atmospheric pressure. This expansion causes the temperature to drop sharply, interrupting the reactions that cause pentose to decompose to furfural and hexoses to hydroxymethylfurfural. Fiber bundles separate, causing particle size to decrease, which in turn increases the exposed area between them, ultimately favoring bagasse digestibility.

Table 17.3 presents the typical composition and properties of prehydrolyzed bagasse.

Feedlot tests have confirmed the efficiency of prehydrolyzed bagasse in cattle feed. There was no reduction in rumen cellulolytic bacteria activity. The performance of feed made

Table 17.3: Composition of prehydrolyzed bagasse.

Parameter	Amount
Moisture %	60–65
pH	3–4
Reducing sugars % s/ms	17.6
Acetic acid % s/ms	1–2
Furfural % s/ms	Less than 0.2
Cellulose % s/ms	35.8
Hemicellulose % s/ms	16.4
Lignin % s/ms	20.2
Ash % s/ms	2.2
Digestibility % s/ms	50–60

**Table 17.4: Typical composition of animal feed made
with byproducts from the mill.**

Dry Matter %	90.00
Crude protein % s/ms	5.98
Ester Extract % s/ms	0.47
Raw fiber % s/ms	27.59
Mineral matter % s/ms	5.09
Non-nitrogenated extractives % s/ms	50.87
Calcium % s/ms	0.17
Phosphorus % s/ms	0.10
Potassium % s/ms	0.71
Methionine mg/kg	0.09
Cysteine mg/kg	0.04
Lysine mg/kg	0.27
Tryptophan % s/ms	0.08
Arginine mg/kg	0.20
B-complex vitamins mg/kg	0.68

with prehydrolyzed bagasse proved to be satisfactory. Remaining levels of furfural and phenolic residues did not compromise feed digestibility.

An evolution of prehydrolyzed bagasse is its use in the formulation of a cattle feed based on byproducts of the sugar and alcohol industry: prehydrolyzed bagasse, dry yeast, and final molasses, at 80:15:5, respectively.

The prehydrolyzed bagasse is dried first (moisture content: 20−25%), then the three components are mixed and extruded as *pellets*. This animal feed is stable, and can be kept for a long period of time. Density is 600−700 kg/m^3, which makes transportation and storage easy.

Table 17.4 presents the typical composition of feed obtained by processing the byproducts of the sugar and alcohol mills (Magnani et al., 1985). This feed is highly digestible and has proteins, vitamins and minerals that are essential for animal growth. It adds value to byproducts such as bagasse and yeast, and diversity to mill production, thereby creating alternatives for feedlot systems integrated to ethanol, sugar, and energy.

17.2 Yeast and Molasses Recovery

Among ethanol industry byproducts, *Saccharomyces cerevisiae* calls our attention for its high nutritional value and protein content. This yeast is also very important for the industry and can be used in several ways.

In its active form, used in bread-making, alcohol fermentation, and in other industrial processes, the yeast must be produced by well-controlled processes, in fermenters, to get a high purity product. In ethanol distilleries, surplus yeast cells, generated by fermentation, can be deactivated, by thermal or other processes and used directly (integral yeast cells) or processed to produce several byproducts.

17.2.1 Chemical Composition of Yeast

The chemical composition of yeast can vary as a function of a series of factors, such as the substrate used, the type of yeast, fermentation method, cell age, and drying conditions (Desmonts, 1968). Moreover, during the process of obtaining recovery yeast, successive washings, to eliminate impurities in the yeast milk or the residue found at the bottom of the vats, can lead to significant variations in yeast composition.

Table 17.5 presents the approximate chemical and mineral composition of dry recovery yeast (Lahr Filho et al., 1996). In addition to the high protein content, yeast products are rich in B-complex vitamins (B_1, B_2, B_6, pantothenic acid, niacin, folic acid, and biotin), minerals, and in macro and micro elements, particularly selenium and dietary fiber.

17.2.2 Yeast Recovery for Animal Feed

During the alcohol fermentation process there is a concomitant generation of yeast, which is necessary to replace and renew viable cells, so as to ensure process quality and stability. Yeast loss during this process takes place during the wine centrifugation stage (yeast milk tapping), at the bottom of the vats and during the acid treatment of the vats (Butollo, 1996).

The amount of yeast that can be recovered from the alcohol fermentation process is of the order of 20 kg of yeast per cubic meter of alcohol produced. However, in order to obtain this amount, fermentation conditions must be strictly controlled, so that alcohol yield is not affected by the partial removal of yeast during the industrial process.

Table 17.5: Chemical and mineral composition of dry yeast (*Saccharomyces cerevisiae*) resulting from alcohol fermentation.

Components (%)	Yeast (*Saccharomyces cerevisiae*)
Dry matter	86.71%
Crude protein	26.95%
Ester extract	0.96%
Mineral matter	4.92%
Calcium	0.57%
Phosphorus	0.52%
Magnesium	0.17%
Manganese	26 mg/kg
Iron	518 mg/kg
Zinc	173 mg/kg
Potassium	2 mg/kg

Yeast recovery from fermented wine centrifugation is carried out immediately after the end of the fermentation, and this stage is necessary to ensure continuity to the ethanol production process, through yeast reactivation and recycling to a new cycle of alcohol fermentation. The tapped cream is washed with water and centrifuged again, after which it is sent, at an 11% concentration (dry base) to the thermolysis tank and kept for 15−30 min at temperatures ranging from 70°C to 85°C. Finally, the yeast is dried in a drum drier to obtain a product that can be used for feed. This product is shaped like scales and has an 8% moisture content.

Chemical composition of yeast resulting from alcohol fermentation can be changed by means of a starvation process. It takes place at 37°C, in approximately 12 h, under constant agitation and in the absence of an energy source. This process causes mass to be lost, metabolizes carbohydrates, generates ethanol and increases protein content.

Figure 17.4 presents a diagram illustrating the production of active dry yeast.

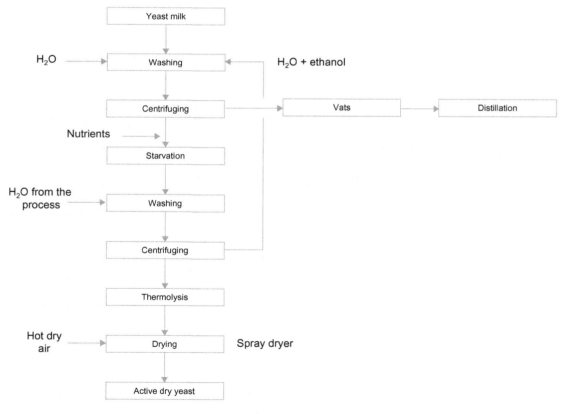

Figure 17.4
Yeast drying.

Currently, the dry yeast resulting from this practice is used mostly in feed for the fish farming industry (shrimps, fish), piggeries (piglets, sows), the dairy cattle raising sector, and by companies that raise small animals, mostly cats and dogs. Dry yeast production has been growing in almost every market in the world, especially because of its properties. It provides flavor, promotes growth, serves as an immune stabilizing agent, as a source of proteins and of the B-vitamin complex. It also has toxin sequestering properties (Desmonts, 1966).

17.2.3 Use of Yeast Byproducts in Food

In producing yeast byproducts for human nutrition, it has been found that cells from the distillery tapping do not have the necessary quality. For this reason, an option is to perform a pure yeast culture. The culture is prepared in compliance with sanitary standards, specifically for the purpose of developing a cell mass by means of aerobic fermentation (air bubbling). The substrate used to prepare the yeast must be a source of sugar. Molasses, a byproduct from the sugar manufacturing process is most of the time used for this purpose. In producing alcohol, it must undergo purification and sterilization.

The potential of the yeast species *S. cerevisiae* to provide important elements for human metabolism, such as proteins, nucleic acids, polysaccharides, and vitamins is significant. It also has the potential to generate biochemical products that are commercially interesting for the food industry.

Depending on the type of process used, yeast cells can give rise to total auto lysates, extracts, protein concentrates, and cell wall, such as is presented in the diagram in Figure 17.5.

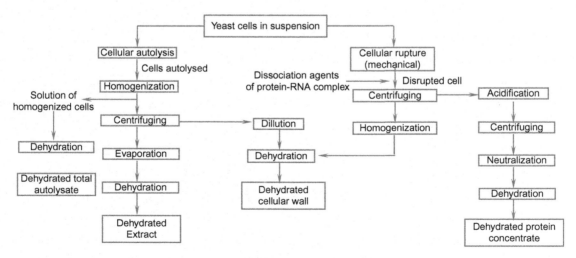

Figure 17.5
Obtaining products and food ingredients from yeast.

Controlled autolysis is the source of total auto lysates, yeast extracts and part of the cell wall. Autolysis is a cellular auto digestion process which is triggered by the internal cell enzymes, such as proteases, nucleases, glucanases, phopholipases, estearases, aminopeptidases, and other lysosomal enzymes. This operation is controlled by plasmolytic agents and the time/temperature interaction of the cell suspension in the equipment. This cell suspension is then centrifuged, to separate the soluble fraction (extract) from the insoluble (cell wall), and both are thermally treated and dried. The resulting products are dry yeast extract, and dry cell wall, respectively.

An option for enzyme lysis is mechanical lysis. This operation is performed by equipment that is similar to a homogenizer, and the cells are broken by mechanical pressure. This process causes more than 90% of the cells to rupture and further treatments result in a protein concentrate. Intermediary operations include centrifuging the ruptured cell suspension with the consequent separation of the insoluble fraction – the cell wall, which is dried and sold in this format – from the soluble fraction, which contains the proteins. Then comes the process of acidification, which serves to precipitate the proteins, a new centrifugation, to isolate them, neutralization of the acid solution that was added, and drying. The result will be a dried protein concentrate which is ready for sale.

Autolyzes and yeast extracts have organoleptic characteristics that bring to mind the flavor of meat, cheese, or mushroom, and can be used with sensory and nutritional benefits in sauces, canned food, soups, meat sauces and meat stock, sausages and chemically enhanced pasta (Sgarbieri et al., 1999).

In the case of the protein concentrate from yeast, studies aimed at finding applications for human nutrition are not as advanced, since this process requires expensive and sophisticated technology. Additionally, other byproducts that are more easily obtained have already been found to give satisfactory results.

17.3 Filter Cake Recovery

Filter cake can be defined as the residue that is eliminated during the cane juice decantation process, during the treatment stage, in sugar and/or alcohol production. After the juice treatment decantation stage, the resulting slurry is sent to the filtration sector where residual sugar is removed, resulting in the filter cake. Cake composition depends on several factors: type of soil, sugarcane variety, harvesting method (manual or mechanized), juice extraction, lime dosage, and other products used for clarification, filtration methods, among others. *Unprocessed* filters contain approximately 75% water and mean composition is presented in Table 17.6.

**Table 17.6: Typical composition of sugarcane
filter cakes (% of dry matter).**

Components	Content (%)
Wax, lipids, and resin	4−14
Fiber	15−30
Sugars	1−15
Protein	5−15
Total ash	9−20
SiO_2	4−10
CaO	1−4
P_2O_5	1−3
MgO	0.5−1.5

Source: Paturau (1969).

In most of the mills, filter cake is used as fertilizer in sugarcane fields. This is the most widely disseminated use of this byproduct, especially because of the significant amount of nitrogen, phosphorus, calcium and organic matter available. Recently, this use has become very popular among farmers and results have been satisfactory for the production of banana, eucalyptus, guava, and vegetable seedlings.

A nobler destination for filter cake is using it to produce wax (Graille et al., 2003). Sugarcane wax is the general term used when referring to the lipids found in sugarcane. These lipids represent, approximately, 0.18% of plant weight and consist of two fractions, the wax and the oil fractions (Paturau, 1969). During the agroindustrial process, only 40% of the lipid material is dispersed in the juice, while the remaining portion remains in the bagasse. Of the portion found in the juice, approximately 95% is concentrated in the filter cake (Laguna et al., 1996).

Cane wax, extracted from the cake filter can be used as an alternative for vegetable, animal, and synthetic waxes, since it meets the quality criteria established by the industry and is an important input for the food, pharmaceutical, chemical, cosmetic, and cleaning and polishing product industries (Graille et al., 2003). This wax can be recovered from the filter cake along with the total lipids by an extraction process that uses a solvent. Yield will depend on the procedure used (type of solvent, temperature, extraction time, relationship between solvent and solute) (Villar et al., 2005) and the characteristics of the matrix that is being extracted (particle size and moisture).

Currently, there are no cane wax producers and suppliers in Brazil, but as a large sugarcane producing country, Brazil has a promising future as a renewable raw material supplier and as a source of a new type of natural wax (Graille et al., 2003).

17.4 Vinasse Recovery

Vinasse is the final residue from distillation of the fermentation wine, used to obtain ethanol. Vinasse composition will depend, among other factors on:

- raw material used in the fermentation process: molasses, juice or a mixture of these materials;
- type of soil, sugarcane varieties, harvesting methods; and
- industrial process used in the mill.

Table 17.7 presents the typical composition of vinasse from Brazilian distilleries.

Vinasse has a high content of oxygen, a low pH and is high in mineral salts. For that reason, it represents a potential danger to the environment if not treated correctly.

Technologies commonly used for treating vinasse are:

- use in fertirrigation in the field, *unprocessed*;
- thermal concentration and use in the crop; and
- biodigestion.

17.4.1 Fertirrigation with Unprocessed Vinasse

In Brazil, practically all the vinasse that is generated is used for soil treatment, for the purpose of using its content, mainly potassium, to meet the nutritional requirements of sugarcane. The phosphorus and nitrogen found in vinasse also play a smaller contributing role. The fertirrigation process also meets the water needs of sugarcane crops, since the harvest season coincides with the dry season. In comparison with other soil treatments, fertirrigation requires less investment.

Transportation and distances are critical parameters, because of the cost of diesel oil. The usual methods of application are:

- transportation in water trucks and direct application in the field;
- transportation in water trucks to harvest sites and distribution by spray guns or water reels and distribution tubes/arms; and
- transportation via channels or pipes and application with spray guns or water reels and distribution tubes/arms.

Studies have shown that the application of vinasse increases productivity by 5−10%. Dosage is close to 100−150 m^3 of vinasse per hectare, and the result is strongly associated to such factors as: vinasse composition, type of soil and depth of the water table. Dosage above recommended values can contaminate underground water. In the State of São Paulo, Cetesb established safe application criteria for vinasse (CETESB, 2005). The standard limits dosage as

Table 17.7: Typical composition of sugarcane vinasse.

Description	Concentrations—Standard			
	Minimum	Mean	Maximum	Per litre of Alcohol
Process Data				
Must Brix (°B)	12.00	18.65	23.65	
Wine alcohol content (°GL)	5.73	8.58	11.30	
Vinasse rates (L/L of alcohol)	5.11	10.85	16.43	10.851
Reference flow (m³/day)	530.00	1908.86	4128.00	
Characteristics of Vinasse				
pH	3.50	4.15	4.90	
Temperature (°C)	65.00	89.16	110.50	
Biochemical oxygen demand (BDO) (mg/L)	6680.00	16,949.76	75,330.00	175.13 g
Chemical oxygen demand (CDO) (mg/L)	9200.00	28,450.00	97,400.00	297.60 g
Total solids (TS) (mg/L)	10,780.00	25,154.61	38,680.00	268.90 g
Total suspended solids (TSS) (mg/L)	260.00	3966.84	9500.00	45.71 g
Fixed suspended solids (FSS) (mg/L)	40.00	294.38	1500.00	2.69 g
Suspended volatile solids (SVS) (mg/L)	40.00	3632.16	9070.00	43.02 g
Total dissolved solids (TDS) (mg/L)	1509.00	18,420.06	33,680.00	223.19 g
Volatile dissolved solids (VDS) (mg/L)	588.00	6579.58	15000.00	77.98 g
Fixed dissolved solids (FDS) (mg/ L)	921.00	11,872.36	24,020.00	145.21 g
Sedimentable waste (RS) 1 hour (ml/L)	0.20	2.29	20.00	24.81 ml
Calcium (236 mg/L CaO)	71.00	515.25	1096.00	5.38 g
Chloride (mg/L Cl)	480.00	1218.91	2300.00	12.91 g
Copper (236 mg/L CuO)	0.50	1.20	3.00	0.01 g
Iron (mg/L Fe_2O_3)	2.00	25.17	200.00	0.27 g
Total phosphorus (mg/L P_2O_4)	18.00	60.41	188.00	0.65 g
Magnesium (mg/L MgO)	97.00	225.64	456.00	2.39 g
Manganese (mg/L MnO)	1.00	4.82	12.00	0.05 g
Nitrogen (mg/L N)	90.00	356.63	885.00	3.84 g
Ammonia nitrogen (mg/L N)	1.00	10.94	65.00	0.12 g
Total potassium (236 mg/L K_2O)	814.00	2034.89	3852.00	21.21 g
Sodium (mg/L Na)	8.00	51.55	220.00	0.56 g
Sulfate (mg/L SO_4)	790.00	1537.66	2800.00	16.17 g
Sulfite (mg/L SO_4)	5.00	35.90	153.00	0.37 g
Zinc (236 mg/L ZnO)	0.70	1.70	4.60	0.02 g
Ethanol-GC (ml/L)	0.10	0.88	119.00	9.1 ml
Glycerol (ml/L)	2.60	5.89	25.00	62.1 ml
Yeast (dry base) (mg/L)	114.01	403.56	1500.15	

a function of the cationic exchange capacity of the soil and its potassium content, potassium content of vinasse and the ability to assimilate potassium per hectare/year of the sugarcane.

17.4.2 Vinasse Concentration

Thermal concentration is one of the commercially available technologies to reduce the volume of vinasse.

While thermal concentration using multi-effect evaporators has been widely disseminated in ethanol production from corn and other grains, it is not used to produce ethanol from sugarcane. The main reason for this is the high nutritional value of the vinasse resulting from ethanol production from corn, which adds commercial value to the product. However, the nutritional value of vinasse from sugarcane ethanol is not as significant, and the product's usefulness is limited to fertirrigation due to the fact that it is rich in potassium.

Another obstacle is the energy consumption required for the process, which requires more vapor to effect the concentration.

17.4.3 The Cost of using Vinasse

Data from the sugar and ethanol industry have made it possible to estimate the costs of using vinasse for fertirrigation:

- transportation via channels or pipes and application with spray guns or water reels: R$2.35 per m^3;
- transportation in water trucks (10 km) and the use of water reels: R$2.83/ m^3; and
- concentrate at 40°Brix (R$67.00 to concentrate), transportation and use of a water truck for distribution: R$69.59 per m^3.

17.4.4 Biodigestion for the Production of Biogas

The advantage of biodigestion is that it can be carried out at the production site. Additionally, it generates fuel gas with energy recovery and results in a fertilizer whose pH is closer to neutral, while maintaining the potassium and the other mineral nutrients found in the vinasse. The chemical demand for oxygen of the resulting effluent is a great deal smaller than that for vinasse (of the order of 5000 ppm).

In Brazil, UASB digester (anaerobic biological mud bed with an upward flow), which operate within the mesophilic (35−37°C) and thermophilic (50−55°C) ranges were implemented for demonstration purposes. It was demonstrated that this technology generates a gas containing 60% methane, 40% carbon dioxide, and generates 21,500 kcal/m^3. Productivity is low is these biodigestors, which causes a strong impact in investments made in industrial facilities.

Although this technology has been demonstrated technically, it has not been transferred to the industrial arena, because of high costs. As a consequence, pioneer units have been closed.

17.5 Carbon Dioxide Recovery

Carbon dioxide produced during alcohol fermentation is almost completely disregarded by existing mills. Recovery is possible by means of capture in closed fermentation vats. The alcohol is recovered and the gas is stored in liquid form. CO_2 from alcohol fermentation is highly pure (approximately 99.9%) and originates from a biological process. Approximately 1 kg of CO_2 is generated per liter of ethanol produced.

Gas recovery is a physical process which depends, essentially, on yeast activity. Good fermentation ensures an excellent recovery.

The gas produced during fermentation is collected in a polyethylene gas pipeline and goes through a primary purification and deodorization system which comprises washing with water, pre-compression, and adsorption in activated carbon columns and zeolites, where alcohol, residual water, and water soluble volatile impurities are removed. After it is purified, the gas is subjected to high pressures (20 kg/cm^2), and cooled to 37°C. During the cooling process, the condensate is separated and the cooled gas goes through an activated carbon column to remove aromatic compounds and then undergoes drying. The dried gas which is 99.99% pure is finally condensed in the liquefied carbon gas tubes. The low condensation temperature is made possible by Freon 22 (monochlorodifluoromethane, $CHClF_2$). Under constant temperature and pressure conditions during the liquefying process, the bottled gas contains very few impurities, which translates into a high quality product.

The main applications for carbon gas are: beverage carbonation, rendering the atmosphere inert for welding, casting, fire extinguishers, refrigeration, aerosol propellant, tertiary recovery of oil wells, and transportation of solids through pipelines.

Bibliography

Butollo, J.E., 1996. Uso de biomassa de levedura em alimentação animal: propriedades, custo relativo a outras fontes de nutrientes. Workshop − Produção de biomassa de levedura: utilização em alimentação humana e animal. Instituto de Tecnologia de Alimentos, Anais, Campinas, pp. 70−89.

CETESB, 2005. Norma Técnica P4.231 − Vinhaça − Critérios e Procedimentos para Aplicação no Solo Agrícola.

Desmonts, R., 1966. Tecnologia da produção dos fermentos secos de destilaria. Boletim Informativo da APM. Piracicaba. 8 (2), 1.

Desmonts, R., 1968. Utilização do levedo na alimentação da criança. Pediatria Prática. 39 (7).

Lahr Filho, D., Ghiraldini, J.A., Rossell, C.E.V., 1996. Estudos de otimização da recuperação de biomassa de levedura de destilarias. Centro de Tecnologia Copersucar.

Graille, J., Barea, B., Barrera-Arellano, D., Vieira, T.M.F.S., Mahler, B., 2003. Caracterização analítica da cera de cana-de-açúcar para seu uso cosmético. Cirad-Amis Agroalimentaire.

Laguna, A.G., Carvajal, D.H., García, M.M., Magraner, J.H., Arruzazabala, M.L.V., Más, R.F., 1996. Policonasol, una mezcla de alcoholes alifáticos superiores para el tratamiento de complicaciones atereoescleróticas tales como la hiperagregabilidad plaquetaria, loa accidents isquémicos, trombosis e incluso su efectividad contra úlceras gástricas quimicamente inducidas y su proceso de obtención de la caña. Patente Cubana CU 22229A1.

Magnani, J.L., Campanari Netto, J., Vallezi Filho, A., Rossell, C.E.V.R., 1985. Auto-hidrólise de Bagaço (BPH) visando preparo de rações para bovinos. Parte I-O Processo. Boletim Técnico Copersucar, pp. 32−85.

Paturau, J.M., 1969. By-Products of the Cane Sugar Industry. Elsevier Publishing Company, New York, 274p.

Sgarbieri, V.C., Alvim, I.D., Vilela, E.S.D., Baldini, V.L.S., Bragagnolo, N., 1999. Produção piloto de derivados de levedura (*Saccharomyces* sp.) para uso como ingrediente na formulação de alimentos. Braz. J. Food Technol. 2 (5), 119−125.

Villar, J., García, M.A., García, A., Manganelly, E., 2005. Crude wax extraction from filter cake in stirred tank. Int. Sugar J. 107 (1277), 308−311.

CHAPTER 18

Sugarcane Bioenergy

Sizuo Matsuoka[1], José Bressiani[2], Walter Maccheroni[3] and Ivo Fouto[4]
[1]Vignis S.A., Santo Antonio de Posse, State of São Paulo, Brazil [2]Canaviallis, Maceio, AL, Brazil
[3]AGN Bioenergy, São Paulo, SP, Brazil [4]AGN Bioenergy, São Paulo, SP, Brazil

Introduction

The last century was the scene of an extraordinary social and economic development of humanity. This development had as one of its pillars fossil energy. The discovery of oil led society to shape a development highly dependent on this source of energy, which has presented two major problems: first, its finite nature, with a forecast that does not extend beyond half of this century and, second, all the imbalance of greenhouse gases that this platform has led to, with unforeseeable consequences for the planet. Abundant literature has addressed this issue, so here, it will be abdicated to quote them.

Malthus, 200 years ago, raised a controversial issue: how food production that grows arithmetically can sustain a population that grows in geometric progression? His reflections helped establish several connections between cause and effect in the political, economic, and social development related to demographic trends. One issue says that population pressure would stimulate increase in productivity (Boserup, 1965; quoted by Conway, 2003). And was precisely what happened with the Green Revolution, i.e., the increase in productivity of food, made possible by genetic and agronomic technology, did not allow that threat to sustain itself, despite the increased population having actually occurred in geometric ratio. This statement does not mean to disregard that hunger affects a significant part of humanity, but this occurs not due to the lack of food, but for social, economic, and political inequality (Conway, 2003). Also, specifically in the case of agriculture, we must recognize that oil was essential for this achievement, because it is a source of energy relatively abundant and easy to transport, handle, and process (Bauen et al., 2004).

Now, at the beginning of the twenty-first century, the challenge posed by Malthus returns; and to an even greater degree: the world's population has grown to almost seven billion, with the possibility of reaching nine or even 12 billion after 2050 (UN, 2007; Conway,

2003). There is the threat of depletion of many natural resources that man explored in an immeasurable way, including oil and, finally, the environmental imbalance caused due to their demand for food, energy and welfare. The expansion of agricultural land without being the least threatening to the nature reserve on the planet is approaching its limit, with the added complication that many of the areas in use for centuries are now reaching the stage of serious degradation (WRI, 2000; Gore, 1993; Brown, 1990). Will man, facing this menacing scenario, be able to once again overcome the challenge? More than a worrying issue, the solution of this equation is imperative.

Apart from the extensive debate on sustainable development, including especially the agricultural sector, which intensified in the last two decades of the last century, in the media and on the international forum, the highlight of the discussions on this topic in recent years has been the competition between food and biomass energy. This is because, among the various mitigation alternatives of energy supply in order to meet the current requirement of human civilization, biomass energy has been one of the elected, for several reasons, among them because it recycles carbon and thus helps to mitigate the greenhouse effect (Rubin, 2008; USDOE, 2008; Johnson et al., 2007; Sticklen, 2007; Hill et al., 2006). However, the *boom* of the Brazilian production of ethanol fuel from sugarcane and the United States from corn, on one hand, and the great demand for food arising from population growth and the economic development of nations, on the other, emphasized the possibility of that competition. However, if solar energy is what drives life on Earth, it is expected that itself should be the great motor to replace oil in the new era of civilization (Lewis and Nocera, 2006; Brown, 1990). And in this case, the utilization of energy available from biomass, which is produced by the use of solar radiation by plants, is being taken as an alternative, recognized to be a valuable supplemental solution during a transitional phase to a more long lasting and definitive solution (ACS and AIChE, 2008; Lewis and Nocera, 2006; Rifkin, 2003). Therefore, it is up to agriculture itself, to give its contribution and solution to the current conflict of food versus renewable energy.

18.1 Bioenergy, the Energy from Biomass

The mastering of fire by humans, and the gradual development of its use, was essential for the first revolution of human civilization in its beginning. The discovery that, through the burning of vegetable biomass, man could warm and also illuminate as well as cook food, was indeed essential for the evolution of humanity. Biomass burning was, for the longest time in the history of human civilization, the only source of heat energy used and, in modern civilization, until the advent of oil and coal, was the predominant form. With the depletion of natural resources of biomass, both by direct consumption such as the conversion of forest areas into agricultural lands, and greater use of oil, coal, and natural gas as main energy sources, the share of biomass in the global energy matrix of the last century has become a minority in most developed countries, unlike the LDCs, where it followed with high importance. The Brazilian energy matrix, for example, in 2007 was as follows: the total biomass participated with 31% and only products of sugarcane made up 16%, while the share of hydropower was only 15% (MME, 2008). Therefore, the biomass of sugarcane has had an important role and will have even more of a role in the future.

A point was reached where the technological support of petrochemicals showed instability, both by its nature of exhaustible commodity and by geopolitical issues, which created energy insecurity in the world. The result is an expectation of great changes in world energy supply. The big challenge is how to develop a new source of energy that meets the rapidly growing developing world without compromising the sustainability of the planet. Several innovative technologies are being studied (Lovelock, 2006) which seek to make the energy matrix of each country according to available resources. This new platform will require major social and economic changes, and therefore changes in national regulatory systems, and international technological innovations in production and consumption, creating new demands for governance of the economy and society. At the base of all this, major investments in science and technology will be required (ACS and AIChE, 2008; Sticklen, 2008, Lewis and Nocera, 2006).

The use of the energy contained in plant biomass (bioenergy) will thus return to be one of the most important alternatives for addressing the issue discussed, especially in tropical countries, coincidentally those most poor and therefore without the resources or knowledge to develop high-tech alternatives. But even in developed countries large amounts of resources are applied in research to develop a platform based on conversion of biomass into solid fuel, liquid, electricity, biogas, bio-oil, and chemical products (NREL, 2008; Rubin, 2008; Sticklen, 2008; USDOE, 2008; Johnson et al., 2007). For those countries, reducing the dependence on oil is an urgent need not only for economic reasons, but mainly for strategic and environmental reasons (e.g., NREL, 2008; Sticklen, 2008; USDOE, 2008; Lewis and Nocera, 2006; Grassi and Palz, 1994). The European Union, for example, has a plan to supply 20% in biomass energy of total energy needs by 2020 and 15% as electricity (Bauen et al., 2004); and some of the EU countries want to have complete dominance of the energy biomass in their energy matrix (Business Insights, s/d). In the United States, in addition to ethanol production from corn which obviously has serious limitations, the use of biomass, not only for ethanol production, but also to generate electricity, is considered an important alternative to reduce the use of oil (Haq, s/d).

In terms of alternative sources of biomass, the developed countries of the northern hemisphere have researched the use of agricultural and urban wastes. However, these sources, besides the limited quantity, still have other problems, either being dispersed in the form they occur or their diversity of form and composition. The most abundant source are forest residues, but these also are insufficient to meet demand. The most appropriate and economical energy would be crops especially dedicated to produce energy and, unlike those mentioned, they present spatial concentration, uniformity and density.

In relation to dedicated crops, the issue is to elect the biomass that can contribute more efficiently, especially not competing with food (which depends on the region of the world or country that is considered) and providing energy gain when considered for all relation *input − output* (Schmer et al., 2008; Yuan et al., 2008; Johnson et al., 2007; Hill et al., 2006; Macedo, 1998; Coombs, 1984). And in this context, the tropical and subtropical regions are privileged, because of greater availability of solar energy as well as arable land and water, essential elements in agriculture. These conditions are met by Brazil, which alone has 27% of the new arable land potential in the world (Conway, 2003), which places it in an unique position at the same time as presenting a challenging and high responsibility.

To produce biomass in order to meet the energy needs of humanity without competing with food production, one should prioritize the production of fibrous plants instead of starch and oilseed plants (Sticklen, 2008). Fibrous plants bring several advantages and fit well within the requirements deemed important to be elected as producers of biomass. The work of Rubin (2008), Schmer et al. (2008), Sticklen (2008), Hill et al. (2006), and Coombs (1984) listed the following requirements:

(i) plants of high efficiency, that is, high processing capacity of solar energy into biomass without requiring a lot of water, nutrients, and other *inputs* (C4 plants);
(ii) perennial growth and long-term canopy to allow harvest during most of the year;
(iii) possibility of application of agricultural technology in large-scale production;
(iv) be easily and efficiently processed into usable forms of energy; and
(v) sustainable economically and environmentally.

Two types of crops for the tropics fit the bill: artificial forests, especially eucalyptus, and grasses like sugarcane. They are two plants that grow well in the tropics and subtropics, have high photosynthetic efficiency, being coincidentally plants possessing the C4 photosynthetic process, a process for greater efficiency in fixing C at higher temperature in those areas (El Bassam, 1998). Another grass with high productivity also is elephant grass (El Bassam, 1998; Woodard and Prine, 1993), but the aim of this approach is the sugarcane, because, in addition to potential high yield, as can be seen in Table 18.1 (El Bassam, 1998;

Table 18.1: Productivity of some "high" grasses.

Cultivar	t ha^{-1} Year^{-1}*	
	1987−90	1992−94
Sugarcane (US78-1009)	50	32
Energy cane (US59-6)	53	36
Erianthus (IK7647)	50	18
Elephant grass (N51)	45	19

*Dry mass.
Source: Prine et al. (1997) cited by El Bassam (1998) and Woodard and Prine (1993).

Woodard and Prine, 1993), it provides greater opportunity for a breeding program, as discussed below, and thus greater future efficiency gain.

The use of biomass fiber as raw material for energy purposes can be carried out according to four basic platforms:

  (i) direct combustion to produce thermal energy (steam) and electric power (cogeneration);
 (ii) chemical or enzymatic hydrolysis of fiber (cellulose and hemicellulose) to obtain fermentable sugars and production of liquid fuels;
(iii) gasification to produce synthesis gas (carbon monoxide and hydrogen) or generation of biogas; and
 (iv) pyrolysis to produce bio-oil or coal/coke.

18.2 Sugarcane as a Source of Biomass

The sugarcane, as explored for thousands of years, produces the industrial stalks, which represent 80–85% of its total biomass, the untapped remainder consisting of leaves and top (top of stalk) (Figure 18.1). Part of the leaves and tops are burned when sugarcsne is burned prior to harvest, or left on the field when harvested without burning. In the industry, after extraction of the juice, the remaining residue, named bagasse, once was an undesirable residue. However, with the scarcity of wood, bagasse has become an important source of energy (steam) for the mills. Subsequently, the residue has also been used to generate the electricity necessary in the process and as a natural evolution, surplus electricity came to be produced to add to the grid (cogeneration) (Balbo and Padovani Neto, 1987; Campos, 1987). Also, the bagasse was sold to third parties, especially to feed boilers in the food industries, and other minor uses, such as animal feed, production of organic compound, manufacture of pellets, paper, etc. (Campos, 1987).

The agro-industry of sugarcane has been exploited for thousands of years to produce sugar (sucrose). In 1975, Brazil opened the way for a new exploration, the production of ethanol fuel on a large scale (Natale Netto, 2007; Xavier, 2007; Vidal and Vasconcellos, 1998; Nemir, 1983; Hammond, 1977). However, in addition to ethanol and thermal and electrical energy which is obtained from sugarcane, hundreds of other by-products can be developed from this raw material (ICIDCA, 1999), even though today they are only marginally explored. Figure 18.2 summarizes some of these possibilities.

Since its inception, the Brazilian *Proálcool* was the object of observation and analysis, thanks to their daring, ownership, and originality (Coombs, 1984; Nemir, 1983; Hammond, 1977). In 1977, Hammond noted that the program had "the possibility of making Brazil not only a world leader in renewable energy, but also the first country in development stage... to find their own path of energy independence − an original path",

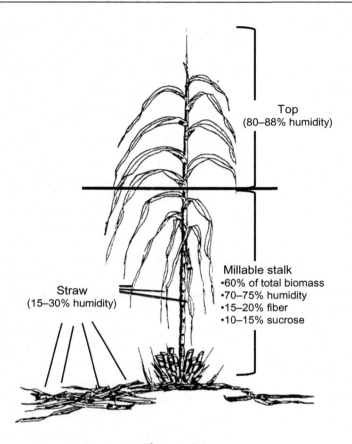

Top
(80–88% humidity)

Millable stalk
•60% of total biomass
•70–75% humidity
•15–20% fiber
•10–15% sucrose

Straw
(15–30% humidity)

Figure 18.1
Schematic drawing of the sugarcane, its component parts (stalk, top and straw) and their
compositions in terms of moisture and biomass (fiber and sugar).
Source: Adapted from Alexander (1985).

and besides considered that, although the main motivation was the balance of payments,
the program would result in a profound and positive effect on environmental issues,
creating jobs, developing the industrial park and that "the history of Brazilian ethanol,
produced in large scale, which is just beginning, has enormous potential and could
become a model and example for a world increasingly hungry for energy and
increasingly poor in oil". Indeed, all this is confirmed, and today, ethanol is already an
important component of the national energy matrix, with more than 17 billion liters
produced annually, contributing to a substantial saving of foreign exchange, reduction of
environmental pollution, mitigation of the greenhouse effect, strengthening and growth of
Brazilian industrial park, sustaining the population inland, and socioeconomic
improvement of the great mass of Brazilian heartlanders.

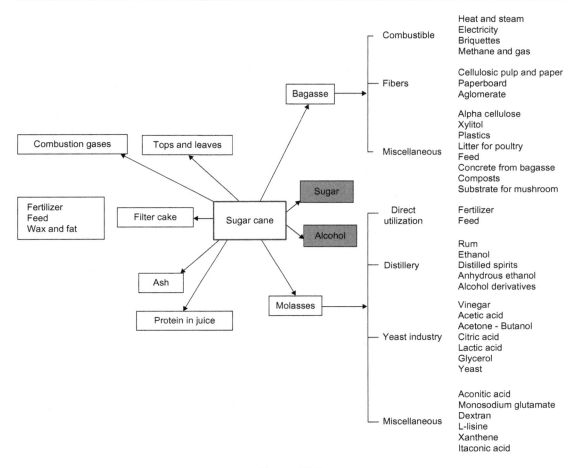

Figure 18.2
Some possible products and by-products in the sugarcane agro industry.
Source: Adapted from Paturau (1982).

Although the sugarcane agro-industry has been traditionally criticized, today there is growing awareness in society of its value to the country. In recent years, the sector has grown at a rate higher than many others, with a prospect to continue so for many years. Of the many projections about ethanol production in Brazil in the coming decades, one of them foresees for 2015 the production of 47 billion liters and the production of 11,500 MW of bioelectricity, or 15% participation in the Brazilian energy matrix (Jank, 2007). Therefore, we would be planting 11.4 million hectares of sugarcane, including therein the production of sugar, whose area would still be less than half the current area of soybeans. The authors argue that already in 2012/13 it would be possible to produce bioelectricity equivalent to the Itaipu plant output (9,699 Wm) if 75% of bagasse and 50% more trash were utilized, and of course, if an appropriate policy will be established and some of the restrictive factors eliminated.

What is commonly called trash (leaves and top) of adult sugar cane represents about 15% in weight of the stalks at harvest, or 12% when dry (Abramo Filho and Matsuoka, 1993). From a purely energy-focussed perspective, this trash is almost 40% of non-utilized energy (Hassuani et al., 2005; Ripoli et al., 2000). When the burning of sugar cane to process the crop became recognized as a major environmental problem and the no-burn harvest began to increase, so too increased the idea of promoting the usage of this major wasted energy source (Pinto, 1992; Ripoli, 1991). Even if the trash is not collected at harvest, it can be collected after drying in the field and utilized for both direct combustion and conversion as cogeneration of energy into heat or electricity and for conversion into liquid fuel, when the technology of cellulose digestion is mastered (Rubin, 2008; Sticklen, 2008, 2007). This alternative production of cellulosic ethanol is receiving substantial technological investment in rich countries such as the United States, Canada, and the European Union, since its efficiency will be even greater when compared with the current technology of transformation of sucrose (Rubin, 2008; Schmer et al., 2008; Johnson et al., 2007; Sticklen, 2008, 2007).

Figure 18.3 shows balances of power generation in the form of sucrose, ethanol and electricity, for a sugar mill in conventional or optimal operation mode (state-of-art machinery and high-performance, already available on the market), taking into account two panoramas of raw material: with standard sugarcane or energy cane cultivars. In these calculations, values of agricultural and industrial productivity were used. Considering that the total biomass of the sugarcane crop is 110 t, we have 85 t of manufacturable stalks reaching the mill. Fifteen tons of trash (25% moisture) will remain in the field, as per the usual process. There is, then, 12.8 t sucrose (TRS — total recoverable sugar) extracted from the stalks, which will result in 6 t of sugar, and the rest of the sucrose, once fermented,will result in 3500 liters of ethanol. The by-product bagasse (23.8 t, 50% humidity) will be burned and will generate 6.1 MWh, one part of which (62 t of steam or 5.18 MWh for existing mills and 36 t of steam or 3 MWh for optimized mills) will be used by the mill itself to feed the whole process and the rest will be a surplus of electricity of 0.92 MWh (conventional power plants) or 3.1 MWh (optimized power).

The final result is 6 t of sucrose, 3500 liters of ethanol and 3.1 MWh of electricity surplus in the case of optimized plants. If the energy cane is used as raw material we would have higher productivity (137 t of industrial stalks and 24 t of trash) and by using 60% of the trash, considering a TRS equal to the current cultivars, a final balance of the same 6 t sucrose, 3500 liters of ethanol and a electricity surplus coming from burning the bagasse (9.8 MWh) and trash (4.9 MWh) of 9.6 MWh or 11.8 MWh in the case of power-optimized plants, with a power gain of about four times in relation to the use of current cultivars.

Figure 18.4 shows the balance of power generation in the form of sucrose, ethanol and electricity, for a bio-refinery operating optimally and able to hydrolyze cellulose and hemicellulose and fermentation of the resulting sugars into ethanol, considering also those

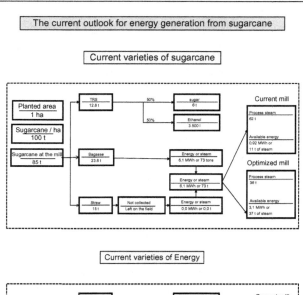

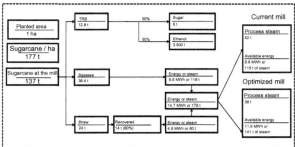

Figure 18.3

Projection of the potential for power generation (sucrose, ethanol and electricity) from ethanol and sugar conventional mill or from optimal operation in two panoramas of raw materials, with cultivars of current sugarcane (upper scheme) and cultivars of energy cane (lower scheme).
*Co-generation of electricity with 10% recovery of total energy from biomass.
Source: Isaiah Macedo; Faep/Consecana; Clippings; VNN Analysis.

same two panoramas of raw material. What we can see about changes with the introduction of hydrolysis technology in this case is that half of the bagasse can be burned to produce electricity (3.9 MWh) and the other half will be hydrolyzed and produce 3.9 t of fermentable sugars that will give 2200 liters of ethanol, and left over 1.2 t of lignin, which is burned to produce 0.8 MWh of energy. More than half of the trash (60%) will be used to burn and will generate 3.1 MWh of electricity. The bottom line is 6 t of sucrose, 5700 liters of ethanol, increasing ethanol production by more than 60% as a direct consequence of the technology of hydrolysis, and 3.9 MWh of electricity surplus. When the cane for energy is the raw material used, it will be because of their increased productivity in terms of bagasse and trash and by the introduction of the technology of hydrolysis, the total

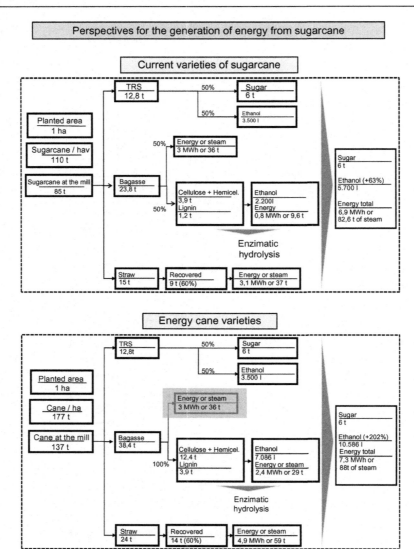

Figure 18.4
Future projection of the potential power generation (sucrose, ethanol, and electricity) from a sugarcane mill with hydrolysis operation of cellulose and hemicellulose in two panoramas of raw materials, with cultivars of current sugarcane (upper scheme) and cultivars of energy cane (lower diagram). *Co-generation of electricity with 10% recovery of total energy from biomass.
Source: Isaias Macedo; Faep/Consecana; Clippings; VNN Analysis.

use of bagasse in hydrolysis process, which will provide 12.4 t of fermentable sugars, which will turn into 7086 liters of ethanol, with surplus of 3.9 t of lignin, which will be burned and will produce 2.4 MWh of energy. Again, the trash should be used to burn and generate 4.9 MWh of electricity. Therefore, the final balance will be 6 t sucrose, 10,586 liters of ethanol and 4.3 MWh of electricity surplus, an increase of ethanol production by more than 200% as a direct result of the use of hydrolysis technology and of energy cane cultivars.

Furthermore, the biomass of energy cane can be used, as described, directly as a source of thermal energy in thermoelectricity or even thermo-chemically transformed into charcoal, with the release of biogas and bio-oils, all of great economic value and especially environmentally-friendly. However, because of the physical characteristic of humidity and granulation of the bagasse, its main current application is limited to procedures of direct combustion in boilers to generate heat, steam, and power (cogeneration). Other features of the bagasse, as homogeneous ligno-cellulosic composition (cellulose, hemicellulose and lignin), processing (crushed and washed) and availability (abundance and storage), enable this biomass as one of the best raw materials for use in future processes of chemical or enzymatic hydrolysis. Because of these characteristics, the bagasse can also be used in more sophisticated processes such as gasification, since it is processed into briquettes or pellets, to increase density and its proper application and use in gasifiers. The gas generated from gasification can be used to feed engines in direct combustion processes or as synthesis gas in thermochemical processes or fermentation. The energy cane in its *in natura* and dried form can also be used directly in combustion processes and gasification, since the cutting of the stalks is done in the size of pellets or directly with whole stalk in processes of pyrolysis to produce bio-oil or coal/coke.

The centuries-old sugar mills, only in the last century turned into sugar agro-industries, most recently in Brazil. Also in the ethanol agro-industry, there will evolve complex bio-refineries to produce food, electricity, liquid fuels and a whole range of substitute byproducts from oil, besides various chemicals and drugs. More recently technology has been developed to produce different types of hydrocarbons other than just ethanol (Lakshmanan et al., 2005; Rubin, 2008; Sticklen, 2008). Manufacturing projects of poly-3-hydroxybutyrate (PHB), biodegradable plastics are also now becoming a reality.

18.3 Sugarcane Breeding Program: Tradition and Future

Historically, the agro-industry of sugarcane was one of the main economic activities during the colonial period of the New World and heavily influenced the social, cultural and economic shaping of each country. After the independence of the colonies, it remained as the main source of economy and still today is a significant business if not essential for many countries. These same considerations apply perfectly to Brazil, today the largest producer and exporter of sugar and also pioneer in the production of a renewable liquid fuel from the sucrose of sugarcane that has become the trend of the moment, ethanol.

The raw material that enabled man to discover the possibility of production of sugar was a plant containing sweet juice and placed in the humid region of New Guinea and later to be called "sugarcane", botanically classified as species *Saccharum officinarum* (Stevenson, 1965). Over time, various natural forms of this species or natural hybrids with other species of the same gender were selected and planted at the same time as the development of procedures for the extraction of the juice and sugar manufacturing. However, as happens with any plant that man removes from nature and begins to cultivate in large areas and for a long time, problems started to affect crops of this plant, especially health problems.

Pressed by these issues, visionaries who worked in the Dutch colony of Java, now Indonesia, recognized that it would be necessary and possible to create artificial forms resistant to diseases. Persevering, they tried to breed a plant that was juicy and relatively gentle with other very rustic species, especially *S. spontaneum*, with high fiber content, but with low sugar content. This tireless work began in the last decade of the nineteenth century, so even before the discovery of Mendel's genetic laws. However, only in the beginning of the second decade of the twentieth century were the commercially valuable results of that first process of plant breeding noticed. Following this successful example, several other sugarcane centers in the world have established their own breeding programs in the decades that followed, including Brazil.

The creation of cultivars adapted to different environments and farm management on a large scale, and disease resistant, allowed the sugarcane agro-industry to expand throughout the world and establish itself as an important source of wealth throughout the last century. A comprehensive review of sugarcane breeding programs at the international level can be found in Heinz (1987) and Ming et al. (2006), and with reference specifically to Brazil in Matsuoka et al. (2005) and Landell and Bressiani (2008).

In the process of photosynthesis, the sugarcane produced on the one hand, the very structure of the plant, basically the fiber (cellulose, hemicellulose and lignin) and on the other, the substance of energy reserves, sucrose. The stalk of a modern hybrid of sugarcane has proportionally from 10−14% fiber and 12−17% sucrose, on average, the rest being water. Industries established that fiber limit as the ideal to ensure good extraction of the juice according to the mechanical process (Gravois and Milligan, 1992; Coombs, 1984). As mentioned above, in the breeding program, *S. officinarum* contributed with the sweet juice broth and *S. spontaneum* with the fiber. Successive backcrosses to the first reached out to the current hybrid fiber with that range mentioned.

On the centenary of the sugarcane breeding program, much effort has been directed to increase the productivity of sucrose, so that the conciliation of sucrose − fiber eventually dictated the limits of the physical−biological breeding programs, i.e., the osmotic concentration of sucrose supported by reserve cells was always the final barrier. The search for high sucrose content, which means recurrence for *S. officinarum*, always ended up making the plant less rustic, and also imposing an agronomic limit. Thus, in modern cultivars the high sucrose content and the high productivity of stalks always acted inversely, so the final productivity gain in sucrose occurs more because of biomass gain than because of sucrose content (Jackson, 2005).

Over the last century, there was great effort in breeding programs and conventional agricultural research to increase the yield of sugarcane and sugar to reach the current level. However, there has been analysis considering that it has reached a difficult level to be outweighed when considering the current conditions of agronomic management (Moore, 2005), especially because it seems occur a difficult limit to be surpassed in the partition of assimilated between sucrose accumulation and growth (Jackson, 2005). For this reason, much effort has been made to deepen their understanding of the synthesis, transport and accumulation of sucrose (Moore, 2005; Rae et al., 2005; Lingle, 2004; Zhu et al., 1997) and genes that regulate this process, in order to, through genetic modification, break that barrier (Ming et al., 2006; Lakshmanan et al., 2005; Moore, 2005; Singels et al., 2005; Watt et al., 2005; Zhu et al., 2000). Notwithstanding the merit and value of these works, a paradigm shift is under way and the new way is indicating that if the contrary route is taken, i.e., an increase of fiber instead of sucrose, the benefits are greater, because then the increased productivity of biomass will be greater and less expensive in effort and resources.

Gravois and Milligan (1992) showed high heritability for fiber and, consequently, the high potential of selection for this character. Other researchers, who reported 79% of genetic variation for fiber as additive, considered that the assessment in one place and at any cycle of the crop would be effective for the selection of fiber clones, allowing speed and resource saving. Gravois and Milligan (1992) argue further that breeders always selected against clones with little stalk diameter (<1.9 cm) due to the negative correlation between diameter and fiber, that is, the smaller the diameter, the greater the fiber. As fibers are more concentrated in the rind, the smaller the diameter the higher the ratio of surface area to volume. In turn, Milligan et al. (1990) found that stalk diameter and number of stalks are inversely related. As the number of stalks is the dominant component of sugarcane (Hogarth, 1987; Mariotti, 1972), it is concluded that selection for thin stalks and the highest number of stalks will lead to higher biomass productivity. Add that the smaller diameter is related to *S. spontaneum*, mainly, and it follows that these materials should also be more rustic. The positive consequences of this fact will be addressed next.

Sugarcane breeding programs are, therefore, facing a new path in the beginning of this century: after 100 years of looking for greater productivity of sugar, now the new type of cane should be directed to high yield of fiber. The advantage of producing more fiber instead of sugar is that the plants will be more rustic, which brings a series of economic and environmental advantages: the plants will be less demanding in soil, climate, water and nutrients and more resistant to pests and diseases, resulting in greater efficiency in its cultivation, that is, higher units of energy produced per energy expended, if we consider all the production chain (output / input). This is a final and essential parameter that will determine the energy options to be considered if the ultimate goals are environmental preservation and sustainability (Johnson et al., 2007; Hill et al., 2006).

The energy cane, therefore, should be planted in areas of soil and climate worse than those reserved for the production of food and requiring less use of fertilizers and pesticides, which are one of the biggest offenders regarding the environment and human health. Because it produces more stalks, energy cane allows a higher multiplication ratio (1:30 or more, against the 1:10 common sugarcane cultivars), which turns out to be another great economic advantage. In addition, these plants will be important for the containment of soil erosion and to assist in rescuing those degraded, given the known ability of grass to do that, because of its strong and abundant fasciculate roots (Johnson et al., 2007) and also, due to the force of ratoon, it will allow more cuts. Alexander (1985) found that the energy cane harvested in annual cycles increases productivity in subsequent ratoons or remains stable at worst, at least for six cuts, or up to eight cuts, as reported by Giamalva et al. (1984). Thanks to the vigorous rhizome of *S. spontaneum*, precisely the characteristic that the pioneer breeders of sugarcane sought as a complement to the vulnerability of *S. officinarum* for this feature, we can predict 10−12 ratoons, or even more.

Thus, curiously, those visionary scientists crossed succulent plants with fibrous plants to take advantage of the rusticity genes in these fibrous plants and subsequently, all sugarcane breeding programs in the world held divergent selection for sucrose content for a century. Now, those fibrous plants should be used in a new process of introgression, this time directing the selection for hardy plants, less juicy or even with no juice and high productivity of fiber. For this to be done, nature will bring back its essential contribution: the ancestral species and genera, that in millions of years led to the development of forms which constitute the basis for the divergent selection of the new type of cane, the energetic plant. We can get an idea of the potential existing in the Saccharum complex by looking at the simple analysis of some data from Bull and Glasziou (1963), among other data in the literature, as presented in Table 18.2.

18.4 Energy Cane: Potential and Collection

The idea of the use of sugarcane as energy plant rather than just a source of sucrose started in the late 1970s in the United States, because of the oil crisis and the harbinger of more problems

Table 18.2: Levels of sucrose, reducing sugars, and fiber in access of ancestral genera and species of sugarcane.

Genera/ Species	Sucrose	Reducing Sugar	Fiber
Erianthus maximus (3)	2.24 ± 0.44	0.73 ± 0.23	26.4 ± 0.9
Erianthus arundinaceus (2)	0.62 ± 0.16	0.61 ± 0.17	30.3 ± 0.3
Miscanthus floridulus (5)	3.03 ± 0.56	0.79 ± 0.24	51.0 ± 2.0
Saccharum spontaneum (30)	5.35 ± 0.38	1.66 ± 0.06	31.8 ± 0.9
Saccharum robustum (10)	7.73 ± 0.83	0.27 ± 0.02	24.8 ± 1.6
Saccharum sinense (2)	13.45 ± 0.02	0.38 ± 0.08	12.8 ± 2.0
Saccharum officinarum (25)	17.48 ± 0.35	0.32 ± 0.02	9.8 ± 0.4

(n) = Number of evaluated accesses.
Source: Bull and Glasziou (1963).

Table 18.3: Results of an evaluation of second generation "energy cane" compared to the usual commercial hybrid, in Puerto Rico.

Cultivar	Brix (B)	Fiber (F)	B + F	Brix	Fiber	B + F	Stalk
	$g\ kg^{-1}$			$Mg\ ha^{-1}$			
L79−1002*	94	255	349	6.6	18	24.6	70.7
LCP85−384**	148	160	308	9.6	10.7	20.3	67.6

*energy cane from second generation.
**commercial cultivar.
Source: Adapted from Samuels et al. (1984).

ahead (Bischoff et al., 2008; Alexander, 1985, 1988). At that time it was demonstrated that in addition to use for the production of fuel ethanol, as Brazil was doing, one should look at the sugarcane as a major biomass producer plant, because until then only the stalk was the target and from this, only sucrose produced (Alexander, 1985). Once the fiber is considered the most important carbon compound, there would be the possibility to increase productivity to a greater extent than with traditional sugarcane, even at the expense of a reduction in sucrose content (Alexander, 1985; Giamalva et al., 1984). This new type of cane was known as energy cane in Louisiana and Puerto Rico, where the first improvement works occurred directed to this objective (Giamalva et al., 1984; Samuels et al., 1984). At that time, Alexander (1985) demonstrated that a waiver of 25−35% in sucrose, would increase the total biomass of up to 100% if the sugarcane harvest was complete. In another result, Samuels et al. (1984) showed (Table 18.3) that, although the production of soluble solids (Brix) was 37% lower (column 2), the production of dry matter (Brix + fiber) was 13% higher (column 4), due to higher production of fiber (column 6), resulting in a final yield of dry matter per hectare 20% higher (column 7). These programs have been discontinued and now with the renewed interest in biofuels and biomass, energy cane recovers interest in the United States with regard to cultivating L79-1002, reported in Table 18.3 (Bischoff et al., 2008).

Table 18.4: Results of evaluation of energy cane compared to the usual commercial hybrid in Barbados.

CLONE	Mg cane ha^{-1}	Brix % cane	Pol % cane	Fiber % cane	Mg MS ha^{-1}	% gain MS
WI81456	125	12.2	8.9	23.9	45.1	169
WI79460	112	14.2	9.3	26.9	46.0	173
WI79458	111	14.0	9.8	21.6	39.5	88
B77602*	78	19.4	17.3	14.8	26.7	—

*Commercial hybrid.
Source: Adapted from Rao et al. (2007).

In Barbados, where a program for introgression in energy cane also has been conducted since the 1980s there were hybrids (Table 18.4) that in terms of productivity of dry matter (DM) showed gains of 72% over a conventional commercial range (Rao et al., 2007). However, this energy cane does not lend itself to the traditional industries of sugar production, because its juice has a low purity (between 70–73% versus 89%). It serves rather for the production of ethanol, but is of interest in the discussion on how to make improvements in industrial technology for the extraction of juice due to the high fiber content these plants contain.

There is much viable genetic diversity for biomass production in germplasm commonly used to create new hybrids of sugarcane. These hybrid genes share basically the species *S. officinarum* and *S. spontaneum*, which are complex aneuploid with formation 2n + n, with total fixation of the chromosomes of *S. officinarum* and {1/2} of the *S. spontaneum*, and having yet combined chromosomes (D'Hont et al., 2008; Grivet and Arruda, 2001). Due to this gene complexity, they are common, in hybrid progeny, closest forms of *S. spontaneum*, small plants, thin stalks with low Brix and low sucrose purity, as well as plants of high production capacity, more rustic than those selected in cultivars, but usually discarded because the sucrose content is below the current standard, i.e., less than 12%. If these plants were selected, there would be a gain in total yield, soluble solids and fiber, of up to 20% compared to the main commercial hybrid, if considering only the industrialized stalks, or more than 30% of heat energy per unit area, if using the leaves and stalks (Matsuoka and Arizono, 1987). Therefore, in the usual population of genetic breeding it would be possible to select plants of the first stage of energy cane, i.e. plants with higher biomass productivity. With these plants, in the current ethanol agro-industry, with an annexed ethanol distillery and also an electricity cogeneration unit, there could be a final economic return greater than that afforded by a variety of high sucrose content. However, this type of material has always been discarded by traditional programs because of those pre-established parameters mentioned above.

With the paradigm shift, it would be possible to add a considerable gain with no additional effort in the genetic breeding. Since the selection of its first series (2003), the Canavialis breeding program has conducted a subprogram in which that kind of clone follows a parallel selection process. Moreover, these clones have returned to the active germplasm bank to be part of a recurrent selection program for increased biomass production.

18.5 Introgression Programs

Especially in the second half of the last century, some breeding programs of sugarcane led introgression programs in order to explore a broader genetic base than that originally used, especially with the focus on the crossbreeding between *S. officinarum* and *S. spontaneum* (Wang et al., 2008; Ming et al., 2006). All of them, however, were looking for traditional types of sugarcane and some were successful (Wang et al., 2008).

The existing indications are that the introgression conducted specifically to obtain biomass production, i.e., high content of fiber, would have much greater success, as well as permitting to acquire material of commercial value in a much shorter time and lower cost. Supporting these assertions are the pioneering work done in Louisiana and Puerto Rico, previously cited. This work may have obtained materials with high productivity in the F1 crossbred generation of commercial hybrids with *S. spontaneum*, whose force was not maintained in successive generations of backcrossing (Ming et al., 2006; Legendre and Burner, 1995; Alexander, 1985). In those pioneering work in the United States, production has been reported as high as 240–265 mg ha^{-1} (Giamalva et al., 1984) and up to 307 Mg ha^{-1} in terms of total green mass (Legendre and Burner, 1995). Another cultivar presented fiber content of 25% on average of four cuts, while the commercial variety presented 16% (Bischoff et al., 2008).

In the introgression program started in Canavialis, the result, although preliminary, is as promising as those. Data from clones that stood out among hundreds of F1clones from a crossbreeding between a commercial hybrid and *S. spontaneum* are presented in Table 18.5. The number of stalks per linear meter ranged from 35 to 40, the fiber content ranged from 15.35 to 19.90 against 12.05 of the commercial variety, the stalks productivity from 155 to

Table 18.5: Preliminary results from five energy cane clones compared to a commercial hybrid.

Clone	Nostalks.m^{-1}	Pol cane	Fiber%	t stalks ha^{-1}	t fiber ha^{-1}
1	40	6.40	19.9	205	40.25
2	36	5.29	15.35	236	36.74
3	36	7.23	19.55	175	34.20
4	35	9.23	17.96	173	30.98
5	39	8.74	19.8	155	30.63
RB72454	14	14.60	12.05	148	17.08

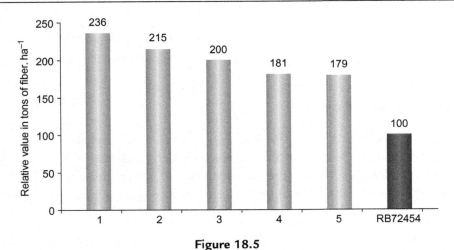

Figure 18.5

Preliminary results in productivity of dry mass (DM) and fiber per hectare of energy cane clones with improved performance in sugarcane plant.

236 tons, against 148 t of commercial variety, and the productivity of fiber from 30.63 to 40.25 t. Figure 18.5 shows the relative fiber productivity of the same clones expressed in percentage compared to the commercial variety: the clone of higher productivity of fiber surpassed the standard in more than 130% and the 5th, in 79%. Considering the leaves and stalks, the advantage would be even greater, because if in the commercial variety they represented 15%, in the energy cane they exceeded 25%. In Figure 18.6, can be seen the morphology of this type of plant: when *S. officinarum* is intercrossed with *S. spontaneum*, the resulting hybrid has intermediate aspect between the first and a commercial hybrid, while a commercial hybrid is intercrossed with *S. spontaneum*, the plant is morphologically presented more similar to this.

Introgression of *S. spontaneum* has proven to be a fruitful path if the target is the total production of dry matter, predominantly fiber. And in this species there are still hundreds of accesses that were not evaluated, which creates the expectation of a very wide path to be followed. However, many other ways also exist. Introgression can be done with several other genera of the Saccharum complex (Daniels and Roach, 1987), as Erianthus and Miscanthus, and Sclerostachya and Narenga. During the last century, several collecting expeditions were conducted in centers of origin and the diversity of this complex and one part of this collection has been maintained by the International Society of Sugar Cane Techonologists (ISSCT) in Florida, USA, and in Kannur, India (Berding and Roach, 1987).

All commercial cultivars now in cultivation in the world originated from a few interspecific crossbreeds performed at the beginning of the last century. This pioneer crossbreeding used a few dozen clones of *S. officinarum* and only two to three clones of *S. spontaneum* (Daniels

Figure 18.6
Examples of energy cane: to the left, F1 of *S. officinarum* × *S. spontaneum*; to the right, F1 commercial hybrid × *S. spontaneum*.

and Roach, 1987). However, when considering important agronomic traits, for example in the collections of *S. spontaneum* and other species and genera of the Saccharum complex maintained by ISSCT (Tai and Miller, 2002), we can see a wide variation among the various types and forms of unexplored germplasm. These collections are therefore real banks of new traits, which may be incorporated through conventional breeding programs or sugarcane genetic modification, especially in the development of energy cane. More evidences that exotic germplasm collections contain large proportions of genetic variability have not yet been exploited by breeding programs of sugarcane come from a study conducted recently by Canavialis. In this study, a worldwide collection of germplasm of sugarcane ISSCT, held in Miami (USA), was imported to Brazil and a portion of this collection was analyzed for genetic variability by means of microsatellite molecular markers (Maccheroni et al., 2007). The existence of at least 40% of new markers or alleles was observed, besides those that had already been observed in a collection of more than 1000 conventional hybrids.

18.6 Energy Cane: Characteristics, Quality, and Utility Value

To conclude and summarize, there follows a list of the characteristics, quality, and value in use of energy cane:

- renewable energy generation, which helps to mitigate the greenhouse effect;
- has a high conversion of atmospheric carbon into organic carbon, i.e., high biomass production;
- may be one of the main crops in the paradigm shift of the civilization of oil to the civilization of multifaceted energy, including multiple renewable sources of energy;
- has high energy density, i.e., in terms of energy and cost this raw material is more efficient than that of edible plants;
- has high resistance to biotic and abiotic stresses, so that can be produced with less input (less fertilizers, pesticides, and energy), and is grown on less valuable agronomic land, i.e., less fertile, less availability of water, more extreme temperatures (both low and high) and more saline;
- offers less competition with food production as a consequence of these above factors;
- has the power to control soil erosion greater than herbaceous plants, fixes more carbon in the soil due to its fasciculate, abundant and vigorous roots, adding to it its semi-durable characteristic, has great ability to protect the soil and, consequently, recovery of degraded soils;
- improves water quality and habitat for wildlife, as compared to other crops, due to the factors already mentioned;
- its cultivation, management, harvesting, and transport are procedures already dominated; it is possible to develop cultivars with high productivity for each geographic region for process improvement, it is already dominated and relatively simple and fast;
- unlike sugarcane that has limitation of harvesting time due to the the requirement of sugar concentration (ripening), energy cane harvesting can endure almost year round and also the fiber, after extraction of juice, or *in natura* in the case of the fiber cane, can be stored for extended period of use; forms of non-producing seeds can be produced so that the multiplication is vegetative and thus preventing it from becoming a weed.

It is important to emphasize the last mentioned characteristic, because there are cases of plants dedicated to the production of energy which constitute environmental threat, due to the possibility of becoming highly invasive weeds (DiTomaso et al., 2007). It follows that energy cane offers an unparalleled opportunity for Brazil to further increase its competitiveness in the field of renewable energy from biomass, with immeasurable benefits to Brazilian society, as well as of strategic value.

Bibliography

Abramo Filho, J., Matsuoka, S., 1993. Resíduo da colheita mecanizada de cana crua. Álcool e Açúcar. 67, 23–25.

Alexander, A.G., 1985. The Energy Cane Alternative. Elsevier, Amsterdan, 509p.

Alexander, A.G., 1988. Sugarcane as a source of biomass. In: Sansoucy, R., Aarts, G., Preston, T.R. (Eds.), Sugarcane as Feed. FAO Expert Consultation, Proc., Santo Domingo, Dominican Republic, July, 1986. FAO Corporate Document Repository, Roma, p. 11. <www.fao.org/docrep/003/s8850e/S8850E04.htm>. (Last access 29.07.08.).

American Society of Chemistry and American Institute of Chemical Engineers, 2008. Science and Technology to Meet our Energy Needs. <http://portal.acs.org/portal/fileFetch/C/CSTA_015269/pdf/CSTA_015269. pdf> (Last access 01.09.08.).

Balbo, J.M., Padovani Neto, A., 1987. Excedentes de energia elétrica e sobra de bagaço para diferentes concepções e sistemas de conversão e utilização de energia aplicáveis a indústria sucro-alcooleira. STAB, Açúcar, Álcool e Sub-produtos. 692, 52–58.

Bauen, A., Woods, J., Hailes, R., 2004. Bioelectricity Vision: Achieving 15% of Electricity from Biomass in OECD Countries by 2020. Imperial College, London. <www.wwf.de/fileadmin/fm-wwf/pdf-misc_alt/ klima/biomassreport.pdf>.

Berding, N., Roach, B.T., 1987. Germplasm collection, maintenance, and use. In: Heinz, D.J. (Ed.), Sugarcane Improvement through Breeding. Elsevier, New York, pp. 143–210.

Bischoff, K.P., Gravois, K.A., Eagan, T.E., Hoy, J.W., Kimbeng, C.A., LaBorde, C.M., et al., 2008. Registration of "L79-1002" sugarcane. J. Plant Regist. 2, 211–217.

Brown, L.R., (Org.). 1990. Salve o Planeta! Qualidade de Vida-1990. Globo, São Paulo, 308p.

Bull, T.A., Glasziou, K.T., 1963. The evolutionary significance of sugar accumulation in *Saccharum*. Aust. J. Biol. Sci. 16, 737–741.

Business Insights, s/d. Electricity from biomass in the Netherlands. <www.globalbusinessinsights.com/rbi/ content/rben0160m.pdf> (Last access 11.09.08.).

Campos, R.M., 1987. Valor econômico do bagaço. Brasil Açúcar. 55 (105), 20–24.

Carmo, A.T., 1977. O Proálcool e a economia agrocanavieira. Brasil Açúcar. 90 (3), 32–36.

Conway, G., 2003. Produção de Alimentos no Século XXI: Biotecnologia e Meio Ambiente. Estação Liberdade, São Paulo, 375p.

Coombs, J., 1984. Sugar-cane as an energy crop. Biotechnol. Genet. Eng. Rev. 1, 311–345.

Daniels, J., Roach, B.T., 1987. Taxonomy and evolution in sugarcane. In: Heinz, D.J. (Ed.), Sugarcane Improvement through Breeding. Elsevier, Amsterdam, pp. 7–84.

D'Hont, A., Souza, G.M., Menossi, M., Vincentz, M., Van-Sluys, M.A., Glaszmann, J.C., et al., 2008. Sugarcane: a major source of sweetness, alcohol, and bio-energy. In: Moore, P., Ming, R. (Eds.), Genomic of Tropical Crop Plants. Springer, Berlim, pp. 483–513.

DiTomaso, J.M., Barney, J.N., Fox, A.M., 2007. Biofuel Feedstocks: The Risk of Future Invasions. The Council for Agricultural Science and Technology, Commentary QTA-2007-1. Ames, Iowa, 7p.

El Bassam, N., 1998. Energy Plant Species. Science Publishers Ltd, London, 321p.

Giamalva, M.J., Clarke, S.J., Stein, J.M., 1984. Sugarcane hybrids of biomass. Biomass. 6, 61–68.

Gore, A.A., 1993. A Terra em Balanço. Augustus, São Paulo, 447p.

Grassi, G., Palz, W., 1994. O futuro da biomassa na União Européia. Álcool e Açúcar. 76 (Out. /Nov.), 28–34.

Gravois, K.A., Milligan, S.B., 1992. Genetic relationship between fiber and sugarcane yield components. Crop Sci. 32, 62–67.

Grivet, L., Arruda, P., 2001. Sugarcane genomics: depicting the complex genome of an important tropical crop. Curr. Opin. Plant Biol. 5, 122–127.

Hammond, A.L., 1977. Photosynthetic solar energy: rediscovering biomass fuels. Science. 197 (4305), 745–746.

Haq, Z., Biomass for Electricity Generation. Washington: Energy Information Administration, USDOE. 18p. <www.eia.doe.gov/oiaf/analysispaper/biomass/>Acessado em (10.09.08.).

Hassuani, S.J., Leal, M.R.L.V., Macedo, I.C., 2005. Biomass Power Generation: Sugar Cane Bagasse and Trash. PNUD-CTC, Série Caminhos para a Sustentabilidade, Piracicaba, 216p.

Heinz, D.J. (Ed.), 1987. Sugarcane Improvement through Breeding. Elsevier, Amsterdam, 603p.

Hill, J., Nelson, E., Tilman, D., Polasky, S., Tiffany, D., 2006. Environmental, economic, and energetic costs and benefits of biodiesel and ethanol biofuels. Proc. Natl. Acad. Sci. 103 (43), 11206−11210.

Hogarth, D.M., 1987. Genetics of sugarcane. In: Heinz, D.J. (Ed.), Sugarcane Improvement through Breeding. Elsevier Press, Amsterdam, pp. 255−271.

ICIDCA, 1999. Manual de derivados da cana-de-açúcar. ABIPTI, Brasil, 474p.

Jackson, P.A., 2005. Breeding for improved sugar content in sugarcane. Field Crops Res. 92, 277−290.

Jank, M.S., 2007. A velha cana-de-açúcar. Opiniões out.-dez., 12−16.

Johnson, J.M.F., Coleman, M.D., Gesh, R., Jaradat, A., Mitchell, R., Reicosky, D., et al., 2007. Biomass-bioenergy crops in the United States: a changing paradigm. Am. J. Plant Sci. Biotechnol. 1 (1), 1−28.

Lakshmanan, P., Geijskes, R.J., Aitken, K.S., Grof, C.L.P., Bonnett, G.D., Smith, G.R., 2005. Sugarcane biotechnology: the challenge and opportunities. In Vitro Cell. Dev. Biol. -Plant. 41, 345−363.

Landell, M.G.A., Bressiani, J.A., 2008. Melhoramento genético, caracterização e manejo varietal. In: Dinardo-Miranda, L.L., Vasconcelos, A.C.M., Landell, M.A.G. (Eds.), Cana-de-açúcar. Instituto Agronômico, Campinas, pp. 101−155.

Legendre, B.L., Burner, D.M., 1995. Biomass production of sugarcane cultivars and early-generation hybrids. Biomass Bioenergy. 8 (2), 55−61.

Lewis, N.S., Nocera, D.G., 2006. Powering the planet: chemical CHALLENGES in solar energy utilization. Proc. Natl. Acad. Sci. 103, 15729−15735.

Lingle, S.E., 2004. Effect of transient temperature change on sucrose metabolism in sugarcane stalks. J. Am. Soc. Sugar Cane Technol. 24, 132−141.

Lovelock, J., 2006. A Vingança de Gaia. Intrínseca, Rio de Janeiro, 159p.

Maccheroni, W., Jordão, H., Degaspari, R., Matsuoka, S., 2007. Development of a dependable microsatellite-based fingerprinting system for sugarcane. Proc. Congr. Int. Soc. Sugar Cane Technol. 26, 889−899.

Macedo, I.C., 1998. Greenhouse gas emissions and energy balance in bioethanol production and utilization in Brazil. Biomass Bioenergy. 14 (1), 77−81.

Mariotti, J.A., 1972. Associations among yield and quality components in sugarcane hybrid progênies. In: XIV International Society of Sugar Cane Technologists Congress, Anais New Orleans. pp. 297−302.

Matsuoka, S., Arizono, H., 1987. Avaliação de variedades pela capacidade de produção de biomassa e pelo valor energético. STAB, Açúcar, Álcool e Subprodutos. 6 (2), 39−46.

Matsuoka, S., Garcia, A.A.F., Arizono, H., 2005. Melhoramento da cana-de-açúcar. In: Borém, A. (Ed.), Melhoramento de espécies cultivadas. UFV, Viçosa, pp. 225−274.

Milligan, S.B., Gravois, K.A., Bischoff, K.P., Martin, F.A., 1990. Crops effects on broad-sense heritabilities and genetic variances of sugarcane yield components. Crop Sci. 30, 344−349.

Ming, R., Moore, P., Wu, K.K., D'Hont, A., Glaszmann, J.C., Tew, T.L., et al. 2006. Sugarcane improvement through breeding and biotechnology. In: Janick, J. (Ed.). John Wiley e Sons, Inc., Plant Breeding Reviews 27, 15−118.

MME, 2008. Matriz energética brasileira. <www.mme.gob.br/site/menu/select_main_menu_item.do? channelId = 14328&pageId = 15043> (Acessado em 06.11.08.).

Moore, P.H., 2005. Integration of sucrose accumulation processes across hierarchical scales: towards developing and understanding of the gene-to-crop continuum. Field Crops Res. 92, 119−135.

Natale Netto, J., 2007. A Saga do Álcool. Novo Século Editora, SP, Osasco, 343p.

Nemir, A.S., 1983. Alcohol fuels: the Brazilian experience and its implications for the United States. Sugar J.10−13, May.

NREL, 2008. Biorefinaries. National Renewable Energy Laboratory. <www.nrel.gov/learning/re_biomass. html> (Acessado em 12.09.08.).

Paturau, J.M., 1982. Byproducts of the Sugarcane Industry. Elsevier, Amsterdam, 2. ed.

Pinto, L.A.R., 1992. Aproveitamento de bagaço e da palha de cana-de-açúcar em programas energéticos comunitários. Álcool e Açúcar. 63, 34−37.

Rae, A.L., Grof, C.P.L., Casu, R.E., Bonnett, G.D., 2005. Sucrose accumulation in the sugarcane stem: pathways and control points for transport and compartmentation. Field Crops Res. 92, 159–168.

Rao, P.S., Davis, H., Simpson, C., 2007. New sugarcane cultivars and year round sugar and ethanol production with bagasse-based cogeneration in Barbados and Guyana. In: 26 Cong. International Society Sugar Cane Technologists. Proc., Durban, AS, pp. 245–246 (Abstracts).

Rifkin, J., 2003. A economia do hidrogênio. M. Books do Brasil Editora Ltda, São Paulo, 300p.

Ripoli, T.C.C., 1991. Utilização do material remanescente da colheita da cana-de-açúcar (*Saccharum* spp.): equacionamento dos balanços energético e econômico. ESALQ/USP, Piracicaba, 1991. 150p. (Tese de Livre Docência).

Ripoli, T.C.C., Molina Jr., W.F., Ripoli, M.L.C., 2000. Energy potential of sugarcane biomass in Brazil. Scientia Agricola. 57, 677–681.

Rubin, E.M., 2008. Genomics of cellulosic biofuels. Nature. 454 (14), 841–845.

Samuels, G., Alelxander, A.G., Rios, C.E., Garcia, H., 1984. The production of energy cane in Puerto Rico: the Hatillo project. J. Am. Soc. Sugar Cane Technol. 3, 14–17.

Schmer, M.R., Vogel, K.P., Mitchell, R.B., Perrin, R.K., 2008. Net energy of cellulosic ethanol from switchgrass. Proc. Natl. Acad. Sci. 105, 464–469.

Singels, A., Donaldson, R.A., Smit, M.A., 2005. Improving biomass production and partioning in sugarcane: theory and practice. Field Crops Res. 92, 291–303.

Stevenson, G.C., 1965. Genetics and Breeding of Sugarcane. Longmans, London, 284p.

Sticklen, M.B., 2007. Feedstock crop genetic engineering for alcohol fuels. Crop Sci. 47, 2.238–2.248.

Sticklen, M.B., 2008. Plant genetic engineering for biofuel production: towards affordable cellulosic ethanol. Nature Rev. 9, 433–443.

Tai, P.Y.P., Miller, J.D., 2002. Germplasm diversity among four sugarcane species for sugar composition. Crop Sci. 42, 958–964.

UNITED NATIONS, 2007. World Population Prospects. The 2006 Revision Population Database. 20 Sept. 2007. 2007.

USDOE, 2008. Biomass. Multi-Year Program Plan, United States Department Of Energy. 192p. <http://www1.eere.energy.gov/biomass/pdfs/biomass_program_mypp.pdf> (Acessado em 30.08.08.).

Vidal, J.W.B., Vasconcellos, G.F., 1998. O poder dos trópicos. Editora Casa Amarela, São Paulo.

Wang, L.P., Jackson, P.A., Lu, X., Fan, Y.H., Foreman, J.W., Chen, X.K., et al., 2008. Evaluation of sugarcane × *Saccharum spontaneum* progeny for biomass composition and yield components. Crop Sci. 48, 951–961.

Watt, D.A., McCormick, A.J., Govender, C., Carson, D.L., Cramer, M.D., Huckett, B.I., et al., 2005. Increasing the utility of genomics in unraveling sucrose accumulation. Field Crops Res. 92, 149–152.

Woodard, K.R., Prine, G.M., 1993. Dry matter accumulation of elephantgrass energycane and elephantmillet in a subtropical climate. Crop Sci. 33, 818–824.

WRI, 2000. World Resources 2000–2001: People and Ecosystems: The Fraying Web of Life. World Resources Institute, Washington, DC, 389p. <www.wri.orgwr2000>.

Xavier, M.R., 2007. The Brazilian Sugarcane Ethanol Experience. Issue Analysis, Competitive Enterprise Institute, Washington DC, N 3. 12p.

Yuan, J.S., Tiller, K.H., Al-Ahmad, H., Stewart, N.R., Stewart Jr., C.N., 2008. Plants to power: bioenergy to fuel the future. Trends Plant Sci. 13, 421–429.

Zhu, Y.J., Komor, E., Moore, P.H., 1997. Sucrose accumulation in the sugarcane stem is regulated by the difference between the activities of soluble acid invertase and sucrose phosphate synthase. Plant Physiol. 115, 609–616.

Zhu, Y.J., Albert, H.H., Moore, P.H., 2000. Differential expression of soluble acid invertase genes in the shoots of high-sucrose and low-sucrose species of *Saccharum* and their hybrids. Aust. J. Plant Physiol. 27, 193–199.

CHAPTER 19

Remuneration System of Sugarcane

Francisco de Assis Dutra Melo

Universidade Federal Rural de Pernambuco, Recife, Pernambuco, PE, Brazil

Introduction

The regulatory framework of the sugarcane payment system to suppliers began with the adoption, by the then Executive Committee of the Institute of Sugar and Alcohol — IAA, at its meeting on 27 June 1945, Resolution 109/45. Later in the 1960s, the IAA appointed a committee with the specific aim to investigate the implemented system and propose suggestions to the current model. However, this Commission was extinguished without carrying out its work and goals. It was renewed in early 1968 and then stopped, mainly due to disagreements between the Resolution 109/45 — IAA and Law No. 4870 of 01/12/65, which provided the legal basis for the implementation of a sugarcane payment system by its sucrose content. However, a number of obstacles prevented the adoption of a system of payment for sugarcane sucrose content, among which we can highlight the appropriate methods for direct sampling of sugarcane shipments (Almeida, 1975). These obstacles were then solved with the experimental results of the "Core Sampler" System (CONSECANA—PE, 2007). With the establishment of the IAA/Planalsucar and the Central Cooperative of Sugar and Alcohol of the State of São Paulo — Copersucar, the work could be accelerated, given the close mutual collaboration that exists.

The system of direct sampling by means of mechanical probe was designed in Hawaii and is used in several sugarcane-producing regions in the world. This system has the great advantage of the practicality and security of the identity of each shipment, taking into account the impurities and the heterogeneity of the shipment (Sturion and Gemente, 1981; Sturion and Parazzi, 1985).

The sugarcane payment system for quality contains some basic rules, as to be neutral, independent of the concerned sectors, and flexible enough to allow changes and improvements over time. Thus, the implemented method entered the concept of improving the quality of raw material and encouraged the industry administrators to seek better returns.

Sugarcane. DOI: http://dx.doi.org/10.1016/B978-0-12-802239-9.00019-0

Basically, the payment system can be divided into several interrelated steps:

- determining the net weight of the raw material;
- sampling of delivered raw materials;
- analysis of individual samples;
- calculation of tradable products; and
- calculation of the value per ton of sugarcane.

The creation of Planalsucar — National Sugarcane Breeding Program, of the Institute of Sugar and Alcohol in 1971, was crucial to the sector, nationally, for the definition of payment for sugarcane. Research aimed at finding a technology to pay sugarcane was placed in the highest Planalsucar priorities in harmony with the program to raise more productive sugarcane cultivars, richer in sugar and more suitable for different sugarcane regions in the country (PLANALSUCAR, 1983). Thanks to the results obtained, it was possible for the government to inaugurate the PSSC system — Payment of Sugarcane by the Sucrose Content — in 1978, in the State of Alagoas. In 1983, the system was deployed in the State of São Paulo and in 1984, in the states of Rio de Janeiro and Pernambuco (PLANALSUCAR, 1983). So its introduction was completed in the sugarcane states with the most production and which were more traditional at that time, and its use throughout the national territory was consolidated in the 1986/1987 season (Oliveira et al., 1985; Pagamento de cana pelo teor de sacarose: o sistema implantado em São Paulo, 1985).

As for the results obtained with the PSSC system, it is emphasized that the system led to the adoption of new technologies, both in agriculture and industry, culminating with the improvement of the entire ethanol and sugar sector. Therefore, the criterion that quantifies the value of the raw material has fundamental importance for the development of the ethanol and sugar agro-industry.

After the release of prices in the ethanol and sugar agribusiness, on 1 February 1999, and subsequent extinction of the Institute of Sugar and Alcohol by the federal government, the producers of the sugarcane industry, based on successful experience gained from PSSC in the State of São Paulo, decided to accept the existing analytical model and adapt it to their situation, creating a new pattern of management relationship of the sugarcane agro-industry, the State Council of the Producers of Sugarcane, Sugar and Alcohol — Consecana, composed of a Board of Directors, Technical Board and Arbitration Board (which in practice requires government official rules) with the following purposes:

- take care of the relationship of the production chain of sugarcane agro-industry in the state, combining efforts of all those who participate, from the planting of the cane to the sale of final products, aiming for its support and prosperity;
- ensure the improvement of sugarcane quality evaluation system, performing studies, developing research, and promoting the systematic and constant updating of technological criteria and negotiation in the sector;

- develop and disseminate technical analysis on the quality of sugarcane and its verification as well as about the structure and evolution of the sugarcane agro-industry, including those related to conditions of employment and bargaining in the sector;
- foment the reconciliation of disputes between members of the system, for those who come to appeal to Consecana; and
- forward to the Environmental Chamber the conflicts that were not reconciled among the members of the system, to do that, they come to appeal on Consecana.

With Law 9307/96, arbitration has gained momentum in Brazil, aiming to resolve disputes between individuals or entities through the trial of a third in a matter of economic and financial nature, promoting the settlement of disputes between the parties.

19.1 Model Assignments of Self Management — Consecana

The model of Self Management Relationship between Sugarcane and Industry Executives — Consecana is operated by a Board of Directors comprising of representatives of the Trade Associations and a Technical Board.

The Board of Directors is responsible for:

I. consolidating, systematizing and disseminating the results of the tests and studies undertaken by the Technical Board — Canatec, in the areas of their allocation, targeting system integrators, in order to improve the conditions for hiring and evaluating the quality of cane in the market;
II. issuing acts aimed at outlining the rules and regulations;
III. solving any doubts and promoting reconciliation of disputes between members of the system who appeal to Consecana, when required;
IV. setting the annual budget for operation of the entity, including the operation of Canatec;
V. issuing resolutions and publications from CONSECANA, previously approved by the Board; and
VI. defining the constitution of the Arbitration Judgment, according to relevant legislation.

The Technical Board is responsible for:

I. conducting studies and developing research aimed at persistent improvement and upgrading of the technological criteria for evaluating the quality of sugarcane, and techniques of negotiation and contracting in the market for the sugarcane agro-industry;
II. informing and updating the producers of sugarcane, sugar and ethanol on the evolution of the criteria used for evaluating the quality of cane and techniques in this trading sector;
III. guiding producers of sugarcane, sugar, and ethanol in order to seek and maintain the best performance and continuity of economic activity they undertake;

IV.   participating in technical committees of other bodies and agencies, aiming at standardization and development of technical standards regarding the quality of sugarcane;

V.   monitoring the evolution of prices and costs of products in the industry, defending, via the federal government, the achievement of regional agricultural policies; and

VI.   preparing technical reports, clarifing doubts and reconciling conflicts among members of the system when they deal with the issues of systematic evaluation of the quality of sugarcane or the contracting and trading sector.

19.1.1 Steps of the Consecana System

- Sale and purchase agreement for raw material.
- Delivery and quality assessment of raw material and determination of the TRS.
- Participation of raw material in the production cost of sugar and ethanol.
- Ascertain the prices of sugar and ethanol:
  - Sugar price on the domestic market through the ESALQ / BM & F.
  - Sugar price in the foreign market through the Stock Exchange.
  - Price of anhydrous ethanol and hydrated ethanol through ESALQ.
- Definition of the production mix.
- Price per ton of sugarcane.

The Consecana model calculates the involvement of the supplier according to production cost spreadsheets of sugarcane, sugar, and ethanol, in the total recoverable sugar, and by the final price that will be calculated from the prices of manufactured products — sugar in the domestic and external markets, and ethanol of all types in the domestic and external markets in the region, free of taxes or freight, i.e., in the PVP (Price Vehicle Plant)/PVM(Price Vehicle Mill) condition, and taking into account the production mix.

19.2 Quality Evaluation of the Raw Material

In simple terms, the analytical process of the payment system, as discriminated in Figure 19.1, starts with the weighing of the shipment, mechanical probe sampling, and sample preparation in disintegrators/crushers. The next phase of analysis of samples begins with the extraction of the juice and includes weighing the residual bagasse called wet cake and the determinations of soluble solids, percentage of Brix and apparent sucrose (Pol).

With these laboratory results and the net weight of the sugarcane delivered, the data processing that will define the remuneration of the raw material provided is developed.

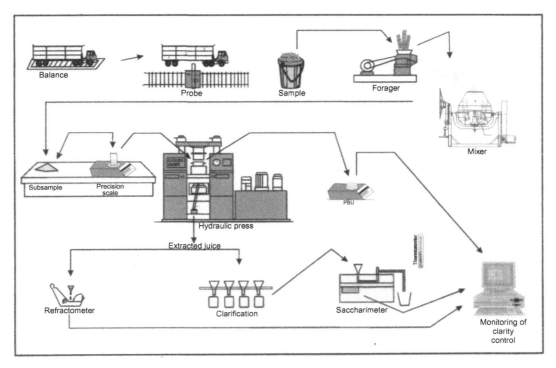

Figure 19.1
Technical and operational flowchart of the Sugarcane Payment System.

It is important to emphasize the importance of standardization of operational Canatec techniques of aspects related to methodologies, equipment, reagents, and supplies used in the process of ascertaining the quality of raw material.

19.2.1 Technical Standards for Operational Control

The Laboratory of Sugarcane Payment is built after the sugarcane balance and projected so as to allow, as a rule, the analysis of at least 50% of the shipments of sugarcane.

The quality of sugarcane is measured through technological analysis on collected samples at the time of delivery to the processing plant, aiming to quantify the content of total recoverable sugars and subsequent payment for the raw material.

The framework for assessing the quality of sugarcane supplied, including the balance for weighing the shipments, the probe sampler, the laboratory and its equipment, the system of automatic capture and processing of analytical data will be the responsibility of the processing plant, being permitted to supervise the entire process by representatives of the sugarcane suppliers duly accredited by the Trade Associations.

The weighing of sugarcane, held in balance for weighing shipments is performed by the National Institute of Metrology, Standardization and Industrial Quality — Inmetro— with at least two measurements during the milling season, the first one at the beginning and the second halfway through the season. The selection of shipments that will be sampled, as a rule, is performed while weighing the shipment, randomly, by lottery, including an indication of the sampling points.

In the sugarcane payment system, the collection of samples can be done through the drilling of the shipment with mechanical horizontal or oblique probe samplers. To the mechanical horizontal probe sampler, simple random samples are taken at three different points of the shipments, which may not have matching vertical and/or horizontal alignment, as established in the combinations shown in Figure 19.2.

With respect to the oblique probe for sampling the shipment, the samples cannot be collected at only one point of shipment. It is important to note that the vehicles are with their wagons suitable for mechanical horizontal or oblique sampling probe, allowing the processing plant access to the samples taken in the entire area of the wagon. When the sugarcane is transported in vehicles which are coupled to one or more trucks, the samples can be taken according to one of the options listed below, linking to the weighing certificate:

- treat each trailer as an individual shipment, taking the samples in each unit;
- carry out the withdrawal of the sample in every trailer; or
- take the sample only from one trailer, in this case, the choice of shipment must be carried out randomly in the balance.

In all such cases, the sampling criterion should be obeyed. For purposes of sampling, in any event, the position of the holes determined by lottery cannot be disobeyed.

The sampling of sugarcane collected may wait up to 6 h to the steps of sample preparation, otherwise it will be neglected and the shipment considered not analyzed. The sample of cane to be analyzed will result of homogeneous mixing of the material collected by the probe, prepared in disintegrating devices that meet the technical specifications and in perfect mechanical and operating conditions, since poor preparation will influence the result of evaluating the quality of raw material. Accordingly, we recommend the use of a mixer coupled to the disintegrator as a way to ensure adequate homogeneous mixing of the disintegrated sample.

In the cane payment system, the minimum rate of preparation of the disintegrators is less than 85% and the disintegrated sample can wait for the step of extracting the juice, to determine the percentage of fiber, pol and soluble solids in a maximum of 60 min. At the end of that period, the sample should be disregarded and referred shipment not analyzed.

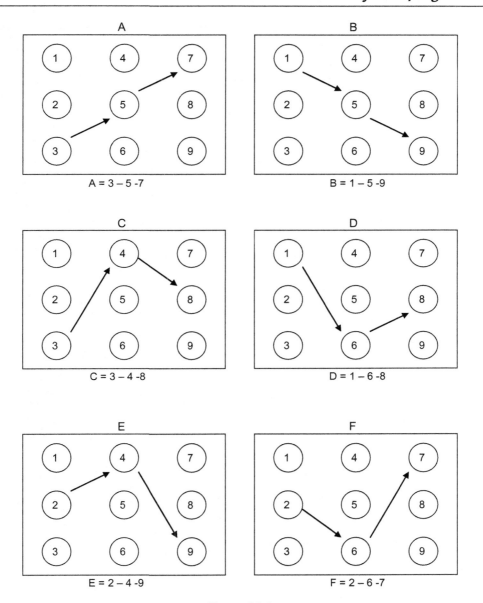

Figure 19.2
Lottery system to collect samples of sugarcane from the shipments using horizontal probe.

19.3 Determination of Technological Parameters of Sugarcane

The disintegrated and homogenized sample forwarded to the final 500 g sample weighing, after adequate dispersion is performed in a precision balance with readability of 0.5 g. After this step, the juice is extracted by hydraulic press machine at a pressure of 250 ± 5 kgf/cm^2 during the period of 1 min, watching the conditions for full drainage of the juice.

The weight of the moist bagasse (cake) used to calculate the percentage of sugarcane fiber is obtained with the aid of a precision scale.

After extraction of the juice, the processing plant will process the sample in a maximum of 30 min; after that period, the sample should be disregarded and considered as not analyzed.

The extracted juice should be homogenized to collect material for analysis of Brix and Pol.

The calculation of total soluble solids — Brix % will be made in a digital refractometer with automatic adjustment of the field, equipped with automatic temperature correction, and the final value is expressed at 20°C.

When there are mineral impurities in the juice, the Brix may be determined in filtered juice on qualitative paper filter, from the fifth drop of the filtrate.

The calculation of Saccharimeter reading — Pol percentage of juice — will be determined in a digital automatic saccharimeter, with normal weight equal to 26 g, resolution of 0.01°Z (one hundredth of a degree of sugar) and calibrated at 20°C, at a wavelength of 587 and 589.4 nm, equipped with continuous streaming polarimetric tube and output to printer and/ or magnetic recording of data.

The Pol will be determined after juice clarification with lead sub-acetate (Horne salt) and/or octapol — Quimatec in automatic digital saccharimeter calibrated with standard quartz tube and PA sucrose solutions.

The lead subacetate used in juice clarification should be mandatorily within the international quality standards as specified below:

- Total lead (as PbO) — minimum 75 g/100 g.
- Basic lead (as PbO) — minimum 33 g/100 g.
- Loss on drying (2 h at 105°C) — maximum 1.5 g/100 g.
- Insoluble Material in water — more than 2.0 g/100 g.
- Insoluble material in acetic acid — maximum 0.05 g/100 g.
- Chloride (as Cl) — 0.0005 g/100 g maximum.
- Not precipitated Substances by H_2S — maximum 0.03 g/100 g.
- Copper (as Cu) — maximum 0.005 g/100 g.

- Iron (as Fe) — maximum 0.005 g/100 g.
- Nitrate (as NO_3) to the limit test — maximum 0.003 g/100 g.
- Particle size: 75% must pass through 115 *mesh tyler* sieve, and 100% must pass through 35 *mesh tyler* sieve.

To measure the temperature of the laboratory in order to correct the reading, the thermometer should be posted near the saccharimeter.

In the case of not being able to clarify the juice with the use of recommended amounts, one of the following procedures should be adopted:

- re-filter the clarified juice;
- repetition of the analysis, new procedure to clarify the juice still available, or new extraction of juice in the presence of an accredited representative of suppliers; or
- dilution of the extracted juice at a ratio of one part of distilled water with one of juice, volume/volume, and subsequent clarification, multiply in this case the value of saccharimetric reading by two.

Characteristically, in the case of failure to clarify the juice, it will be attributed the purity of 65% for the sample and the percentage value of the juice Pol calculated.

19.4 Consecana System Calculations

The direct analysis of sugarcane by the method of hydraulic press requires the use of coefficients of transformation of the percentage of extracted juice in percentage of absolute juice, to calculate the percentage of industrial fiber.

The calculation of the percentage of industrial fiber of sugarcane is based on the correlation between the fibrous residue and the percentage of industrial fiber sugarcane, experimentally established according to the following formulas:

(1) Industrial Fiber % Sugarcane

$$\text{Industrial Fiber \% Sugarcane} = \frac{(100.PS) - (PU.b)}{5(100 - b)}$$

where PS = weight of dry cake in the oven at 105°C; PU = weight of wet cake (fibrous residue); b = Brix do caldo extraído; b = Brix of extracted juice.

(2) Fiber Factor

$$\text{Fiber Factor} = \frac{\text{Industrial Fiber \% Sugarcane}}{\text{Wet cake (fibrous residue)}}$$

or.

Industrial Fiber % Sugarcane = Wet cake × Fiber Factor.

Note: the fiber factor is found through statistical surveys.

Example: the equation for calculating the Fiber % Sugarcane from State of Pernambuco

$$\text{Fiber \%} = 0.379 + 0.0919 \, \text{PBU}$$

The calculations of Pol % Juice and Purity of the juice (Pza) of sugarcane are initially made by correcting the temperature to 20°C, as in the following equation:

$$LC = LS.LC = LS.(1 + 0.000255 \, (T - 20))$$
$$\text{Pol\% Juice} = LC \, (0.2605 - 0.0009882 \times B)$$

where LC = corrected saccharimetric reading; LS = saccharimetric reading; T = laboratory temperature; B = Brix% juice; Purity of juice (Pza); Pza = Pol × 100% juice/Brix% juice.

The coefficient "C", which is used for the transformation of Pol percentage of juice extracted by press in percentage of cane Pol is determined statistically by means of research, using the methodology of the digester — South Africa for the Pol percentage of the absolute juice.

Example: equation used in the State of Pernambuco.

$$C = 1.0313 \text{ to } 0.00575 \times \text{Fiber \% Sugarcane}$$

The Pol% cane (PC) will be calculated according to the expression:

$$PC = \text{Pol \% juice} \times (1 - 0.01 \times \text{Fiber \% sugarcane}) \times C \, \text{Pol}$$

Aiming to enhance the quality of sugarcane in some regions, is the employed purity factor that relates the purity of the extracted juice and the standard purity of 83.28 in the following expression:

$$FPza = Pza/83.28$$

The calculation of Free Sugars, AR percent by weight, can be determined analytically or be calculated by correction equation with the purity of sugarcane juice.

For the state of Pernambuco, the expression used is as follows:

$$RS = 9.9408 - 0.1049 \times \text{purity of the juice}$$

The juice of Reducing Sugars (RS) % sugarcane is held by the equation:

$$ARC = RS \times (1 - 0.01 \text{ Fiber} \% \text{ sugarcane}) \times C$$

The Total Recoverable Sugar (TRS) from sugarcane of the parameters of Pol% sugarcane (PC) of Reducing Sugars % sugarcane will be calculated by the equation:

$$TRS = 10 \times PC \times (1 - \% \text{ industrial losses}) + 1.0526 \times 10 \times TRS \times (1 - \% \text{ industrial losses})$$

Considering the industrial losses of 11%, we have:

$$TRS = 10 \times PC \times (1 - 0.11) \times 1.0526 + 10 \times ARC \times (1 - 0.11)$$

$$TRS = 10 \times PC \times 0.89 \times 1.0526 + 10 \times ARC \times 0.89$$

$$TRS = PC \times 9.36814 + ARC \times 8.9$$

where Stoichiometric Relation: Sucrose/ARL = 360/342 = 1.0526; TRS = Reducing Sugars % sugarcane; PC = Pol % sugarcane.

The remuneration of the raw material placed in the treadmill of the processing plant will be defined by the expression:

$$R = QATR \times P$$

where R = revenue of provider in R$; QTRS = kg of TRS delivered by the supplier; P = monthly average value in R$ of (1) kg of TRS by the mix of products, adjusted with the participation of the raw material.

19.5 Methodology Employed in the Final Price of Sugarcane

The price of sugarcane will be calculated using the following parameters:

I. the quality of sugarcane expressed in kilograms of TRS (total recoverable sugar);
II. the average price of finished products, sugar and ethanol, free of taxes and freight, provided PVP / PVM (Price Vehicle Plant / Price Vehicle Mill) by producers in the state in relation to external and internal market;
III. share of the cost of sugarcane (raw material) in the cost of sugar and ethanol, statewide, based on technical studies, which should be reviewed each season; and
IV. value to pay the liquidation of the crop year, should be used the production MIX and marketing effectively accomplished. The determination of the mix is performed using as a basis the preparation of sugar and ethanol and their technical specifications, with data from beginning to the end of the harvest.

19.6 Conversion Factors

19.6.1 Direct Ethanol

Anhydrous ethanol (99.3° INPM)

According to the stoichiometric yield of Gay Lussac, 1 kg of TRS (Total Recoverable Sugar) leads to 0.6503 liters of anhydrous ethanol to 99.3° INPM.

$$0.6503 = (0.6475 \times 0.7893)/(0.993 \times 0.7915)$$

where 0.6475 is the efficiency of Gay Lussac; 0.7893 is the density at 20°C of absolute ethanol; and 0.7915 is the density at 20°C of anhydrous ethanol to 99.3° INPM.

Considering an efficiency of 85.5% of fermentation and distillation of 99%, the efficiency of the distillery in liters of anhydrous ethanol per kg of TRS will be:

$$E = 0.6503 \times 0.855 \times 0.99 = 0.5504 \text{ liters/kg of TRS}$$

To produce one liter of anhydrous ethanol the following is necessary:

$$\text{Anhydrous Ethanol} = 1/0.5504 = 1.816860 \text{ kg of TRS/liter}$$

Hydrated ethanol (93.0° INPM)

Conversion factor of anhydrous ethanol (99.3° INPM) to hydrated ethanol INPM 93.0° depending on the specific masses and their ethanol content = 1.04361.

$$1.04361 = (0.993 \times 0.7915)/(0.93 \times 0.8098)$$

where 0.8098 is the density at 20°C to 93.0° hydrated ethanol and INPM; hydrated ethanol = 0.5504 × 1.04361 = 0.5744 liters/kg TRS.

To produce one liter of hydrated ethanol, the following is required:

$$\text{Hydrated ethanol} = 1/0.5744 = 1.740947 \text{ kg TRS/liter.}$$

19.6.2 Sugar

Sugar = Pol sugar × 1.0526/100 = kg of TRS.
Refined sugar = 99.8 × 1.0526/100 = 1.050495.
Special sugar = 99.7 × 1.0526/100 = 1.049442.
Superior sugar = 99.5 × 1.0526/100 = 1.047337.
Standard sugar = 99.3 × 1.0526/100 = 1.0452318.
Demerara sugar = 98.4 × 1.0526/100 = 1.035758.
VHP (very high polarization) sugar = 99.1 × 1.0526/100 = 1.043127.
Amorphous sugar = 99.0 × 1.0526/100 = 1.042074.

19.6.3 Honeys

HTM $= 0.763 \times 1$ (HTM with 76.3% of TRS).
Molasses $= 0.55 \times 1$ (molasses with 55.0% of TRS).

Based on surveys about production costs, the participation of cane sugar is determined in the cost of manufactured products, sugar, and ethanol. This participation, however, should be reviewed each season, which in practice does not happen, as well as adding new expression products in the production chain.

In determining the weighted average price of the TRS, it takes into account the mix of products in the region and/or product mix of the plant and sugar prices in domestic and foreign markets and the anhydrous and hydrated ethanol in the condition PVP and PVM.

In the Consecana system, the residual ethanol produced from molasses that comes from the process of sugar production is calculated by the classic expression of recovery SJM, where:

$$\text{Recovery SJM} = (\text{Sugar purity}/\text{sugar purity} - \text{Molasses purity})$$
$$\times [(1 - (\text{molasses purity}/\text{purity of the extracted juice} - 1)]$$

$$\text{Purity of sugar} = \text{sugar Pol}/100 = \text{unit of sugar}$$

19.7 Methodology for Evaluation of Product Prices

The standard of the product whose prices will be analyzed to meet predetermined specifications in terms of parameters that allows the specification of the most common sugar in the negotiations for the domestic market (to be determined in the market over the first stage of labor). The leading of the work to obtain indicators of the type of fuel anhydrous ethanol and hydrated ethanol, respectively, shall follow the equivalent procedure for obtaining the specified indicator of sugar in the domestic market detailed in Section 19.1.

19.7.1 Calculation Procedure

Sources and characteristics of primary information

Information will be collected from a representative sample of the market segments that make up the market in the state of Pernambuco, including processing plants, distilleries, and consumer industries and large wholesalers. The prices reported by consumers (demand) will be used to monitor the prices reported by the production sector (supply).

The prices to be researched concern the value of sales, charged on the spot market, or made with the appropriate financial adjustments when done through installment negotiations.

The amount collected in the market should express the "value placed vehicle plant" (or PVP), or in storage, in currency (Brazilian Real), related to transactions in the physical market with cash payment. Transactions based on contracts with fixed prices for the crop year will not be included in calculating the average. However, transactions in which contracts only set the volume to be delivered to the customer, with price being fixed at the time of billing will be considered.

Employees should inform if the negotiated price corresponds to transactions for cash payment or installment. In this latter case, a discount rate will be applied based on rates of Rural Promissory Notes: NPR. This discount rate will correspond to an arithmetic average of rates on NPR informed daily by commercial banks that operate the largest volumes of these Promissory Notes.

19.7.2 Data Collection

The collection of price information will be made through phone calls by technicians involved in the project, ensuring the proper treatment of confidential information.

19.7.3 Treatment of Collected Information

Pricing information will be processed and organized to form a database, after being subjected to appropriate statistical tests. Such testing requires the exclusion of the values exceeding the threshold of two standard deviations from the overall average. Additionally, the information about the product origin will be added to the information collected from consumers in the region to monitor and control the quality of information gathered.

19.7.4 Indicator Prices for Sugar Export

The price indicators for export sugar are obtained by taking as basis the prices set by stock exchanges and/or invoices.

19.7.5 Technical Terms

Sugar — nnd product of the sugar mill consists of granules of sucrose crystallized, involved or not in a film of honey.

Fermentable sugars — name of sugars which can undergo alcoholic fermentation.

Reducing sugars — name of sugars which reduce cupric ion to cuprous oxide from Fehling's solution. The vast majority of these sugars are monosaccharides, especially glucose and fructose, present in both sugarcane and commercial sugar as in intermediate products and process fluids in the sugar and ethanol industry.

Total reducing sugars (TRS) — corresponds to the reducing sugar content of a sweetened product, in which all its sucrose was split into glucose and fructose.

Granulated sugar — generic name of all white sugar produced directly by the sugar mill. They are made from a sulfited juice. The crystals are washed in the centrifuge and dried in a dryer.

TRS — total recoverable sugar, expressed as kg/ton of cane.

Bagasse — the fibrous residue from the operation of extracting the juice of sugarcane, comprising the fiber of the sugarcane and the residual juice.

Brix (degree) — the percentage of total dissolved solids in a sugar solution.

Refractometric Brix — unit of scale of a refractometer, which, through the index of refraction of light, expresses the percentage in weight of dissolved solids in a sugar solution at 20°C.

C — coefficient of extraction of the juice that correlates the extracted and absolute juice.

Fiber — insoluble matter in water contained in sugarcane. For purposes of industrial control, fiber includes the so-called foreign matter (trash).

Foreign matter (impurity) — any material that is not stalk of sugarcane present in the raw material as it is delivered to the industry.

Pbu — weight of wet cake extracted by the press.

PC — Pol percentage of the corrected sugarcane.

Polarization — quantity, in weight of sucrose in 100 ml of solution, measured by optical deviation caused by the solution in the plane of polarized light.

Purity — the relationship between the sucrose content of this solution (% dissolved sucrose) and the concentration of total dissolved material, expressed as a percentage. It can be defined as the percentage fraction of sucrose in the total solids of a sugar solution. The purity may be real, Clerget, apparent, or refractive, depending on how it is analyzed.

Optimal yield of an alcoholic fermentation — also known as stoichiometric yield or Gay—Lussac, is calculated according to the possible stoichiometrically amount of ethanol to be obtained, if the reaction of conversion of glucose into ethanol was 100%. According to Gay—Lussac, 100 kg of glucose or levulose produce 51.11 kg of pure ethanol at 100°GL.

Saccharimeter — instrument used to determine the amount of sugars through the deviation of polarized light. This instrument is similar to the polarimeter, differing only by scale, because the polarimeter presents the results in angular degrees and saccharimeter in degrees of sugar.

Sucrose — disaccharide resulted from the condensation of a glucose molecule with a molecule of fructose and has the empirical formula $C_{12}H_{22}O_{11}$ (molecular weight = 342.30). The sucrose crystals exhibit optical activity along its three axes and have the crystallized shape of a transparent prism in the monoclinic sphenoidal system.

Bibliography

Almeida, D., 1975. O pagamento de cana fornecida. Brasil Açucareiro, n. 49.

CONSECANA – PE, 2007. Manual de Instruções Técnicas.

Fernandes, A.C., 1982. Princípios básicos de um sistema de pagamento da cana-de-açúcar pela qualidade. Ano V (19).

Oliveira, E.R., Sturion, A.C., Gemente, A.C., Parazzi, C., Valsechi, O.C., 1985. Modelo de autogestão da agroindústria canavieira do Nordeste do Brasil. CONSECANA – NE.

Pagamento de cana pelo teor de sacarose: o sistema implantado em São Paulo, 1985. Brasil Açucareiro, ano VIII, n. 40.

PLANALSUCAR, 1983. Manual de orientação – Pagamento de cana pelo teor de sacarose. Piracicaba.

PLANALSUCAR, Tecnologia, pagamento de cana pelo teor de sacarose. Relatório sobre o sistema de pagamento de cana pelo teor de sacarose, preparado pela ComisSão Especial designada pelo Planalsucar.

Sturion, A.C., Gemente, A.C., 1981. A agroindústria do açúcar no Brasil e o pagamento da cana-de-açúcar de fornecedores pelo teor de sacarose. Planejamento e Pesquisa, Ano IV, n 14.

Sturion, A.C., Parazzi, C., 1985. O método da prensa hidráulica e a utilização do fator de transformação (coeficiente C) da Pol% de caldo extraído em Pol% de caldo absoluto. Coordenadoria Regional Sul do IAA/PLANALSUCAR, STAB.

CHAPTER 20

Theoretical Background of Sugarcane/Ethanol Analyses

Celso Caldas

Central Analítica LTDA, Maceio, AL, Brazil

Introduction

The development of analytical equipment industries and the constant evolution of the world of information technology have catered for increasingly simple operations in the laboratories, which is good. Conversely, this ease of operation induces the great majority of analysts to become accommodated, unwilling to know the theoretical basics regarding quality control tests, which are extremely important for obtaining reliable and precise data. Moreover, the wide diversity of theoretical contents underlying the analyses performed also hinders the acquisition of this knowledge.

Other factors may also be considered in this assessment of the level of knowledge of the vast majority of the technicians of sugar/ethanol laboratories, such as the lack of a condensed and specific literature which is totally focused on theoretical analyses and involves theoretical information on the methods used in daily work activities.

This chapter endeavors to, in a concise, clear, and objective way, bring light to some of the theories concerning the main methods implemented in the industries of sugar and ethanol. It is practically impossible to comment on the theoretical backgrounds of all the analyses performed in a sugar/ethanol laboratory. Therefore, we will consider only the theories involved in the main determinations of the productive processes (which are Brix, Pol, and reducing sugars), starting, however, with the concept of solutions, essential in any laboratory.

20.1 Solutions

We begin this chapter setting out, in general terms, the concept of a solution: a homogeneous mixture in which a substance called solute is dissolved in another called solvent. It is precisely the relationship between the amount of solute and the amount of solvent or of solution that we call concentration, which can be expressed in various ways.

Sugarcane. DOI: http://dx.doi.org/10.1016/B978-0-12-802239-9.00020-7

Some of the expressions involve mol and gram equivalents (or equivalent weight), and for that reason examples of these calculations will be outlined.

The mole of a chemical element or a substance is calculated from the atomic mass units (u) of the element(s) involved. The calculation of the chemical equivalent depends on the valences of the elements and functions of the substances involved.

As described above, the concentration of a solution is the ratio of solute dissolved in a given amount of solvent or solution. Some authors use the term "title" to define concentration, hence the term "title" or concentration of the solution. The amount of solute can refer either to mass or volume (physical quantities). This is implicated in the classification of solution concentrations in three ratios:

- Mass versus mass
- Volume versus volume
- Mass versus volume

In the mass versus mass ratio, one finds the percentage in weight (% w/w) and molality, which is defined as the number of moles dissolved in 1 kg of pure solvent. In volume/volume, the most common form of expression is the percentage of volume/volume (% v/v). Finally, for the mass/volume ratio, there are expressions in physical units, such as the percent composition in mass, which is the ratio of the amount in grams of solute per liter of solution, and the mass/volume percentage (% m/v). Other expressions are based on chemical units, such as molality, defined as the number of moles of solute dissolved in 1 liter of solution, and the normality, established as the existing number of gram equivalents in 1 liter of solution.

Richet's law is known as the Principle of Equivalence: in any chemical reaction, the number of moles or equivalents of the substances that participate in the reaction is always the same. Through its mathematical equation, this law can be used in the calculations of neutralization and dilution of solutions, a very common procedure in laboratories of the most varied types. Generalizing, one can write the law of Richter as:

$$C_1 V_1 = C_2 V_2$$

where C_i = concentration of the solutions; V_i = volume of the solutions.

In case the concentration is expressed in normality or molality, the expression will take the following forms, respectively:

$$N_1 V_1 = N_2 V_2$$

$$M_1 V_1 = M_2 V_2$$

where N_i = normality of the solutions; M_i = molality of the solutions.

For chemistry technicians involved in the work of a quality-control laboratory, one of the points on which they must have good knowledge is the standardization of solutions. This term "standardization" is replaced in some cases with "factoring," but they refer to different procedures: standardization refers to the act of making a solution of unknown concentration react with a standard solution (prepared from a standard reagent), therefore one of exactly defined concentration, aimed at understanding their true concentration. This becomes possible in quantitative chemistry, since, theoretically, it is assumed that the true concentration of a standard solution is identical to the theoretical concentration. Conversely, factoring refers to the procedure for calculating the ratio of the solution according to theoretical and actual concentrations.

There are two types of standardization: direct and indirect. Standardization is called direct when there is a reaction of the solution of which one wishes to know the concentration as compared with a standard solution. Indirect standardization happens when one standardizes the solution of unknown concentration by comparison with a solution that is not standard, but which was previously standardized by direct input. This type of standardization is very common in laboratories which, for example, are in lack of biphthalate potassium, oxalic acid, benzoic acid and/or hydrogen potassium iodate for use in the standardization of bases, but have carbonates in stock to standardize its acids. In these cases, the acid solution which has been standardized with a carbonate standard will serve as "standard" for the standardization of the base. Clearly, indirect standardization does not have the same degree of reliability as direct standardization, but it helps to minimize operational error in an analyses laboratory.

20.2 Densimetry

Densimetry is widely used in the sugar—ethanol sector, in both sugar and ethanol factories. This use is due to correlations between the densities and the mass percentages of total dissolved solids and ethanol content.

The determination of the content of dissolved solids in a solution has its importance in the sugarcane industry not only to calculate the purity of the materials analyzed, but also for their mass balance and in the division of the harvested cane to produce sugar or ethanol, since that calculation is performed from the existing solids in the juice that is sent for the production of sugar and ethanol. This determination is known as Brix, in honor of the researcher who conducted studies on this topic.

The density of a substance at a given temperature is defined as the relationship between mass and volume of the substance at the same temperature and the mass of the same volume of water at that reference temperature. This reference ratio of the density of a substance in comparison with the density of water is also known as relative density. Thus, the density at 20°C/4°C of a given substance indicates the relationship between the mass of a given volume of the substance at 20°C and the mass of the same volume of water at 4°C. Thus, it is obviously concluded that the relative density is a dimensionless measure. Taking into consideration that these measures relate to the masses of certain volumes, they are also synonymous with specific weights.

While studying densimetry, one also needs to differentiate between the concepts of true density, absolute density, bulk density and specific mass. Density is called "true density" or "absolute density" when the measurement is made in a vacuum, eliminating the effect of air buoyancy. "Apparent density" is used when the measurement is made in the presence of air. The difference is negligible. On the other hand, we can have another relationship between the mass at atmospheric pressure of 1 cm^3 of a homogeneous substance, at a given temperature. This relationship is defined as density, or apparent density. Thus we can conclude that the density of a homogeneous substance is the mass of unit volume at a given temperature, expressed in g/cm^3 or kg/m^3.

It is observed that, as happens with relative density, the specific mass or apparent density is also a function of the temperature at which measures of mass and volume were taken. Thus, one can have different values for the same substance, due to different temperatures. The following examples illustrate this difference:

Specific mass of water at 20°C = 0.998200 g/cm^3
Specific mass of water at 4°C = 0.999973 g/cm^3

Bulk density or specific mass has a great importance in the identification of substances, as well as other determinations, such as melting point, freezing point, etc. It is thanks to Archimedes, who wished to clarify if there was silver in a crown of gold, that the discovery of another important property of matter has been made. While bathing in a bathtub, Archimedes realized that, when dipped, people feel the thrust of the water, and concluded that the weight loss of a body immersed in water equals the weight of the water he relocated. Once one knows the weight of the displaced water, he/she can determine its volume. Having the weight and volume, we can calculate the density of the body. Obviously, this could be done with the crown, and thus, knowing the density of the crown besides that of silver and gold, the proportion of silver added to the crown could be obtained. As his question could now be resolved in all its simplicity, Archimedes went screaming *eureka* (found it) and went in search of the crown to immediately determine the nature of its materials.

Among the devices used for the determination of density, there are simpler ones, based on Archimedes' thrust, called aerometers, and the more sophisticated ones, such as electronic densimeters, which make use of the principle of resonance frequency of a glass oscillator, as is the case of the Anton Paar densimeter. Besides these, there are others, like the pycnometer and the Westphal balance.

Figure 20.1
Illustration of a hydrometer (aerometer).

Table 20.1: Densities of pure solutions of sugars.

Sugar Solutions at 10% m/m	Density at 20°C/4°C
Arabinose	1.0379
Glucose	1.0381
Fructose	1.0385
Galactose	1.0379
Sorbose	1.0381
Sucrose	1.0381
Maltose	1.0386
Lactose	1.0376
Raffinose	1.0375
Average	1.0380

The Brix and Gay–Lussac aerometers are the best known devices in the sugar and ethanol production sectors. The scales of these densimeters are based on the concentrations of soluble solids and ethanol in the materials analyzed, respectively, both based on experimental studies. Regardless of type, Figure 20.1 shows a densimeter which, when placed in a liquid, immerges up to a certain height, at which point is established the equilibrium observed when Archimedes' buoyancy equals the weight of the aerometer.

In the sugarcane–ethanol industry, densimetry was introduced by Balling (1839) and later by Brix (1854). This occurred because sugar solutions with the same concentrations and prepared with different sugars had very similar densities at 20°C/4°C. Table 20.1 shows the densities for several of these solutions, where it can be noticed that the average densities at

20°C/4°C of these solutions (1.0380) is very similar to the density of sucrose at 20°C/4°C, which justifies the fact that in many bibliographic references, the definition of Brix is found to be the percentage by weight of sucrose.

Measurements were prepared from standard solutions at 20°C. The corrections should be made when the readings occur at different temperatures and are made over Brix (concentration) rather than density.

20.3 Refractometry

The physical phenomenon that explains the deviation of a light ray when it focuses on a transparent surface is called refraction. This deviation occurs because a beam of light falling obliquely changes direction when passing from one transparent medium to another transparent medium which presents different speed of light from the first medium. Schematically, one always uses Figure 20.2 to explain the phenomena of reflection and refraction.

The change of course that the light ray undergoes when crossing from one medium to another depends on the speed of light in the two media involved. The index of refraction (n) is the physical quantity that relates the velocities in both media by the equation:

$$n = v_1/v_2$$

where $n \rightarrow$ refraction index; $v_1 \rightarrow$ speed of light in medium 1; $v_2 \rightarrow$ speed of light in medium 2.

When one of the media is considered to be a vacuum, the refractive index is called the absolute refractive index and relates the speed of light in a vacuum with the speed of light in another medium. Mathematically, we can write:

$$n = c/v$$

where $n \rightarrow$ absolute refraction index; $c \rightarrow$ speed of light in a vacuum; $v \rightarrow$ speed of light in the other medium.

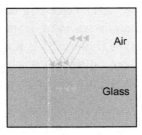

Figure 20.2
Reflection and refraction phenomena.

From the absolute refractive indices of two media, one can determine the relative refractive index of a medium in relation to another, which is equal to the ratio between the refractive indexes of the media considered, whose expression is:

$$N_{21} = n_2/n_1$$

where $N_{21} \rightarrow$ relative refraction index; $n_2 \rightarrow$ refraction index of medium 2; $n_1 \rightarrow$ speed of light in medium 1.

The law that allows us to study the phenomenon of refraction, and to establish the optical properties of lenses, is called Snell−Descartes Law. Consider that an incident ray propagates from medium 1 to medium 2, where 1 is less refractile than 2 (Figure 20.3). Note that the phenomenon of refraction occurs precisely when the incident ray reaches the surface that separates the two media. The refracted ray forms an angle ($\emptyset_2$) with a normal (N) perpendicular to that surface, called angle of refraction. Note also that another angle is formed ($\emptyset_1$) between the normal and the incident ray, now termed angle of incidence.

According to the Snell−Descartes Law, the ratio between the sine of the incident angle ($\emptyset_1$) and the sine of the angle of refraction ($\emptyset_2$) is a constant value equal to the relative refractive index N_{21}, for a given wavelength (λ). The refractive index of a sucrose solution is a measure of the concentration of sucrose in that solution, in the same way that the refraction index of a juice, for example, express the concentration of the total solids dissolved in it. This measure, like the measurement made from density, is also called Brix, but refractometric. For the case of impure solutions, such as broth, pasta and honey, the measures of refractometric Brix are more reliable than the measurements obtained by the densimeter, that is, than densimetric Brix, because the determination of the refractive index is less affected by suspended solids than the determination of density by a Brix densimeter. In addition, the refractive index varies little with the addition of impurities and is not affected by surface tension.

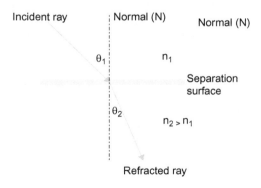

Figure 20.3
Refraction of an incident ray.

The devices that measure the refractive index are called refractometers. Brix has developed tables that relate the concentrations of sugar solutions, expressed as a percentage, with their respective refraction indices. These tables were built into the devices, so that the readings obtained already correspond to the percentages of those solids.

20.4 Polarimetry

This technique is undoubtedly the most important one among the techniques used in the industry, if we take into consideration that, through it, the apparent concentrations of sucrose in the materials analyzed are determined. It is not the most appropriate for these determinations due to interference from other optically active compounds, but it is widely accepted due to their convenience and speed.

Prior to beginning the study of this technique, it is worth remembering that the previous items included mentions of earlier theories about the nature of light, one of which describes light as an electromagnetic phenomenon composed of two oscillating fields, one electric and one magnetic, perpendicular to each other and also perpendicular to the direction of propagation of the beam. As these fields propagate, vibrations occur in infinite planes. To select or filter one of these planes, one uses a polarizer. Ordinary light is made to pass through a polarizer which interacts with its electric field, so that this field remains oscillating in a single plane. Figure 20.4 shows a schematic representation of light and its polarization.

Since a single plane of light is desired, one of those two has to be eliminated. The polarization process established through using a Nicol Prism consists of the elimination of the ordinary ray, making a diagonal cut in it and then pasting it with Canada balsam. This balm, being more refractile than a $CaCO_3$ crystal (due to the fact that its refractive index is equal to 1.530, a value intermediate between 1.658 of the ordinary ray and 1.487 of the extraordinary ray), makes the ordinary ray, upon hitting it, have a higher incidence angle than the so-called limit angle and, thus, the ordinary ray, instead of being refracted, as before, is now reflected. Consequently, only the extraordinary ray traverses the Nicol Prism, being therefore the only polarized light obtained. The whole scheme of polarization of light through the Nicol Prism can be seen in Figure 20.5.

Some substances, whether solid or liquid, have the ability to divert the plane of polarized light, and thus come to be called optically active substances. In crystals, optical activity is due to the fact that they do not have a symmetry element. For liquids, this property is due to asymmetric molecules.

Thus, optical activity is directly related to spatial isomery, which is defined as the ability of two compounds to have the same structural formula, identical physical properties (e.g., melting and boiling points), but different spatial structures. When the deviation from the

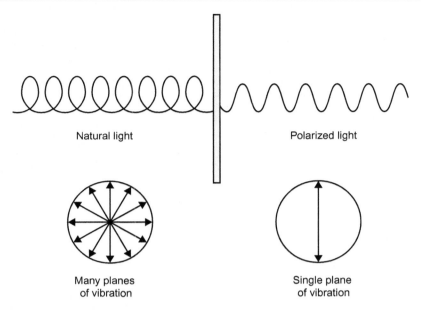

Figure 20.4
Representation of the polarization of light.

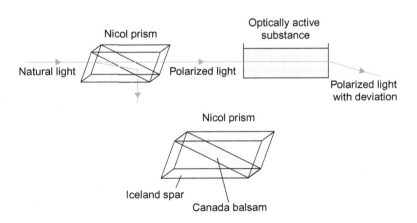

Figure 20.5
Polarization scheme by the Nicol Prism.

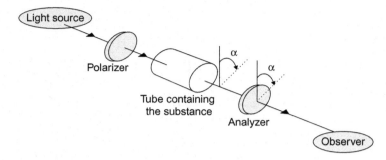

Figure 20.6
Schematic representation of a polarimeter.

plane of polarized light is to the right, or clockwise, the substance is called dextrorotary and is represented by the letter 'd' or the positive sign " + ". When the deviation is to the left, counterclockwise, the substance is called levorotary and is represented by the letter "l" or the negative sign " − ".

In this study, we need to better understand what specific rotation is and how it is measured. Initially, let Figure 20.6 be observed, which shows a schematic representation of a polarimeter, where we highlight the parts that compose it: (a) light source, (b) Nicol polarizer, (c) polarimetric tube, where the substances to be analyzed are placed, (d) Nicol analyzer, and (e) observer.

When the polarimetric tube is empty, or contains water or a saline solution − that is, any optically inactive substance, without asymmetric carbons − the maximum amount of light will occur and the second Nicol Prism of the polarimeter, the analyzer, will mark 0°. In cases where the solution contained in the tube is optically active, the analyzer rotates up to a certain position, which is equal to the number of degrees by which the plane of polarization is rotated. This value is called specific rotation and depends on factors such as:

- concentration of solution in the tube;
- length of the polarimetric tube in which the solution is contained;
- temperature of the solution analyzed;
- wavelength of light used in the polarimeter; and
- solvent.

From these variables, we can now more precisely define what specific rotation is: *It is the rotation in degrees, produced in a plane of polarized light, by 1 g of substance of optical purity, dissolved in 1 cm³ of solution, when the length of the polarimetric tube is 1 dm.*

Experimentally, the three last variables are fixed to determine specific rotation: (a) temperature of the solution − 20°C, (b) wavelength of light − a sodium light is used, with radiation corresponding to the streak, or line D, which has a wavelength of 589.44 nm and is responsible for the yellow flame of sodium; (c) solvent — water is preferably used; if the sample is insoluble in water, another solvent may be used, for example, ethanol, ether, or C_6H_6, but it must be identified.

The polarimeter is the equipment used to measure the deviation of polarized light from the plane caused by an optically active substance. The source used in this equipment is an ordinary monochromatic light, from sodium, which focuses on the first polarizer Nicol Prism. When going through this prism, the light is converted into a plane of polarized light. Thus, it crosses the polarimetric tube, where the sample, a solution whose optical activity one wishes to analyze, is found. After going through the solution, the polarized light reaches another Nicol Prism, called the analyzer, which is mobile. It is important to realize that if the tube is empty, or filled with an optically inactive solution, the polarized light passes through it completely unchanged. In this case, the instrument indicates 0° and the analyst notes the maximum amount of light crossing it, since the polarizer and analyzer prisms are parallel, as mentioned in the previous item.

The saccharimeter is a polarimeter designed, theoretically, only to measure sucrose solutions using quartz wedge compensation, which eliminates the use of the sodium lamp or other source of monochromatic light, therefore allowing it to maintain the polarizer and analyzer Nicol Prisms fixed into position. The quartz wedges were introduced in these devices by Soleil in 1845 and are also called rotary polarizing plates. They should be used whenever measurement accuracy is required. That happens because it is also common to use standard solutions for the calibration of saccharimeters. In such cases, the polarization plane of linearly polarized light beams is set to rotate, so as to obtain angles which are previously known from the equation given above for specific rotation, obviously at temperatures and wavelengths previously known.

The results of these polarizations, much as the technical criteria are met for the preparation of standard solutions, have error margins which are no better than 1%. When using Standardized Reference materials established by the American "National Bureau of Standards", these error margins can reach 0.5%. This error is taken along to routine analyses, since the analyzed samples can only be compared to the standard sample within the limits of accuracy provided by the equipment.

Between 1842 and 1843, Ventzke proposed a method to set the scale of sugar, avoiding calculations and standardizing analyses through saccharimeters. He established that the 100th point in this scale was the rotation of a sucrose solution with relative density of 1.1000 at 17.50°C, with water at the temperature of 17.50°C as the reference. As the results were not entirely satisfactory using as reference relative density, the standard solution then changed to be one prepared from 26.048 g of sucrose to 100 cm^3, being this weight called normal. In 1855, the Mohr balloon was introduced into the sugar industry, forcing polarimeter manufacturers to establish a new 100th point for that scale. This new point of the scale became the rotation of a solution prepared with 26.48 g, because 100 cm^3 on the Mohr balloon amounted to 100.234 ml. Finally, the 100th point on the Ventzke scale was determined by the rotation that produces 26.048 g of sucrose dissolved in 100 cm^3 on the balloon at 17.50°C. Thus, between 1855 and 1890, all saccharimeters, except those of the French scale, were made based on this scale.

Although it has been recognized for many years, the Ventzke scale always brought confusion on the basis of the calibrations of the flasks of Mohr. For that reason, in 1900, when meeting in Paris, the International Commission for Uniform Methods of Sugar Analysis (ICUMSA), recommended a new definition for the 100th of saccharimeter scales, establishing the cm^3 as the unit of volume and a temperature of 20°C, resulting in a new weight of sucrose, changing it to 26.010 g. For simplicity reasons, and to work with integers, a final value 26.000 g of sucrose was established. To avoid further confusion, this scale was renamed the International Sugar Scale.

In 1932, when the ICUMSA assembled in Amsterdam, the consensus was to set the value of 34.620° for the 100th of saccharimeter scales working with sodium light. Currently, the scale of saccharimeters is the one established by ICUMSA in the meeting of 1986, with the change of scale of degrees sucrose (°S) to degrees "Z" (°Z) of azúcar (Spanish), zucchero (Italian), and zucker (German). The international scale °Z has as its reference the specific rotation of a normal pure sucrose solution, prepared from 26.0160 g of sucrose weighed in a vacuum and dissolved in 100 cm^3 of pure water at 20°C. This amounts to a solution made from 26.0000 g (of sucrose weighed in air at standard temperature and pressure STP), that is, 1 atm pressure and 20°C temperature.

The reference value of 100°Z is equivalent to the optical rotation of the normal solution of sucrose, read at the wavelength of the green line of the mercury $_{190}Hg$ isotope, amounting to 546.2271 nm, which is equal to 40.777°. Using the filtered sodium light at wavelength equivalent to 589.44 nm, the value of the 100th point in °S amounts to a rotation equal to 34.626°. In saccharimeters compensated by quartz wedges with a wavelength equal to 587.000 nm, the rotation of section 100°Z is equal to 34.934.

20.5 Redox Titration

All chemical reactions involving electron transfer or change in oxidation number (NOx) among the substances which react with each other are considered part of the so-called redox titration reactions. Hence, these substances are necessarily reducing or oxidizing agents. Obviously, redox titration does not apply to the direct determination of elements that are presented with a single valence state. The most common types are permanganometry, dichromatometry, cerimetry, iodometry, etc., depending on the substances involved.

Stoichiometric knowledge of these reactions allows for the quantitative calculation of the substance being oxidized or reduced. The equivalent of an oxidizing agent or a reducing agent is simply defined as the mass which reacts or contains 1.008 g of hydrogen or 8.000 g of oxygen available, capable of being used in oxidation or reduction. These equivalents can also be calculated from the oxidation number, because oxidation and reduction are accompanied by change of valence, as already mentioned.

In oxi-reducing reactions which are processed through titration, it is necessary to know the equivalence point or endpoint of the titration. As in acid-base titrations (titration of neutralization), where this point is defined by the sudden change in pH during titration (making it necessary to resort to the use of indicators also called neutralization indicators), in oxi-reducing titrations the stoichiometric point is marked by an abrupt change in oxidation potential. The indicator of oxi-reducing reactions is a compound that exhibits different colors in the oxidized and reduced forms.

The use of oxi-reducing titration in the sugarcane industry is based on the property that certain sugars have to reduce the copper of the alkaline solutions of certain metallic salts from the copper to the cuprous state, through its aldehyde and ketone groups. The first researcher to make use of oxi-reduction in redox titration in the determination of aldehyde and ketone groups was Trommer (1841), employing a solution of copper sulfite in alkaline medium. However, it was Fehling (1848) who established in detail the function of copper in the reaction. Soxhlet (1878) also modified the method separately preparing a solution of copper and an alkaline solution of sodium tartrate and potassium. Soxhlet called these solutions respectively Fehling solution A and Fehling solution B, with an equal-part mixture of these two known as Fehling liquor. Soxhlet was also the one who found that the stoichiometric relationship between copper and the reducing sugars varies with the excess of copper during the reaction, and from that discovery is justified the need for standardization of Fehling Solution A before its use through a standard solution of inverted sugar.

Finally, Lane and Eynon (1923), to facilitate detection of the endpoint, introduced methylene blue as an indicator due to its property to become colorless with a slight excess of reducing sugars. The introduction of this indicator has become the fastest and more easily employed method. The endpoint of the titration was also sharper, making the method more reliable and accepted by all segments of analytical chemistry.

Methylene blue is an internal indicator of redox reaction, which has a formal potential of 0.52 V. In the oxidized form, it has a blue color, while the reduced form is colorless. In the presence of high alkalinity, as is the case for determinations of reducing sugars by the

Lane—Eynon method, it re-oxidates with the air. Therefore, this analysis must necessarily be made under heat, in soft boiling, thus avoiding re-oxidation of methylene blue during titration.

The reduction of copper from the cupric state (Cu_{2+}) to cuprous state (Cu_+) by the reducing sugars is a complicated reaction, difficult to define stoichiometrically, mainly due to the existence of intermediate reactions not yet fully clarified. Another factor that hinders the establishment of stoichiometric equivalence is that copper can be reduced by both the aldehydic group (glucose) and the ketone group (fructose).

Comparing the stoichiometric relationships involved in these reactions, it appears that there is a significant difference when the monosaccharide is an aldose or ketose. Due to the different relationships between the copper sulfate and the reducing sugars, the molar equivalence to the reaction was thus established:

5 moles of copper sulfate ($CuSO_4$) for each mole of monosaccharide ($C_6H_{12}O_6$) (reaction 7.2.11).

The standardization of Fehling's liquor should always be made before analysis, so that the true stoichiometric relationship is known. To facilitate the standardization of this solution, one must calculate a factor that will correct the concentration of the solution. This factor, in some laboratories, is overlooked or forgotten. This usually leads to results which do not have the desired reliability. One of the most commonly adopted procedures for this standardization is to prepare a solution of inverted sugar at 1% from sucrose, unheated. From this, prepare another solution at 0.25% and use it as a titrant solution. The titration should be performed using 20 ml of Fehling's liquor. The theoretical amount being spent of the standard solution is 40 ml. That's because 20 ml of liquor equals 0.10 g of reducing sugar, so:

$$100 \text{ ml of the standard solution} \qquad 0.25\text{g of RA}$$
$$x \qquad 0.10 \text{ g of RA}$$
$$x = 40 \text{ ml}$$

The use of the factor is extremely important in this analysis of reducing sugars. Its calculation and proper use allow for more reliable results and higher success rates for the analysis. To illustrate this calculation, one can also consider standardizing the procedure above. Thus, any calculation will be based on the theoretical volume being spent, that is, 40 ml. Being "Vg" the volume spent on standardization, equal to 40 ml, the factor is 1.0000. If different than 40 ml, there are two ways to calculate the factor:

$$F = 40/Vg \text{ and } F = Vg/40$$

The common doubt in laboratories is which of the factors to use, and where to apply it: the volume consumed in the analysis or the outcome? One can easily eliminate such doubts calculating the actual amount of copper sulfate weighed, and from the result, calculate the real molar equivalence. In doing so, we come to this conclusion:

- $F = 40/Vg$ is applied to the amount spent on analysis; and
- $F = Vg/40$ is applied to the result of the analysis.

The standardization factor should have a limit that can be established by the laboratory itself, depending on the degree of accuracy which they desire their results to have. In general, these factors are accepted when they range from 0.9975 to 1.0025.

Bibliography

ABNT. Associação Brasileira de Normas Técnicas, 1989. Xarope Invertido Concentrado-Determinação de Brix Aerométrico e Brix Refratométrico-Projeto 13:10-02-048. Rio de Janeiro.

Allinger, N.L., et al., 1978. Química orgânica. Guanabara Dois, Rio de Janeiro.

Associação Brasileira de Normas Técnicas, 1989. Melaço-Especificação. Rio de Janeiro.

Armentano, E. Açúcares redutores em mel rico invertido. Société Générale de Surveillance.

Basset, J., et al., 1981. Análise inorgânica quantitativa. Guanabara Dois, Rio de Janeiro.

Bureau of Sugar Experiment Stations, 1991. The Standard Laboratory Manual for Australian Sugar Mills. Brisbane.

Caldas, C.S., 2005. Teoria básica das análises sucroalcooleiras. first ed. Central Analítica, Maceió.

Campbell, J.M., Campbell, J.B., 1986. Matemática de laboratório. three ed. Editora Roca, São Paulo.

Cooperativa de Produtores de Cana, Açúcar e álcool do Estado de São Paulo, 1995. Métodos de análises para fabricação de açúcar e álcool. Centro de Tecnologia COPERSUCAR. 108p.

Francisco, W. de, 1982. Estatística. Editora Atlas, São Paulo.

Gonçalves, D., 1978. Física. Ao Livro Técnico, Rio de Janeiro.

International Commission for Uniform Methods os Sugar Analysis, 1990. Proceedings. Colorado Springs.

International Commission for Uniform Methods of Sugar Analysis, 1994. Methods Book. London.

Meade, G.P., Chen, J.C.P., 1977. Cane sugar handbook, tenth ed. John Wiley & Sons, Inc., EUA.

Núcleo de Absorção e Transferência de Tecnologia, 1986. Determinação de brix, açúcares redutores, açúcares redutores totais, açúcares totais e sacarose em high test molasses. Caldas, Maceió.

Ohlweiler, O.A., 1976. Química analítica quantitativa. Livros Técnicos e Científicos Editora S.A, Rio de Janeiro.

Oliveira, A.J., Açúcares redutores-Princípios da oxirredutimetria. Escola Superior de Agronomia Luiz de Queiroz, São Paulo.

Orear, J. F., 1978. São Paulo: Livros Técnico e Científicos Editora.

Solomons, T.W.G., 1982. Química Orgânica. Livros Técnicos e Científicos Editora S.A, Rio de Janeiro.

Sugar Technologist's Association, 1985. Laboratory Manual of South African Sugar Factories, third ed. Pietermaritzburg.

Tipler, P.A.F., 2000. Rio de Janeiro: Ao Livro Técnico.

United Molasses Trading Company, 1971. The analysis of molasses. London.

Voguel, A.I., 1992. Análise química quantitativa. fifth ed. Guanabara Koogan, Rio de Janeiro.

Suggested Websites

www.ecientificocultural.com
www.dqi.ufms.br
www.rbi.fmrp.usp.br

www.livrariapolitecnica.com.br
www.quimilab.com.br
www.geocities.yahoo.com.br
www.jroma.pt/polarimetro.htm
www.fsc.ufsc.br
www.metroogia.br
www.unicam.br
www.buladequimica.com.br/isomeria
www.qca.ibilce.uesp.br
www.quiprocura.he.com.br/isomeria
www.quimica.fe.usp.br
www.enq.ufsc.br
www.ifi.unicamp.br
www.cgs.com.br
www.quansa.com.br
www.ital.or.br
www.allchemy.ig.usp.br
www.chemkeys.com
www.fisica.uel.br
www.educar.sc.usp.br
www.ufpa.br/quimicanalitca
www.debiq.faenquil.br
www.sbq.org.br
www.ualg.pt
www.qt1.iq.usp.br
www.gravidade.hpguip.ig.com.br
www.quimica.com.br
www.ufv.br
www.analiseinstrumental.hpg.ig.com.br
www.perfline.com
www.fc.unesp.br
www.rossetti.eti.br

CHAPTER 21

Managing Costs of Production and Processing

Willians Xavier de Oliveira[1], Angélica Maria Patarroyo[2] and Paulo do Carmo Martins[3]

[1]*Fundação Getúlio Vargas, São Paulo, SP, USA* [2]*Universidade Federal de Minas Gerais, Belo Horizonte, MG, Brazil* [3]*Federal University of Juiz de Fora, Juiz de Fora, MG, Brazil*

Introduction

This chapter addresses the importance of being aware of costs, whether in production, administration, trade, or otherwise. It is essential to master concepts, methodologies, and practices that can be adopted in day-to-day business. In this sense, the approach presented here involves the discussion of concepts and techniques established by management theory coupled with market practices.

Cost has always been a controversial issue, especially regarding allocation. Since there is no prevailing methodology and the dissemination of techniques and success stories is poor, the subject ends up restricted to empiricism, i.e., people do things the way they learned or the way they think should be. However, this approach takes companies with similar structures in similar activities, to produce quite different results, and worse, leads to wrong decisions as to product pricing and costing.

After all, what is the use of adequate knowledge about costs? In order to answer this question, one should understand that there needs to be an alignment between the strategy of the organization and what it practices day-to-day. For Porter (1991), an industry can ensure competitiveness through three generic strategies: cost leadership, differentiation and focus. However, as regards commodities, the obvious strategy is to be price competitive, since they are products with little differentiation, and traded with a basic determinant, which is price, such as ethanol or sugar. In this case, the market sets commodities prices anywhere in the world, so, those with good profit margins earn more. In addition to strategy, it is necessary to differentiate the types of costs that are most important and how they should be appropriated, i.e., allocated to products.

Based on this basic knowledge, it is possible to outline the proper stance of companies vis-á-vis the market and, especially, to make decisions about what, how and at what cost to produce the goods demanded. It is worth mentioning that this kind of decision cannot be supported only by cost calculation and allocation, but other variables should be taken into account, such as market trends, selling prices, demand, and tax incentives, among others. On the other hand, the use of appropriate analytical tools, including cost tools, is also needed due to several factors, including economic globalization, the evolution of markets, the need to be competitive and, above all, ensuring good profit margins for shareholders and investors.

The current scenario of sugarcane, sugar, and ethanol production is promising and very dynamic. For the first time in the history of the country, energy sources are not mutually exclusive but complementary. Ethanol has been gaining strength and takes an important position in the Brazilian energy matrix. Since Brazil has emerged as the largest producer of sugarcane and sugarcane products, matters such as process management, costs, and strategic planning have become habitual in plants and companies involved in the activity.

The constant advancements accomplished in agricultural and industrial research also contribute to the steady increase of Brazilian competitiveness in the sector. Accordingly, cost management takes on a major role since it represents a significant opportunity to streamline production processes, optimizing resources and providing better results. Thus, one can guarantee the country's strategic position in the market, as well as profit for entrepreneurs.

This chapter combines good theoretical framework to best market practices, with illustrations and examples, so as to provide readers with an important tool for decision-making as regards costs.

21.1 Important Concepts

Costs constitute a natural consequence of production. In order to produce, one needs to use production factors such as raw materials, equipment, energy, facilities, and labor, among others. Companies must pay to have access to these items, i.e., they have to incur costs. On the basis of this assumption, the search for efficient production methods constitutes one of the main objectives of well-managed companies, in other words, methods for maximum production at minimum costs, that is, process optimization.

Much of the literature on costs refers to the definition of the term. This text will provide a view of the practice related to costs. Thus, it is expected that readers will find it relatively easy to relate the concepts presented here with the everyday running of their companies.

21.1.1 Understanding Costs

Cost means all expenditure directly or indirectly involved with production, without which production could not happen. Broadly speaking, cost is any spending made to leave products ready to be sold.

Martins (2003) considers cost as an expense on a good or service used in the production of other goods or services. In this perspective, the term cost is whatever is spent during the production process so that the final products can be obtained.

For Megliorini (2001), costs are an expression of the company reflected as attitudes, behaviors, structures, and operation modes, and the better it is structured, the better the results; conversely, the less information there is available, or the lower the quality of such information, the poorer the results. This concept incorporates the relationship between costs and the importance of knowledge about them for the organization's structure.

As an example, consider the following extreme scenario:

(a) an ethanol-producing industry maintains idle as much as 30% of its industrial facilities and manpower available, so that it is able to implement a sudden increase in output in case a market opportunity arises;

(b) the market indicates that at any time the amount of ethanol in gasoline can be raised by a government decree, which would increase demand, since the industry is the direct supplier of leading refineries.

Apparently, the situation of the industry in question is comfortable. However, one must take into consideration that idleness has a positive side to it, but it also has negative effects, such as high spending on equipment maintenance, wasted manpower, etc. This leads to poorer results, and therefore, loss of operational and economic efficiency. Consider, for example, that the increase in demand does not occur, or that it takes longer to occur: was it worth paying that extra to have that structure in place? Would it not have been better for the company to plan the increase in structure according to the increase in fuel consumption statistically expected without considering political variables?

Anyway, one must be aware of information, especially statistical information, and of structural efficiency, as this is the only way companies are able to ensure the best ratio between cost structure and operational results.

Costs should not be confused with expenses. Although there is a fine line between them, there is a difference between the two concepts.

21.1.2 What Is Expense?

Expense is any spending directly or indirectly linked to the management of businesses or to the marketing of products or services produced. Broadly speaking, expense is the effort required to market the product, involving spending with the administrative and commercial structures of the business, i.e., expenditures made after the production process.

21.1.3 Costs versus Expenses

Of course, the border between costs and expenses is not so easily outlined. Some expenditures can be costs or expenses, depending on the activity or stage in the process of production / marketing in which they occur. A clear example is product storage. If storage is part of the production process, to wait for processing for example, then storage is cost. But, if storage is a mere step to wait for marketing, as is the case of ethanol and sugar, once products are ready, then it becomes an expense. Take as an example a plant producing sugar and ethanol, with three sectors: production, management and marketing and storage. All spending incurred in the production sector can be considered as costs, because they are required for making sugar and ethanol ready for commercialization. As for the spending incurred in the fields of management and marketing, as well as in the storage for finished products, those can be considered as expenses. Figure 21.1 shows an example of the difference between costs and expenses, according to this line of thought.

21.2 Classification of Costs

21.2.1 Fixed Costs

When costs do not vary as a function of the quantity produced, they are known as fixed costs. This does not mean to say that they remain the same forever, but that they are fixed within certain limits of fluctuation of the activity to which they relate, being able to go beyond these limits, but not to increase proportionally to the increase in activity. In other words, fixed costs are those that will happen, whatever the volume of production (Leone, 1997). It is important to point out that costs can be fixed regardless of the concept of value, that is, even if the values of a certain cost differ from one period to another, its classification does not change: it continues to be fixed.

21.2.2 Variable Costs

When there are increased costs associated with the production of an additional unit of output, the cost is variable. If there is no quantity produced, the variable cost is zero. It should be pointed out that variable costs should be analyzed with respect to the quantities consumed in production and not to the quantities purchased or stored. The amount of sugarcane processed and other inputs used for sugar production are clear examples of variable costs, because the amount of raw materials used is directly related to the quantity produced.

21.2.3 Total Costs

Total costs refer to the sum of total fixed costs and total variable costs. Figure 21.2 shows an example of the three types of cost, considering the situation of a growing industry.

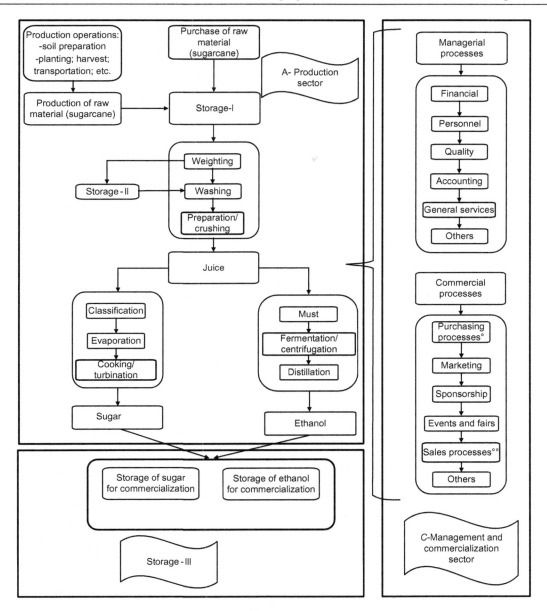

Figure 21.1
Simplified procedures for production, management, marketing, and storage.

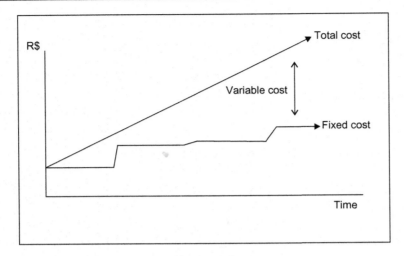

Figure 21.2
Graphic representation of the behavior of costs in a growing industry.

21.2.4 Direct Costs

Direct costs are those that can be directly allocated to each type of product at the time of its occurrence (Martins, 2003). They are related to the actual unit expenditure of a given product and its direct appropriation, i.e., you can allocate the product sugar to the raw materials used in its preparation. Thus, it is possible to know how much money, in sugarcane (raw material), was spent for each kilogram of sugar produced, for example.

Direct costs are those that can be easily identified and allocated to products, without the use of arbitrary apportionment criteria (method of distribution or allocation of costs). Consider the following scenario: (a) the company produces ethanol and sugar and buys a shipment of sugarcane containing 30 tonnes; (b) the company pays $50.00 per tonne; (c) taxes amount to 10% and freight to 5% of the gross value; (d) the raw material is delivered to the plant; (e) employees carry out all stages required to produce sugar; (f) the cycle is repeated, this time for production of ethanol, in the same month. Figure 21.3 illustrates this situation.

Regardless of the amount of sugar produced, or the yield of raw material, it is known that all costs described as direct shall be directly allocated to the final product without any apportionment criterion. Simply divide the total direct cost by the quantity produced. Thus, one obtains the direct variable cost per unit. It is cost because it is involved in the production process, it is variable because it depends on the quantity produced, it is direct because it uses no apportionment criterion, and it is per unit because the total cost is divided by a measurement unit, which can be kilogram, bushel, tonne, or other.

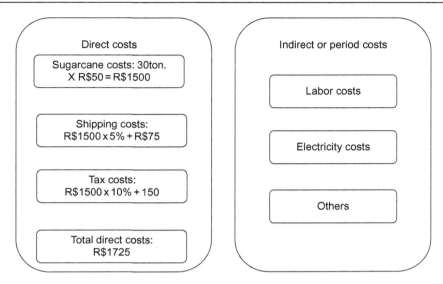

Figure 21.3
Example of difference between direct and indirect costs.

21.2.5 Indirect Costs

Considering the example in Figure 21.2, the costs of labor, electricity, among others, are not directly assigned to the final product. This would only be the case if the industry manufactured only one product, with no by-products, which is not the case in the mentioned example, or the reality for the most part of the domestic sugar and ethanol industry. Therefore, the same labor and electricity used for the production of sugar will be used at another time, but within the same base period, to produce other products such as ethanol, molasses, etc.

Therefore, indirect costs are those whose sum cannot be directly allocated to each product at the time of its occurrence, and that require a distribution criterion so that it can be allocated. Despite numerous attempts to create a methodology for distribution, apportionment or allocation of these costs, despite the vast literature available, and despite the claims by different schools of thought that they have the best methods, there is no efficient criterion for the distribution of such cost. This is because, if there is more than one way to allocate costs, or more than one apportionment criterion for the same expenditure, then the final cost is subject to a decision by the manager, administrator, or accountant on which criteria to use, thus producing an arbitrary result, since the choice of criteria is individual and subjective. It is very likely that competing companies with the same structures and production technology, in the same sector and economic activity, show different results on different costs. This causes distortions in the results presented and prevents companies from comparing like for like, thus hindering the technological development and management of companies involved. The allocation of this type of cost will be further discussed under the topic "Costing Systems" (see Section 21.7).

21.3 Contribution Margin

The concept of contribution margin is very important for companies that want to use cost as a competition strategy. It is a technique easily assimilated and applied in everyday business. It is a simple and very important tool for decision-making. Martins (2003) defines contribution margin as "the difference between the selling price and the variable cost of each product." This difference may be total or per unit: it is total when a company finds the contribution of the total production of one product or mix of products for the company as a whole, and per unit when applied to the sale of a unit compared to the direct variable cost of such unit. The following equations exemplify the concept of contribution margin:

Equation A		Equation B	
Sale of the entire sugar production		Sale per unit (per tonne)	
	Gross Revenue		Gross Revenue
−	Total Direct Variable Cost	−	Direct Variable Cost per Unit
=	Total Contribution Margin	=	Contribution Margin per Unit

There are variable costs that may not be easily allocated to products, for example, electricity. It is considered a cost because it is linked to production, it is variable because it increases or decreases according to industrial activity or volume produced, and it is indirect because it is used without a precise measurement for the production of all products, i.e., one cannot tell exactly how much electricity was used for the production of sugar or ethanol. It could only be considered a direct cost if one could employ individual meters and the electricity used for lighting and for the management of the storage system could also be measured. However, the cost−benefit ratio of such investment would probably not be positive for the company. This type of cost, although variable, can be considered as "fixed" cost, but the best terminology would be period cost.

21.4 Period Costs

Period costs are those which do not have an accurate classification or that are difficult to categorize, but that are not direct variable costs. Typically, period costs are indirect variable costs which, for determination of results, are not directly applied to the product, but are added separately, along with fixed costs, to calculate the contribution margin and the break even point.

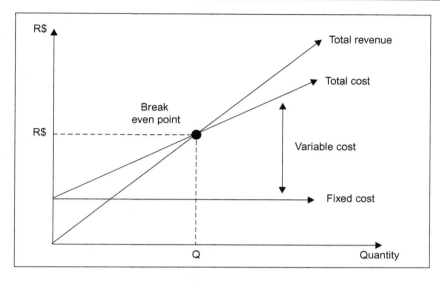

Figure 21.4
Break even point in a period of stability.

21.5 Break Even Point

The break even point corresponds to the volume of transactions in which total revenue equals total cost. Thus, it is the point at which net profit is zero, and it can be expressed in physical or monetary units (Cogan, 1999).

It can also be considered the level reached when the company is able to cover all expenditure (costs and expenses) for the quantities produced. At break even point, the company has already covered or paid all fixed and period costs, in addition to direct variable costs. As of then, the company will have greater flexibility to establish product prices, since each unit will now cost only the value of variable direct costs. Figure 21.4 shows break even point.

21.6 Costs as Strategy

The use of information related to cost for decision-making is not new, but has gained weight due to highly competitive business environments. The transfer of costs to consumers through the prices charged for their products or services is not something that should be done naturally by companies; the right thing to do is to manage costs and make them a great competitive advantage.

Cost management came from a stage when it was used merely as a technique for verification of data on stock value, which used to be part of industry's needs following the Industrial Revolution (18th century), to gain a prominent place as a management tool.

The mere surveying of accounting data no longer makes sense in a competitive environment; the goals of traditional costing systems have evolved to search for information to support the control of operations, results analysis and product costing. In markets with several competitors, some pieces of information are essential to corporate governance, such as pricing policy, marginal gain, analysis of cost structure, timing of purchases and sales, inventory turnover, among others.

Offering competitive costs as a strategy is not as easy as it sounds. It entails considering every detail that may reduce operational, financial, and economic expenses, or the most cost-beneficial investments for the company. This means extreme concern with the quality of raw materials, labor, investment, financing sources, inputs and especially, quality management. Behind the quality lies savings of resources, good products, no duplication of efforts, i.e., the management of costs.

In order to determine cost competitiveness, companies must, of course, check costs accurately. There are several costing methods available, which can be understood as a way to allocate costs to a particular product or service. It is a tool that collects sorts and organizes data on the costs of products or services, thus turning such data into information.

Costing systems are adopted by companies taking into account, for example, their goals, activity, market strategy, and tax legislation.

Among the most commonly known and used costing systems, one can mention absorption costing, direct or variable costing, known as traditional costing methods (Megliorini, 2001), and activity-based costing, or ABC.

21.7 Costing Systems

21.7.1 Absorption Costing Traditional Costing Systems

In absorption costing, also known as the traditional costing method, the value of the costs of goods or services is determined based on all production costs. The procedure is to make each product or production (or service) absorb part of the direct and indirect costs related to manufacturing. Direct costs are allocated by registering them in an objective manner, and indirect costs through apportionment. Among the most used criteria, there is the proportionality to the value of raw materials used in the production process, to the value of direct labor, number of labor hours, and machine hours consumed.

The distinction between cost and expense is essential to the understanding of this costing system. This is because expenses are immediately referenced to the results of the period, as well as the cost of products sold. The costs associated with products underway and unsold finished products, on the other hand, are activated in the stocks of these products. Thus, all production costs are recognized as expenses at the moment of sale, and shown in a manner that is more appropriate for the visualization of the reconciliation between revenues and expenses in the calculation of results (Bornia, 2002).

The critical point of this method is the distribution of indirect costs among products by means of apportionment criteria, which involves subjective and arbitrary aspects. Thus, even if they appear to be accurate, these criteria lead to errors, rendering invalid the calculation of the cost per unit (Megliorini, 2001).

The most obvious advantage of absorption costing is that it observes generally accepted accounting principles and tax legislation. Furthermore, it can be less costly to implement, since it does not require the separation of manufacturing costs in fixed and variable components. However, companies' cost profiles have changed greatly in recent years. Direct costs, especially labor, have been significantly reduced, unlike indirect costs (overheads, costs with new technologies), which have been increasing. As a result, indirect costs have taken on great importance within companies and should be better controlled for efficient management. Therefore, in the competitive environment of the sugar and ethanol sector, one cannot use methods that use arbitrary and imprecise criteria for measuring costs. Cost analysis techniques that allow better visualization and measurement are now required for improved management. Figure 21.5 shows an outline of the dynamics of traditional or absorption costing.

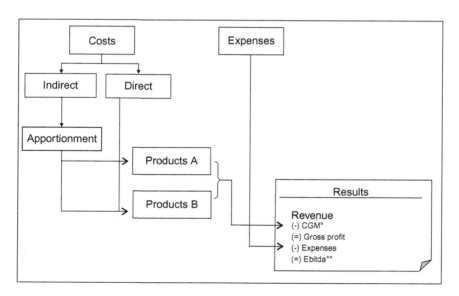

Figure 21.5
Dynamics of traditional or absorption costing.

21.7.2 Contribution Costing System, Direct, or Variable

The contribution costing system seeks to eliminate distortions in the cost apportionment criteria required in absorption costing. In absorption costing, fixed costs are distributed among goods and/or services, while in variable costing these costs are treated as expenses or period costs and, thus, go straight to results.

Direct costing is a kind of costing that only considers as product or production cost the variable costs in the production process that can be easily identified and allocated (without the use of apportionment criteria), thus, fixed costs are considered expenses for the period, since they do not depend on production volume.

The benefits of variable costing are essentially related to the generation of information for decision-making. Padoveze (2000) points out that from variable costs one can extract the contribution margin, which is the difference between sale price and product cost, used to answer several important questions in the decision-making process. Some examples pointed out by the author are:

* What is the contribution margin of a product?
* Should or shouldn't one accept a special order?
* What is the break even point?

Figure 21.6 shows an outline of the dynamics of variable or direct costing.

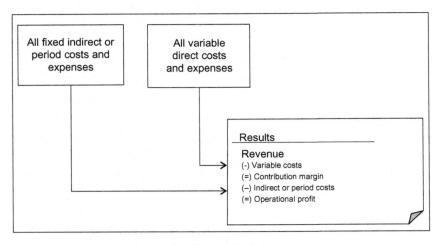

Figure 21.6
Dynamics of variable or direct costing.

21.7.3 Activity–Based Costing

As with contribution costing, ABC seeks to mitigate the distortions caused by the use of apportionment. It could be seen as an evolution of the systems presented, but its direct relation with the activities involved in the process constitutes a mere deepening of the absorption costing system.

The method assumes that the different activities carried out consume resources, therefore, they generate costs. These costs are accumulated in activity centers through resource drivers. Activity means a set of processes comprising people, technologies, materials, methods, and environment, as appropriate. Such activities, used to manufacture products, generate new costs, allocated to products through activity drivers.

The basic idea is to assign basic costs to activities first, and then assign activity costs to products. As in other costing systems, direct costs are allocated directly to products. Figure 21.7 summarizes the dynamics of the ABC.

This costing system has advantages and disadvantages; if well implemented, it certainly represents an evolution vis-á-vis other methods. However, it does not address the issue of distribution of indirect manufacturing costs, since the choice of resource and activity drivers is arbitrary, i.e., depends on managers or organizations. Thus, it goes back to the problem

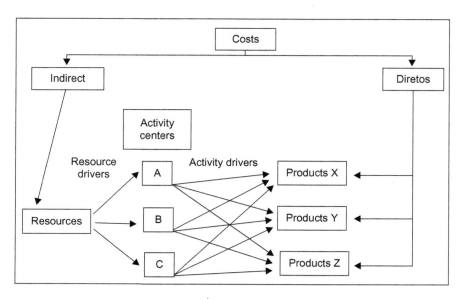

Figure 21.7
Dynamics of ABC.

of differences between companies within the same segment, sector, or activity with respect to results over costs. Another relevant fact is that this type of costing system is recommended for industries with large and diverse product portfolios, which is not the case with the sugarcane industry. Finally, high implementation costs, especially training and adaptation to the system by employees, and high maintenance costs make this costing system unfeasible. The information generated by such system does not justify the effort required for its deployment.

21.8 Cost—Based Decisions

For wise decisions to be made, will, instinct or partial knowledge of the reality of the company and the market will not suffice; one needs to know every detail about the company and possible impacts of the decision. Therefore, information on costs alone, no matter how good the method employed for their measurement and allocation, is not enough to subsidize strategic decisions. Costing systems are tools for operational decision, which must follow organizations' strategic guidelines. This is explained by the fact that costs, expenses and expenditures generally take place in the everyday running of production and marketing, not in a meeting room, where companies' plans are defined.

One needs a system that brings about the best trade-off between implementation costs and the training required for staff to use it. To that effect, two costing systems are recommended for the sugar and ethanol sector. The first, for obvious reasons, is the traditional or direct system, taking into account the tax legislation. However, it deserves special attention as to detailed allocation of costs, because this system is not meant to subsidize decisions; it should be used only for accounting records. The second and most important is the direct costing system, variable or marginal. Easy to use, this system is able to support everyday decision-making, and provides excellent immediate results. However, the decisions to be taken based on the information generated, although they are operational, need to be very well thought out, since they directly impact profit.

Its use should strictly follow the method, so as to enable the determination of the break even point both as regards production volume and monetary values. It also allows calculation of the contribution margin of each product, and thus one can determine the best product to be produced and marketed, according to market variables, such as price, seasonality, and demand and supply fluctuations, for example. It also helps determine whether industries should or should not accept special orders. All one needs to do is to find out whether break even point has been reached in the period and then consider that only direct variable costs and expenses comprise the total cost of the order. It is worth mentioning that tax costs are variable and direct and should be accounted for.

21.9 Final Remarks

All costing methods available have their own peculiarities, advantages and disadvantages. Different information can be generated with the use of different accounting information systems, which will depend on the type of decision to be made. Therefore, managers should be aware of the different accounting and financial management tools and make them useful in the decision-making process. Companies measure costs because they are concerned with overall results, which are directly impacted by costs and by choice of the appropriate costing method. Correct measurement of costs and expenses is of great importance for the profitability analysis of products, chains and the industrial segment. Continuous improvement of processes and activities, and therefore of results, is the path to be followed. One cannot skip steps. Improvements in costing systems processes are gradual, time-consuming, and must be taken seriously and addressed with great dedication. It is not possible to implement efficient costing systems with fragmented work focused on some sectors of the organization. The techniques available must be coupled with overall company planning, so that the whole organization follows the same path, striving for the same ultimate goal.

Bibliography

Bornia, A.C., 2002. Análise gerencial de custos. Bookman, Porto Alegre.

Cogan, S., 1999. Custos e preços: Formação e análise. Pioneira, São Paulo.

Leone, 1997. Curso de contabilidade de custos. Editora Atlas, São Paulo.

Martins, 2003. Contabilidade de custos. Atlas, São Paulo.

Megliorini, 2001. Custos. first ed. Makron Books, São Paulo.

Nakagawa, M., 1994. ABC: Custeio baseado em atividades. Atlas, São Paulo.

Padoveze, C.L., 2000. O Paradoxo da utilização do método de custeio: custeio variável por absorção. Revista CRC-SP 4. 12, 42−58.

Porter, E., 1991. Estratégia competitiva. Editora Campus, Rio Janeiro.

Santos, J.J., 2000. Análise de custos. third ed. Atlas, São Paulo.

Index

Note: Page numbers followed by '*f*' and '*t*' refer to figures and tables, respectively.